Naval Shipboard Communications Systems

John C. Kim
Eugen I. Muehldorf

TRW Inc.

Prentice Hall PTR
Englewood Cliffs, New Jersey 07632

Library of Congress Cataloging-in-Publication Data

Kim, J. C. (John C.)
    Naval shipboard communications systems / by J. C. Kim &
E. I. Muehldorf.
      p. cm.
    Includes bibliographical references and index.
    ISBN 0-13-613498-X
    1. Command and control systems.   2. United States.   Navy-
-Communication systems.   I. Muehldorf, E. I. (Eugen I.)
II. Title.
VB212.K56   1995
359.8′5′0973—dc20   94–16309
                                         CIP

Editorial/production supervision
  and interior design: *Harriet Tellem*
Cover design: *Lundgren Graphics*
Manufacturing manager: *Alexis R. Heydt*
Acquisitions editor: *Michael Hays*
Editorial assistant: *Kim Intindola/Diane Spina*
Sources for cover photographs:
    U.S.S. Arleigh Burke (DDG-51), Courtesy of Bath Iron Works
    AN/URT (D) HF Transmitter, Courtesy of Stewart-Warner Electronics Corporation
    Fleet Satellite Communications Satellite, Courtesy of TRW Inc.
    AN/WSC-3 UHF Transceiver, Courtesy of E-Systems
    AN/WSC-6 Antenna, Courtesy of Electrospace Systems Incorporated

The publisher offers discounts on this book when ordered in bulk quantities. For more
information, contact:

    Corporate Sales Department
    Prentice Hall PTR
    113 Sylvan Avenue
    Englewood Cliffs, NJ 07632

Phone: 201-592-2863, 800–382–3419 Fax: 201-592-2249, e-mail: dan_rush@prenhall.com

Printed in the United States of America

10 9 8 7 6 5 4 3 2 1

ISBN 0-13-613498-X

Contents

Preface   **xv**

Acknowledgments   **xix**

Chapter 1   Introduction   **1**

Chapter 2   Background   **6**

References   10

Chapter 3   Communications in the Navy   **11**

3.1   The Defense Communications System (DCS)   11

3.2   The Naval Telecommunications System (NTS)   12

3.3   Level of Communications Capability as Function
of Conflict Periods   14

3.4   Naval Communications Stations   16

3.4.1   *Fleet Communications, 16*

3.4.2   *Functions of NCTAMSs and NAVCOMMSTAs, 17*

3.4.3 Traffic Volume at a NCTAMS and a NAVCOMMSTA, 20

References   21

Chapter 4   Naval Warfare and Its Communications Connectivities   **22**

4.1   Naval Warfare   22

4.1.1   Anti-Air Warfare (AAW), 23
4.1.2   Anti-Surface Warfare (ASuW), 25
4.1.3   Force Anti-Submarine Warfare, 26
4.1.4   Strike Warfare (STW), 28
4.1.5   Amphibious Warfare, 29
4.1.6   Space and Electronic Warfare, 31

4.2   Strategic Submarine Connectivity   31

4.3   Naval Shipboard EXCOMM Connectivity   34

4.3.1   Shore-to-Shore Connectivity, 34
4.3.2   Shore-to-Ship Connectivity, 35
4.3.3   Shore-to-Aircraft Connectivity, 35
4.3.4   Shore-to-Submarine Connectivity, 35
4.3.5   Ship-to-Shore Connectivity, 35
4.3.6   Ship-to-Ship Connectivity, 36
4.3.7   Ship-to-Aircraft Connectivity, 36
4.3.8   Ship-to-Submarine Connectivity, 36
4.3.9   Aircraft-to-Shore Connectivity, 37
4.3.10   Aircraft-to-Ship Connectivity, 37
4.3.11   Aircraft-to-Aircraft Connectivity, 37
4.3.12   Aircraft-to-Submarine Connectivity, 37
4.3.13   Submarine-to-Shore Connectivity, 37
4.3.14   Submarine-to-Ship Connectivity, 38
4.3.15   Submarine-to-Aircraft Connectivity, 38
4.3.16   Submarine-to-Submarine Connectivity, 38

References   38

Chapter 5   Naval Telecommunications System (NTS) Circuits   **40**

5.1   Tactical Circuits/Links Employed in Naval Warfare   41

5.1.1   Tactical Group (TG) Communications, 44
5.1.2   Anti-Submarine Warfare (ASW) Communications, 44
5.1.3   Anti-Surface Warfare (ASuW) Communications, 45

5.1.4   *Anti-Air Warfare (AAW) Communications, 45*

5.1.5   *Electronic Warfare (EW) Communications, 45*

5.1.6   *Air Operations Communications, 46*

5.1.7   *Tactical Air Communications, 46*

5.1.8   *Amphibious Communications, 46*

5.1.9   *Naval Gunfire Communications, 47*

5.1.10   *Submarine Communications, 48*

5.1.11   *Data Links, 48*

5.1.12   *Distress Communications, 50*

5.1.13   *Mine Countermeasure (MCM) Communications, 50*

5.1.14   *Harbor Communications, 51*

5.2   Long-Haul Communications in Naval Warfare   51

5.2.1   *Wideband SHF Satellite Communications, 51*

5.2.2   *UHF Fleet Satellite Broadcast, 53*

5.2.3   *UHF Fleet Satellite Communications, 53*

5.2.4   *Long-Haul HF Communications, 54*

5.3   Strategic Circuits/Links Employed in Submarine
       Warfare   55

References   56

**Chapter 6   Shipboard Exterior Communication
             (EXCOMM) Requirements   58**

6.1   Voice   60

6.2   Record Message Transmission   62

6.3   Data   64

6.4   Facsimile   65

6.5   Video Transmission (TV)   66

References   68

Chapter 7   Performance Measures   70

7.1   Coverage   71

7.1.1   *Antenna Gain and Directivity, 71*

7.1.2   *Transmitter Power, 72*

7.1.3   *Noise, 73*

7.1.4   *The Signal-to-Noise Ratio, 76*

7.1.5   *Receiver Sensitivity, 78*

7.1.6   *Link Quality, 80*

7.2   Link Capacity, Bandwidth, and Data Rate   80

7.3   Reaction Time   84

7.4   Availability   84

    7.4.1   *Availability Modeling, 86*
    7.4.2   *Improving the Availability of Communications, 86*
    7.4.3   *Circuits, 86*

References   88

Chapter 8   AJ Communications   89

8.1   Introduction   89

8.2   Suppression of Signals in a Repeater by a Stronger
    Signal   100

8.3   Signals for Bandspread Systems   103

8.4   AJ in Naval Communications Systems   108

References   109

Chapter 9   ELF, VLF, and LF Communications   111

9.1   Fundamentals of ELF, VLF, and LF
    Communications   112

9.2   Propagation   116

    9.2.1   *Propagation Over Long Distances, 116*
    9.2.2   *Attenuation in Seawater, 120*
    9.2.3   *The Effects on Sky Wave Interference, 122*
    9.2.4   *Noise and Interference, 122*
    9.2.5   *Long-Wave Propagation Prediction Models, 123*

9.3   Antenna Systems   126

    9.3.1   *Transmitting Antenna, 126*
    9.3.2   *Receiving Antennas, 126*
    9.3.3   *Tuners and Couplers, 132*

9.4   Modulation Systems for Long-Wave Radio
    Transmission   133

    9.4.1   *Long-Wave Modulation System Design
        Considerations, 133*
    9.4.2   *Long-Wave Modulation Techniques, 133*

9.4.3   Waveform Filtering, 136

9.4.4   Examples of Bandwidth Reduction by Filtering for Long-Wave Transmission, 138

9.4.5   Long-Wave Modulation Bandwidths Comparison, 140

9.4.6   Typical Long-Wave Receiving Equipment, 145

9.4.7   Compact VLF (CVLF) Systems, 146

9.5   Other Long-Wave Systems   150

9.5.1   ELF Sanguine Systems, 150

9.5.2   TACAMO/VLF System, 151

References   151

Chapter 10   Communications   154

10.1   Characteristics of HF Communications   155

10.2   HF Propagation, 159

10.2.1   Fading, 160

10.2.2   HF Operating Frequency, 161

10.2.3   HF Sounding, 163

10.3   HF Communications Systems   166

10.4   Typical HF Equipment   170

10.4.1   R-1051 (G) /URR HF Receiver, 171

10.4.2   R-2368/URR HF Receiver, 171

10.4.3   AN/URT-23 HF Transmitter, 172

10.4.4   HF-80 HF Transceivers, 174

10.4.5   AN/USQ-83 Tactical Data Link System, 177

10.4.6   ICS-3 Integrated Communication System (AN/URC-109), 177

10.5   HF Antennas   179

10.5.1   HF Whip Antennas (Single Pole Whips), 180

10.5.2   Twin Whip Antennas, 181

10.5.3   Long-Wire Antennas, 183

10.5.4   Fan-Type Antennas, 184

10.5.5   Near-Vertical Incident Skywave (NVIS) Antennas, 186

10.6   Automatic Link Establishment Techniques   189

10.6.1   Description of ALE Method, 189

10.6.2   HF Radio ALE Standards, 190

10.6.3 *Shipboard ALE Using the AN/URT-23 Radio Equipment, 193*

References   195

Chapter 11   VHF/UHF Line-of-Sight Radio Communications   198

11.1   Characteristics of VHF/UHF Communications   199

11.2   VHF/UHF Propagation   202

11.3   VHF/UHF Transmitters and Receivers   207

11.4   VHF/UHF Transceivers   207

11.4.1   *SINCGARS, 207*

11.4.2   *The AN/ARC-182 VHF/UHF Radio, 211*

11.4.3   *The AN/VRC-40 Series VHF Radios, 212*

11.4.4   *The AN/WSC-3 UHF LOS Radio, 212*

11.4.5   *The AN/URC-93 VHF/UHF LOS Radio, 214*

11.4.6   *The Position Location Reporting System (PLRS) AN/TSQ-129 (V), 214*

11.4.7   *The Joint Tactical Information Distribution System (JTIDS), 216*

11.4.8   *Link 4A, 218*

11.4.9   *HAVE-QUICK II, 220*

11.4.10 *The Common High Bandwidth Data Link (CHBDL) AN/USQ-123, 221*

11.4.11 *The Light Airborne Multi-Purpose System (LAMPS) Data Link AN/SRQ-4, 223*

11.5   VHF/UHF Antennas and Couplers   225

11.5.1   *VHF/UHF Multicouplers, 225*

11.5.2   *The VHF/UHF Antenna Coupler CU-1559, 227*

References   227

Chapter 12   Satellite Communications (SATCOM)—Fundamentals   230

12.1   Satellite Orbits   230

12.1.1   *Geosynchronous Orbits, 231*

12.1.2   *Polar Orbits, 231*

12.2   Tracking   232

12.3   Propagation and Noise   233

12.3.1   Propagation, 233

12.3.2   Noise, 234

12.4   Antenna Parameters   237

12.5   The Range Equation and Link Budgets   239

References   244

Chapter 13   UHF Satellite Communications   245

13.1   FLTSATCOM Satellites   245

13.1.1   The Fleet Satellite (FLTSAT), 246

13.1.2   The LEASAT, 247

13.2   UHF Satellite Subsystems   252

13.2.1   Fleet Satellite Broadcast Subsystem   260

*13.2.2   Common User Digital Exchange Subsystem/Naval
          Modular Automated Communications Subsystem
          (CUDIXS/NAVMACS), 261*

*13.2.3   Officer-in-Tactical Command Information Exchange
          Subsystem (OTCIXS), 263*

*13.2.4   Submarine Satellite Information Exchange
          Subsystem (SSIXS), 265*

13.2.5   Tactical Intelligence Subsystem (TACINTEL)

*13.2.6   Tactical Data Information Exchange Subsystem
          (TADIXS), 267*

13.2.7   Secure Voice Subsystem (SECVOX), 268

13.2.8   Teletype (TTY) Subsystem (ORESTES), 271

13.3   Equipment   271

13.3.1   Receivers/Transceivers, 271

13.3.2   Antennas, 272

13.3.3   Secure Voice (SECVOX) Equipment, 276

13.3.4   Teletypewriter (TTY) Terminal Equipment, 278

13.3.5   Interconnection Groups, 279

13.4   Demand Assigned Multiple Access (DAMA)
       Subsystem   280

13.4.1   The Principle of DAMA, 280

13.4.2   DAMA Operational Considerations, 282

13.4.3   DAMA Equipment, 285

13.5   UHF Satellite Control Subsystem   286

13.6   International Maritime Satellite (INMARSAT)   288

13.7   Currently Planned Navy UFH SATCOM Programs   294

    13.7.1   UHF Follow-On Satellite System, 294
    13.7.2   High-Speed Fleet Broadcast (HSFB), 302
    13.7.3   New Enhanced Terminals, 303

References   303

Chapter 14   SHF Satellite Communications   305

14.1   Satellites   308

14.2   Shipboard Terminal Configuration   312

14.3   Equipment   312

    14.3.1   QUICKSAT SHF SATCOM Terminals, 317

14.4   DSCS Control   318

14.5   An Example of Navy DSCS Ship Communications, 319

References   322

Chapter 15   EHF Satellite Communications   323

15.1   The MILSTAR Satellite   328

15.2   Shipboard Terminals for EHF Satellite
       Communications   334

15.3   MILSTAR Operational Control   338

References   339

Chapter 16   Shipboard Communications Antennas   341

16.1   Shipboard Antennas (Noncommunications)   342

16.2   Overview of Shipboard Communications
       Antennas   348

16.3   A Brief Introduction to Shipboard Communications
       Antenna Fundamentals   348

16.4   Description of Some Shipboard Communications
       Antennas   354

References   359

Chapter 17   Electromagnetic Compatibility (EMC)   360

17.1   Definitions   362

17.2   Descriptions of EMI Sources   363

17.3   EMC and EMC Engineering   365

    17.3.1   *EMI Assessment, 366*

    17.3.2   *EMC Topside Integration, 366*

17.4   Shipboard Interference Analysis   369

    17.4.1   *Shipboard Exterior RF Communications System Design, 370*

    17.4.2   *EMI Control in Shipboard Multichannel HF Transmitter and Receiver System, 370*

    17.4.3   *EMI Control in Shipboard VHF Radios, 377*

17.5   EMC Application to Below-Decks Communications System Design   379

17.6   EMC Engineering During a Ship's Life Cycle   382

References   383

Chapter 18   Shipboard Technical Control and Interior Communications Systems   385

18.1   Shipboard Technical Control Systems   386

    18.1.1   *Technical Control Facility (TCF), 386*

    18.1.2   *The Communications Security Equipment Area, 387*

    18.1.3   *The RF and Baseband Equipment Area, 388*

    18.1.4   *The Surface Ship Exterior Communications Monitor and Controls System (SSECMS) AN/SSQ-33, 388*

    18.1.5   *The Quality Monitoring Control System (QMCS), 389*

    18.1.6   *Frequency Standards AN/URQ-23, 392*

    18.1.7   *Tactical Frequency Management System AN/TRQ-35 and AN/TRQ-42, 394*

18.2   RF Switching Subsystems   394

    18.2.1   *Transmitter and Receiver Antenna RF Distribution, 396*

    18.2.2   *Antenna Multicouplers, 396*

    18.2.3   *RF Patchpanels, 398*

18.3   Communications Switching Systems   402

18.3.1   The Black Communications Switch
         SA-2112A(V)6/STQ, 403

18.3.2   The AS-2112A(V)8/STQ Red Communications
         Switch, 404

18.3.3   The SB-4268 Audio Frequency Patchpanel, 406

18.3.4   The MCS-2000 Interior Communications System, 407

18.4   Data Switching Systems   408

18.4.1   The SA-4176 Data Transfer Switch, 409

18.4.2   The Navy Modular Automated Communications
         System (NAVMACS-V) AS/SYQ-7, 409

18.4.3   The ON-143 (V) 6 Interconnecting Group, 411

18.5   Other Interior Communications Systems   413

18.5.1   The Flight Deck Communications System
         (FDCS), 413

18.5.2   Sound-Powered Telephone Systems, 414

18.5.3   The Announcing System LS-474/U, 417

18.5.4   The TA-970 Telephone Set, 417

References   419

Chapter 19   Shipboard Communications Protocol   421

19.1   Shipboard Combat Systems   422

19.1.1   Description of Typical Shipboard Combat Systems, 422

19.1.2   Shipboard Signals and Interfaces, 423

19.1.3   Requirements for Shipboard System Interconnection, 424

19.1.4   External Shipboard Communications Connectivity, 425

19.2   Overview of Link 11 Message Formats (TADIL-A)   426

19.3   Local Area Networks (LANs)   429

19.3.1   Shipboard LANs, 429

19.3.2   Token Rings, 430

19.3.3   Fiber Distributed Data Interface (FDDI), 431

19.3.4   Use of Fiber Optics in Shipboard Combat Systems, 432

19.4   Open Systems Interconnection (OSI) Standards   432

19.4.1   Open Systems Interconnection (OSI) Model, 433

19.4.2   Government Open System Interconnection Profile, 436

19.4.3   DoD Military Protocol Standards   436

19.5   The Survivable Adaptable Fiber Optic Embedded
       Network (SAFENET)   438

       19.5.1   The SAFENET OSI Suite, 440
       19.5.2   The SAFENET Lightweight Suite, 441

19.6   The MIL-STD-1553B Bus   442

References   443

Chapter 20   Trends in Shipboard Communications   445

20.1   Chronological Review of Naval Telecommunications
       Automation Programs   445

       20.1.1   Naval Tactical Data System (NTDS), 446
       20.1.2   Naval Modular Automated Communications System
                (NAVMACS), 446
       20.1.3   Naval Telecommunications System (NTS) Architecture, 447
       20.1.4   ICS/SCAN, 447
       20.1.5   Survivable Adaptable Fiber Embedded Network
                (SAFENET), 448
       20.1.6   Unified Networking Technology (UNT), 448
       20.1.7   Communication Support System (CSS), 449

20.2   Description of the Communication Support System (CSS)   449

       20.2.1   CSS Concepts, 449
       20.2.2   The Concept of the CSS Communications Services, 451
       20.2.3   Implementing CSS Services with EXCOMM Resources, 453

20.3   Copernicus Architecture   454

References   455

Appendix A   Abbreviations and Acronyms   456

**Appendix B   A Description of Typical Shipboard
               Combat Systems   476**

       B.1   Radar System   476
       B.2   Identification System   477
       B.3   EW System   478
       B.4   Navigation System   478
       B.5   Underwater System   479

B.6   Command and Decision System   479

B.7   Weapon Control System   479

B.8   Weapon System   479

B.9   Telemetry System   480

B.10   Exterior Communications (EXCOMM) System   480

B.11   Interior Communications (INTERCOM) System   480

B.12   Meteorological System   481

B.13   Combat Training System   481

References   481

Appendix C   Joint Electronic Type Designation System   **483**

            Index   **485**

Preface

The mission of the U.S. Navy, stated in Title 10, U.S. Code, is "to be prepared to conduct prompt and sustained combat operations at sea in support of the U.S. national interest." This also means that the U.S. Navy must effectively assure continued maritime superiority for the United States. Derived from that are the Navy's two basic functions: sea control and power projection. Effective communications is vitally important to fulfill this mission, and for this the U.S. Navy depends on a worldwide Command, Control, and Communications (C3) capability.

Naval communications is constantly evolving and adopting modern technology. Satellite communications and modern networking is a part of the repertoire of naval communications technology. The development of communications and data processing technology is fostering significant changes in the manner in which the Navy is communicating among ships and between ship and shore. The chain of operations from sender to receiver includes message format processing, relaying to a communications center, protection to prevent unauthorized intrusion and to guarantee transmission integrity, modulation for optimum sending over the medium, transmission, reception, removal of the protection, and distribution to the recipient.

Naval shipboard communications combines advanced technology with operational procedures, and uses ruggedized equipment that can be maintained with limited resources in the harsh environment of the sea. All aspects of engineering disciplines are applied to provide this essential capability to the U.S. Navy. It includes not only technological aspects but also operational procedures, configuration control, and logistic support considerations. The most advanced technology has little value if it cannot be operated and maintained by naval personnel under stressful conditions.

This book focuses on shipboard communications. Shipboard communications has an exterior component, dealing with antennas, transmitters, and receivers, in short the classical exterior communications (EXCOMM). It also has an intraplatform component dealing with the shipboard distribution system, the communications processors in the communications rooms, and the procedures for processing messages.

In this book we discuss the technologies and circuits of all frequency regions used for naval communications, from ELF to optical frequencies. The shipboard EXCOMM spans a wide spectrum. At the low end of the spectrum, ELF with frequencies from 70 to 80 Hz penetrates seawater and allows very narrowband transmission to submerged submarines. In the VLF region from 10 to 30 kHz, the Navy operates long-range narrowband TTY circuits largely unencumbered by the vagaries of propagation. HF supports long-distance circuits, VHF, UHF, and SHF support line-of-sight communications among ships and between ships and aircraft. EHF, 30 to 300 GHZ, supports the latest satellite communications technology. Optical frequencies, at the high end of the spectrum, penetrate seawater and can be used for communications to submerged submarines.

A modern Navy ship has many communications, radar, guidance, and weapons systems, all of which require antennas. These antennas, many of which are clearly visible when one sees a naval ship, others of which are small, unobtrusive structures blending with the ship's hull, transmit and receive electromagnetic signals that can interfere with each other severely. Electromagnetic compatibility (EMC) is an essential part of shipboard electronic engineering to assure that naval communications are functioning well.

Intraplatform communications provide the interfaces through which the electromagnetic signal is brought from the topside antennas into the ship and are distributed within the ship for processing the communicated information. Protocols, that is, message formats and methods to accomplish the data and message transmission, are essential for naval communications. Message standards are vital to assure Navy-wide and interservice uniformity and understandability of the message context. However, the adherence to standards and formats can also cause problems, which most of us know from working with data processing equipment when we see such messages as: "Incorrect parameter specified." Training to understand the standards applied and message composition processing support systems is essential to naval communications procedures.

This book has grown out of our work: many projects, and immersion into the subject. We have seen the need to bring a lecture series to our junior colleagues, borne out of strictly technical as well as Navy communications practice-related questions which we were asked. Some of the material exists in the open literature and in specialized books, much is in unclassified documents, such as procedures and equipment descriptions. Our contribution was to gather and unify this material and present it in this volume.

This book is intended as a reference textbook for engineers who understand communications and need to become familiar with Navy communications methods and procedures. For example, when one thinks of satellite communications, one envisions broadband and high-speed transmission. Not so in naval satellite communications—here strategic satellites provide narrow-band dedicated circuits, in many cases protected against jamming. We have tried to bridge such gaps of perception and do it without touching on sensitive military aspects.

Acknowledgments

The authors wish to acknowledge the support and assistance we have received in preparing this book. In particular, our thanks go to Dr. John M. Gormally, TRW Inc., for encouragement with this project. Our thanks also go to RADM Ronald Wilgenbusch (USN Ret.), Booz, Allen & Hamilton Inc., for comments that helped focus on some of the key issues confronting Navy communications; Dr. Gunther Brunhardt, Space and Naval Warfare Systems Command, provided us with encouraging comments. We cannot begin to list all persons who helped obtain the photographs for the pictures shown in this book, but we are grateful for the help we received from Mr. John W. Eadie, Naval Sea Systems Command, who provided us with pictures of the shipboard antennas, and Mr. Robert Scruitsky, Naval Electronic Systems Engineering Activities, St. Inigoes, MD, and Mr. Terrence R. Connor, TRACOR INC., who helped us with pictures taken at the Aegis Land Based Test Site.

We also wish to thank all people who helped us in the preparation of the manuscript of this book. EIM wishes to thank his wife Erni for the initial translation of handwritten notes into typed form, and patience during the time it took to prepare the material; JCK wishes to thank his family, Rahn, Janet, William, and Douglas; in particular, Douglas's help in reading the manuscript and pointing out potential improvements is appreciated.

1

Introduction

Naval shipboard communication in pre-electronic days was only feasible at short ranges. It consisted of visual signalling, employed flags or lanterns, and used a semaphore alphabet to relay messages from one ship to another or from ship to shore. It was slow and cumbersome. A ship at sea beyond the visual range or shrouded in fog was virtually cut off and had no way to receive messages, either from the land or another ship.

The invention of radio and radio telegraphy promoted a revolutionary change for shipboard communications. The limitation of communicating at visual range was removed and the speed of communications was drastically increased. Naval radiotelegraphy became of significant military tactical and strategic importance and revolutionized naval communications. Each ship had a radio room and specially trained radio operators performed the vital task of maintaining radio contact with naval authorities and other naval vessels.

As radio and communications electronics developed, the evolving methods were applied to shipboard communications. The use of High Frequency (HF) opened global communications, albeit not always reliable. The use of VHF and UHF provided reliable line-of-sight communications.

Between 1950 and 1990, naval shipboard communications developed very rapidly and became exceedingly complex, spurred by the advent of communication satellites and the pressures of the cold war. A tremendous progress in naval shipboard communications has been accomplished in these decades, providing long-haul reliable global naval command and control communications, data transmission supporting tactical and strategic

1

missions, line-of-sight (LOS) and beyond-line-of-sight (BLOS) ship-to-ship and air-to-ship battleground communications, communications to submerged submarines, and a plenitude of warfighting communications applications.

The whole range of the electromagnetic radiowave spectrum, from ELF to EHF is currently exploited for shipboard communications. In each spectral region electromagnetic waves propagate differently and each frequency region offers a different usable communications bandwidth. Communications technology makes use of the propagation and available bandwidth. Communications are a function of the exploited frequency region, and in order to provide the required communications capability the technical, technological, and operational practices are combined to provide links and circuits for naval shipboard communications.

Figure 1.1 presents an overview of the radio frequency spectrum and shows the key

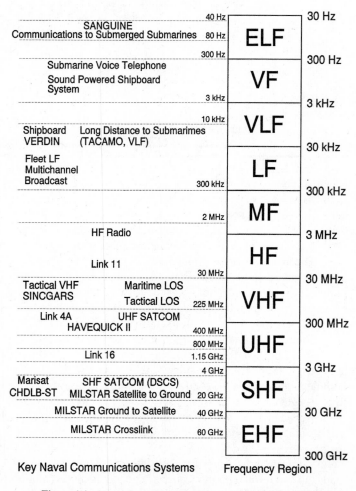

Figure 1.1   Naval Communication Within the Radio Spectrum

TABLE 1.1 SHIPBOARD COMMUNICATIONS IN DIFFERENT FREQUENCY REGIONS

Frequency Region	Primary Usage	Theoretical Considerations	Practical Design Considerations	Limitations
ELF 30–300 Hz	Communication to submerged submarines	• Penetrates seawater, propagation to depths of 25–40 m • Noise from atmospherics and submarine antenna motion • Global coverage	• Shore to submarine • Requires high transmitter power • Very low data rate (< 1 bps) • Special antennas required	• One way • Only short, prerecorded messages
VLF, LF 3–30, 30–300 kHz	Communication to submerged submarines	• Limited penetration into seawater • Strong atmospheric noise • Antenna matching to frequency • Global coverage	• Towed submarine buoys required • High transmitter power and large antennas needed • RF power and insulation critical • Limited bandwidth	• Narrow message bandwidth • Antenna directionality restricts course • Operational limitations (buoy needs to be close to surface)
MF, HF .3–3, 3–30 MHz	Global ship-ship, shore-ship and ship-ship	• Complex propagation, ground wave and sky wave (reflected by the ionosphere) • Noise from atmospherics • Antennas narrowband, matched to frequency • Global coverage	• High transmitter power • Antenna size matched to frequency; matching devices required • Narrowband (3 kHz) • Complex coding and modulation to overcome multipath • For broader band applications (AJ) only ground wave can be used, antenna matching difficult • EMI and EMC need to be carefully considered	• Propagation has strong diurnal variations • Limited circuit operating reliability because of propagation (circuits are interrupted or lost) • Power requirement makes HF source easy target for detection by adversary
VHF, UHF 30–300 and 300–3000 MHz	Line-of-sight ship-ship, ship-air, and shore-ship UHF SATCOM (near global coverage)	• Propagation simple: straight LOS path, signal penetrates the ionosphere • Noise random gaussian: background and receiver front end • Antennas directional, need to be matched to frequency • Coverage: non-SATCOM LOS only, SATCOM near global (between 70°N and 70°S latitudes)	• Transmitter power depends on application: LOS low to medium power, SATCOM high power • Wide bandwidth possible (MHz rather than kHz), used for broadband data transmission for command and control • Multiple antennas matched to frequency and bandwidth • Primarily digital modulation • EMI and EMC very important; frequency region shared with powerful radars • SATCOM antennas directional and steerable	• LOS short distance • SATCOM no polar coverage • Resource may have to be shared • SATCOM is a limited resource

3

TABLE 1.1 *(Cont.)*

Frequency Region	Primary Usage	Theoretical Considerations	Practical Design Considerations	Limitations
SHF 3–30 GHz	SATCOM (near global coverage)	• Propagation straight line through ionosphere • Noise gaussian (background, receiver front end) • Directional steerable narrow beam antennas • Coverage 70°N to 70°S latitudes	• High transmitter power, but concentrated into narrow beam • Antenna steerable parabolic reflector • Two antennas may be needed for full hemispherical coverage • Very broad band • Digital technology • AJ applications • EMI and EMC very important; frequency region shared with powerful radars	• Shipboard antenna small, 1.2 to 2.1 m (4 to 7 ft) diameter, hence the ship is a disadvantaged user • No polar coverage • Complex and costly equipment • Resource shared with other services • Transmitter power needs to be carefully controlled
EHF 30–300 GHz	SATCOM (primarily tactical)	• Propagation straight line • Noise gaussian, increased by absorption (water vapor and rain) • Very directional narrow beam antennas • Coverage 70°N to 70°S latitudes • Very broad bandwidth, used for AJ protection	• 20/40 GHz bands used • Transmitter power less than SHF SATCOM • Reflector antennas (two may be needed for full hemispherical coverage) • Very broad band • Modulation very complex • AJ (strongly reduced message bandwith) • Satellite to satellite links for better global coverage	• No polar coverage • Low data rate • Very costly and complex technology • Resource shared with other services • Currently in redefinition for post–cold war applications

modern naval communications systems it supports. These systems will be described in more detail in the following chapters. In addition, blue-green light frequencies, with a wavelength of about 0.54 μm and a frequency of about 5.5×10^{11} Hz that can be generated by LASERs and propagate into sea water, is considered for communications to submerged submarines.

Since the terms *link* and *circuit* are often used loosely, the following definitions are used:

- A link is a complete facility, encompassing equipment, practices, and procedures by which communications are accomplished; examples are a satellite link, Link 11, Link 16, and so forth. A link is a permanent capability.
- A circuit is one communication path established for message transfer; examples are a teletypewriter circuit, an HF voice circuit, and so forth. A circuit is a temporary communications path.

Recognizing that the frequency of operation is one of the key parameters for determining communications capabilities, Table 1.1 provides a brief overview of shipboard communications capabilities ordered by frequency region. This table lists the type of communications offered by the different frequency bands and highlights the key theoretical and technical design considerations. It also shows the limitations of using a particular frequency band for shipboard communications.

The book has a strong systems orientation. Briefly summarizing its content, the groundwork for this system orientation is laid in the initial chapters, where first the Navy telecommunications system (NTS) is discussed in depth. We then describe, by way of example, the various types of NTS circuits, links, and nets. This is followed by a discussion of operating modes. Next we provide fundamental system essentials: performance measures, shipboard communications requirements, and methods for providing anti-jamming (AJ) protection. The material treated in these chapters will address problems applicable in naval shipboard communications in all frequency regions identified in Table 1.1.

Having laid the groundwork, the book then discusses shipboard communications links and circuits in terms of frequency regions, that is, the chapters cover ELF, VLF, LF, HF, and VHF/UHF LOS communications. This is followed by satellite communications, describing the UHF Fleet satellite (FLTSAT), the SHF Defense Satellite Communications System (DSCS), and the EHF military satellite relay system (MILSTAR).

The discussion of the facilities and circuits opens the need to discuss several specialized naval shipboard technical and operational subjects. They are needed to round out the understanding of shipboard communications. Hence, this book provides chapters with a synopsis of shipboard communications antennas, electromagnetic compatibility (EMC) considerations, followed by command and control and communications protocols.

The book closes with a chapter discussing current Navy communications trends: the move to the new Copernicus communications architecture, where sharing communications capabilities will make optimal use of resources without jeopardizing communications capabilities needed for mission support. In addition, the closing chapter addresses the move toward a more integrated shipboard internal communications capability through the use of buses and fiber optics.

2

Background

In this chapter we address the scope of this book and its background. The naval communications system is worldwide in extent. It encompasses a shipboard and a shore segment as illustrated in Figure 2.1.

The figure presents a top level overview, combining typical shipboard user circuits, for example, teletypewriter (TTY), voice and data, into a ship-shore-ship link. Each link has its own cryptographic protection. The encoded and encrypted information consists of different digital data streams. The data streams are combined for transmission (or decombined at reception) via an internal (baseband) distribution subsystem; for effective transmission some data streams may be combined in multiplexers, for example, data and TTY. The translation from the digital format into a radio frequency (RF) signal is shown in this overview figure as being accomplished by the transmitter; similarly, the receiver will generate a digital data stream from the received radio signal. An antenna coupler and RF signal distribution subsystem will maximize the utilization of the shipboard antennas, that is, one antenna may be used for transmitting and receiving several RF signals.

At the shore station the signals are received (or, for shipbound traffic transmitted) and translated from RF signals into digital signals (or, for shipbound traffic translated into RF signals). The digital signals are formatted for shore communications, usually a multiplexer for transmission (or a demultiplexer for reception), and then reformatted by a data modem for interfacing to the land lines that provide the shore connectivity.

At the shore command the signals are again reformatted. The digital data streams

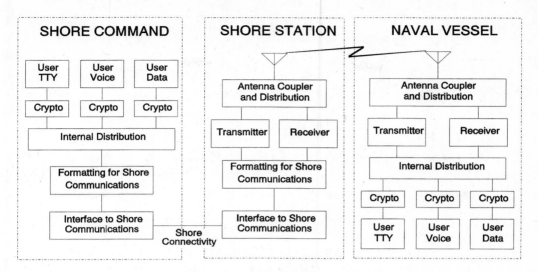

Figure 2.1   Naval Communications Connectivity—Overview

for each user are generated (by a formatting or reformatting process, i. e., multiplexing or demultiplexing) and then connected via cryptographic equipment to the user's end instruments (e. g., TTY, voice or data).

While the figure shows a ship-shore-ship connectivity, there is also a ship-to-ship and ship-to-air connectivity where the elements on the naval vessel are connected (via RF transmission) to another ship or an aircraft. The fundamental translation processes, however, are the same.

The shipboard segment shown in Figure 2.1 is the primary subject of discussion in the chapters to follow. Where necessary the elements of the shore station may also be described for completeness. Some systems diagrams may also show the complete connectivity to promote a full systems understanding.

The communications processes performed by an individual link or circuit shown in Figure 2.1 can be quite complex. Each of the transformations (multiplexing, demultiplexing, modulation, demodulation, transmission, and reception) can be very involved. To illustrate these complexities, a more detailed diagram, Figure 2.2, presents a complete overview of a communications channel.

At the transmitting side of the channel an information source (e. g., a data processor or TTY equipment) inserts the information into the link. The information is formatted into a digital data stream. This is followed by several transformations in order to adapt the information for transmission. At the multiplexer, information from different sources may be combined in this channel. From this point on in the channel there is a multiplexed data stream that will be transmitted. Modulation prepares the purely digital information for transmission. Additional transformations may be used to provide anti-jamming (AJ) protection and efficient access to the transmission medium. The transmitter then generates the radio signal necessary for sending the information to the receiving shore station, ship, or aircraft.

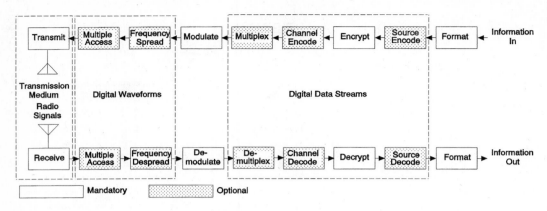

Figure 2.2   Diagram of a Digital Communications Channel

At the receiving side the processes are reversed step by step. In order for a digital transmission system to operate correctly, a synchronization signal is required in order to identify the boundaries of the digital signal elements (bits, bytes, frames, and so forth). The synchronization signal can either be provided by a common clock source, ultra-stable clock sources at each site, or it can be sent with the signal from the transmitter to the receiver. Synchronization is usually imperfect and complex at best: if a common clock for the transmitter and receiver cannot be used because of a large physical separation and delays or distortions along a transmission path, and ultra-stable, but separate, clocks are not usable because of a relative motion between transmitter and receiver, as it is common for communications to and from a ship, then a form of clock adaptation, either through buffering or bit-stuffing and de-stuffing must be provided at the modem or multiplexer level.

Not all links or channels are as complex as shown in Figure 2.2. Some processes are optional and some are mandatory for link operation. The mandatory and optional processes have been indicated in the figure. Examples of optional processes are frequency spreading for AJ protection, source encoding for optimizing efficiency, channel encoding for error protection, multiplexing for combining channels, and multiple access for transmitter and transmission medium (e. g., satellite channel) sharing. Note that since the subject discussed here pertains to naval communications links, encryption is shown as a mandatory process.

A key process is the transmission and reception of the signals. This process is strongly dependent on the frequency region of operation. This has led us to organize this book primarily along the lines of the operating frequency of links, as outlined in Table 1.1. It also left one problem open: Some optional processes shown in Figure 2.2 are highly important in certain naval communications processes, for example AJ protection. Thus, we have included chapters and subsections dealing with such subjects as AJ communications and multiple access. In addition, we saw a need for chapters on selected subjects such as electromagnetic interference and compatibility (EMI/EMC), an introduction to satellite communications, antennas, protocols, and projected developments to round out this book and provide the needed system background.

Next, we present a brief look at the history of naval communications. Figure 2.3 shows the historical development of naval communications [NA-1]. As it can be seen, we are currently in an era of unprecedented technological advance. The last 40 years have brought unparalleled advances that are reflected in extraordinary changes in naval communications. In this book we attempt to describe the naval communications as they exist today, having incorporated these advances, and also project some of the future to come.

A very important precept of naval communications was the availability of certain circuits for certain operational needs under any circumstance, as dictated by the requirements of the cold war strategies. This lead to a fracturing of the communications with incompatibilities among the various naval communications links and with no possibility of sharing circuits. There is now a change in process where circuits may be shared, but with priorities in place that guarantee designated users access to specific assets for mission critical needs. Methods used to implement these changes that will introduce circuit sharing will be discussed toward the end of this book.

Figure 2.4 illustrates the current transition from separate systems to a common system. The early military communications services show a variety of links, circuits, networks, and communications methodologies. The expanded services show a great deal of networking and the establishment of common military nets. It also shows the trend to using nonmilitary government networks as well as commercial networks.

The development of naval communications reflects a similar pattern. First, the Navy has its own established circuits—they will be discussed in the chapters to come.

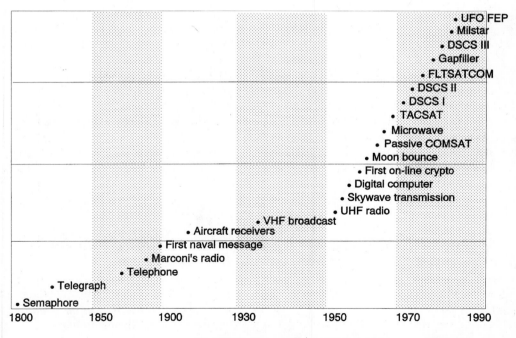

Figure 2.3   History of Naval Communications

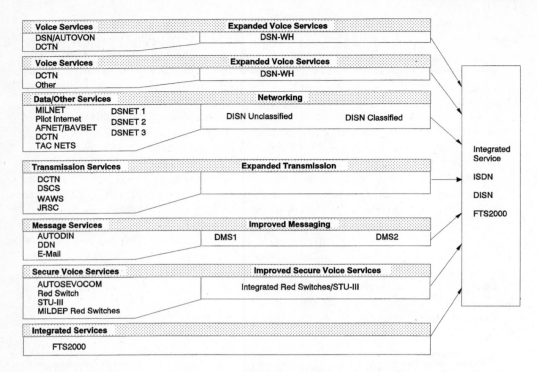

Figure 2.4   Transition of Military Communications Services

Second, the Navy participates in many of the services shown in Figure 2.4 and will continue to participate in them and their future developments. Finally, by introducing the new Copernicus Architecture for naval communications, the Navy plans to streamline its multiplicity of circuits, links, and services into a network-oriented operation.

The considerations discussed above represent the background of this book. While some technologies may be intriguing and could be subjects of books in their own right, the overall system, its shipboard components, and the overall naval communications operations were viewed as the guiding concepts in developing this book.

REFERENCE

[NA–1]   *The Copernicus Architecture—Phase I: Requirements Definition,* Copernicus Project Office, Department of the Navy, Washington, D.C. 20350-2000, August 1991.

3

Communications
in the Navy

The Defense Communications System (DCS) and the Naval Telecommunications System (NTS) support Navy Command, Control, Communications, and Intelligence (C3I) systems, which consist of shore command and support systems and afloat platforms (ships) [BE-1]. The DCS provides worldwide communications for all branches of the U.S. military forces. The NTS is a major component of the U.S. military communications systems; it provides connectivity for various commands and their subordinate elements. These include: (a) the Joint Chiefs of Staff (JCS), (b) Unified/Specified Commanders-in-Chief (CINCs), (c) Fleet CINCs, and (d) other services' support systems. Figure 3.1 shows the structure of NTS and its relationship to the DCS.

3.1 THE DEFENSE COMMUNICATIONS SYSTEM (DCS)

The DCS provides basic communications for worldwide command and control of U.S. military forces: (a) the major unified/specified commands in Europe, the United States, and the Pacific, (b) the land-based Navy command and support systems such as the Ocean Surveillance Information System (OSIS), the Anti-Submarine Warfare Operations Center (ASWOC), the Integrated Undersea Surveillance System (IUSS), and the Fleet Command Center (FCC).

There are four backbone networks in the DCS. The first is the Automatic Digital

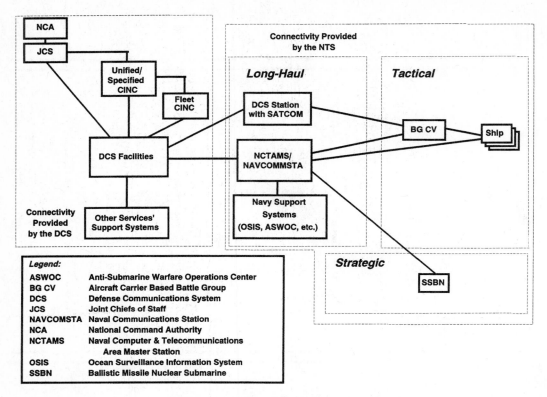

Figure 3.1   Structure of Naval Telecommunications System

Network (AUTODIN), which is a message-switching network, and the Defense Digital Network (DDN), a digital packet-switched network. The second is a circuit-switched voice network that contains the Automatic Voice Network (AUTOVON) and the Defense Switched Network (DSN). The third consists of a secure wideband (56 kbps) and narrowband (2400 bps) voice network; both are called the Automatic Secure Voice Communications System (AUTOSEVOCOM) and support 1500 AUTOVON subscribers. The fourth is a Super High Frequency (SHF) Satellite Communications (SATCOM) network called the Defense Satellite Communications System (DSCS). These worldwide DCS networks use both military and leased commercial satellites and landlines.

3.2 THE NAVAL TELECOMMUNICATIONS SYSTEM (NTS)

There are three components in the NTS: (a) strategic communications, (b) long-haul communications to and from afloat units, and (c) tactical communications among operating afloat units [RI-1].

Strategic communications.   The strategic communications component provides connectivity between the National Command Authority (NCA) and fleet ballistic missile submarine (SSBN) forces. Communications links using Low Frequency (LF) and Very Low Frequency (VLF) radios are employed since the SSBN fleet ballistic missile forces normally are submerged in order to avoid detection.[1]

Long-haul communications.   The long-haul communications component provides connectivity between the shore-based communications stations and afloat units at sea. Long-haul communications are bidirectional: information is sent from the shore to the afloat unit and the afloat units send information to Naval Communications Stations (NAVCOMSTAs) from which the information is distributed to other land-based facilities via the DCS. Figure 3.1 illustrates the long-haul NTS interconnections.

Information originating from a shore command is normally sent to Naval Computer and Telecommunications Area Master Stations (NCTAMSs) or NAVCOMSTAs via the DCS.[2] NAVCOMSTAs serve as hubs for afloat units and as communications relay nodes. NAVCOMSTAs process information received from the DCS for transmission via various circuits to afloat units. SATCOM and High Frequency (HF) links are used for beyond line-of-sight (BLOS) long-haul two-way communications between shore and afloat units. These BLOS long-haul communications are used to send broadcasts from major operational and administrative units and to provide a one-way path for promulgating command decisions. BLOS long-haul ship-to-shore links provide for status reports, information exchanges, and necessary acknowledgments to coordinate fleet operations.

Tactical communications.   The tactical communications components provide the connectivity among afloat units of the battle force/group during naval operations. Various afloat units include major combatants such as carriers, cruisers, amphibious ships, destroyers, carrier-based aircraft, and submarines. Typical transmission media for tactical communications include High Frequency (HF), Very High Frequency (VHF), and Ultra High Frequency (UHF). Extended Line-of-Sight (ELOS) and Line-of-Sight (LOS) communications maintain short-range and medium-range, ship-to-ship, ship-to-air, air-to-ship, and air-to-air links for the information exchange needed to coordinate the movements of individual fleet units and formations.

Computerized combat direction systems in ships and aircraft exchange vital sensor, fire control, and electronic warfare information via a number of high-speed tactical data links. Tactical radio-telephone systems back up these high-speed data links with voice communications, providing an extra measure of flexibility at the tactical level.

---

[1]VLF/LF submarine communications are unidirectional; HF and UHF satellite communications are bidirectional.

[2]Occasionally, a shore command may send time-critical information directly to an afloat node, bypassing NCTAMSs or NAVCOMSTAs.

3.3 LEVEL OF COMMUNICATIONS CAPABILITY AS A FUNCTION OF CONFLICT PERIODS

The naval telecommunications architecture defined by the NTS provides for three levels of communications service: normal standard communications, tactical communications, and minimum essential communications. Figure 3.2 shows the probability of occurrence of levels of conflict and illustrates the activities during the levels of conflict. As shown in Figure 3.3, the levels of communications are supporting four levels of conflict: peace, crisis, regional conflict, and global war [NA-5]. The correspondence is as follows: (a) normal standard communications support peacetime operations and crisis situations, (b) tactical communications support regional conflicts and some global war situations, and (c) minimum essential communications support all-out global war.

The NTS is designed to provide normal standard communications capabilities during peacetime routine information exchange for long-haul ship-to-shore communications. These capabilities are provided by UHF SATCOM, long-haul HF circuits, and LF/VLF communications. For Intra Battle Group communications, these capabilities are provided by HF ground wave, VHF circuits and nets, and UHF circuits and nets. Figure 3.3 shows the type of information transferred by the three levels of communications services in the NTS for both long-haul ship-to-shore and Intra Battle Group communications.

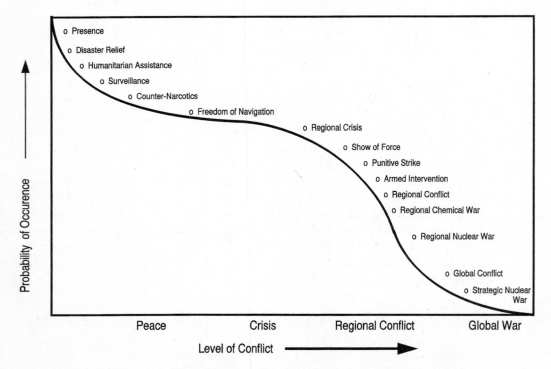

Figure 3.2   Probability of Occurrence of Level of Conflict

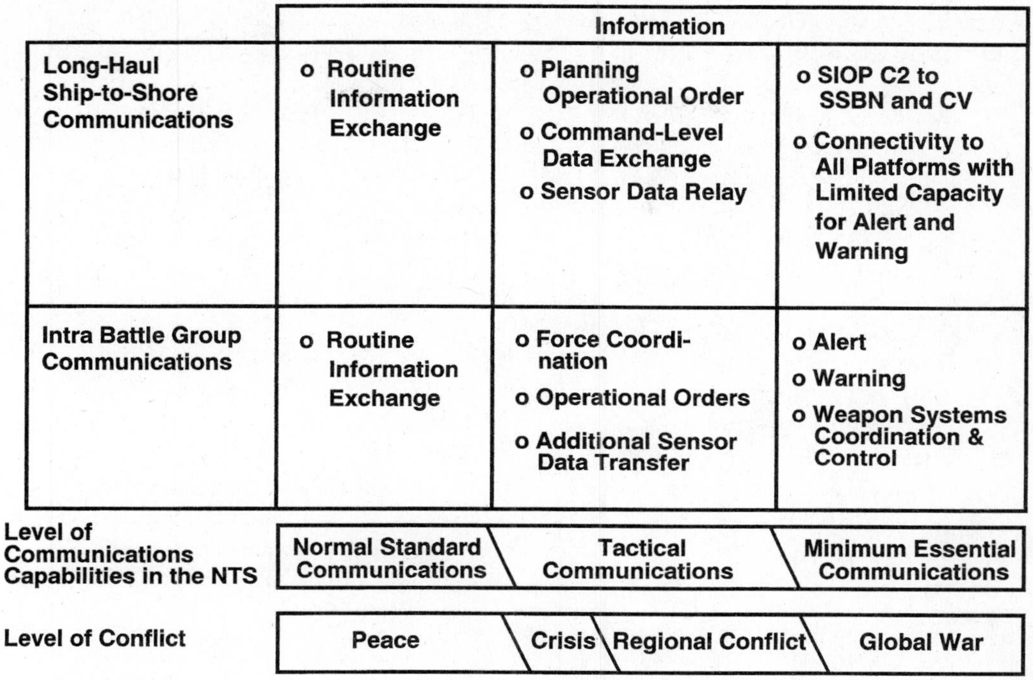

	Information		
Long-Haul Ship-to-Shore Communications	o Routine Information Exchange	o Planning Operational Order o Command-Level Data Exchange o Sensor Data Relay	o SIOP C2 to SSBN and CV o Connectivity to All Platforms with Limited Capacity for Alert and Warning
Intra Battle Group Communications	o Routine Information Exchange	o Force Coordination o Operational Orders o Additional Sensor Data Transfer	o Alert o Warning o Weapon Systems Coordination & Control
Level of Communications Capabilities in the NTS	**Normal Standard Communications**	**Tactical Communications**	**Minimum Essential Communications**
Level of Conflict	**Peace**	**Crisis Regional Conflict**	**Global War**

Figure 3.3 Levels of Communications Capabilities in NTS

The NTS is also designed to provide tactical communications during crises, regional conflicts, and global wars. For long-haul ship-to-shore communications, the NTS supports the information transfer of planning and operational orders, command-level data exchange, and sensor data relay. For Intra Battle Group communications, the NTS supports the information transfer of force coordination, operational orders, and additional sensor data.

With battle damage and reduced communications capabilities during global war, the NTS provides minimum essential communications to maintain traffic necessary to effectively support all levels of command and control of forces at sea reduced by losses. These circuits must provide connectivities for all platforms with limited capability for alert and warning. This capability is provided by the Minimum Essential Emergency Communications Network (MEECN). The minimum essential communications circuits provide information transfer of the Single Integrated Operational Plan (SIOP) [RI-1; p. 73] for strategic ballistic missile nuclear-powered submarines (SSBN).

For Intra Battle Group communications, minimum essential communications are provided to transfer alert warning, and weapon systems coordination and control information. As an illustration, the tactical alert warning weapon systems coordination and control information may be transferred via secure voice conferencing in normal standard communications usually operating via SATCOM at a rate of 2400 bps. With reduced

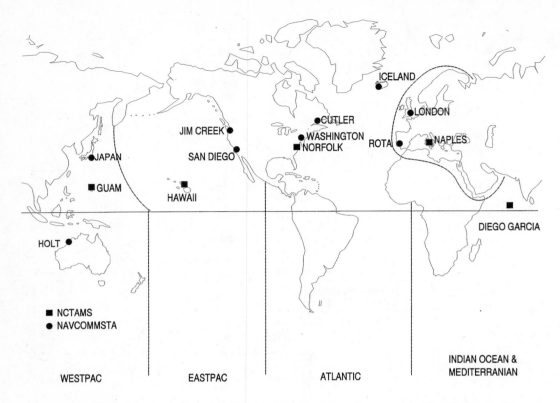

Figure 3.4   Areas Covered by Naval Telecommunications System and Its Navy Communications Stations (Adapted from Bell and Conley, © 1980, IEEE; Reprinted with permission of IEEE.)

communications capabilities, secure voice conferencing falls back to teletypewriter conferencing via SATCOM operated at a rate of 75 bps.

3.4 NAVAL COMMUNICATIONS STATIONS

Naval communications stations provide the on-shore capability needed to support the full range of naval communications.

3.4.1 Fleet Communications

Fleet communications are supported by four NCTAMSs. The primary function of the NCTAMSs is to support naval operational command and control and intelligence functions. The division of Fleet Communications into four regions is intended to streamline this function. The NCTAMSs provide area coordination in the Western Pacific, Eastern Pacific, Atlantic, and Mediterranean regions. In addition to the NCTAMS, the Navy maintains several subordinate NAVCOMSTAs, that are controlled by the NCTAMS in its area. Figure 3.4 shows the location of the various sites [BE-1].[3]

As shown in Figure 3.1, NCTAMSs are gateways for information transfer between shore command facilities and afloat units. Within this structure, long-haul communications are provided by a geostationary satellite system; footprints of the satellites cover each region, and each region has a satellite control station.[4] Other long-haul communications are provided by HF and LF that do not depend on footprint regions. NCTAMSs also have a major role in providing HF communications worldwide that are controlled by regional communications stations.

Navy fleet command centers use the DCS worldwide strategic network for information transfer to and from the fleets via NCTAMS and NAVCOMMSTA. For example, fleet broadcast messages originating from a fleet command center are sent to an NCTAMS via the DDN. They are formatted and processed at the NCTAMSs by the Naval Communications Processing and Reporting System NAVCOMPARS and then transmitted to afloat units either via Fleet Satellite Communications (FLTSATCOM) channels or HF circuits.

Subordinate naval communications stations are also associated with fleet activities. These stations are smaller, and act as a backup for the NCTAMS in that region for terrestrial networks and satellite communications.

3.4.2 Functions of the NCTAMSs and NAVCOMSTAs

The major functions of the NCTAMSs and NAVCOMSTAs are: (a) radio transmitter and receiver operations, (b) satellite transmitter and receiver operations, (c) communications security (COMSEC) operations, (d) network connection, (e) network management, (f) message processing, and (g) technical control [NA-2]. Table 3.1 represents an overview of these functions. Figure 3.5 is a functional block diagram of an NCTAMS; this block diagram structure also applies to NAVCOMSTAs.

Radio transmitter and receiver operations. An NCTAMS and NAVCOMSTA is responsible for to operating radio transmitters and receivers in Medium Frequency (MF), HF, VHF, and UHF bands. LF/VLF stations for submarine communications are operated separately by Submarine Operational Authorities (SUBOPAUTHs).

HF transmitters and receivers are usually at locations remote from NCTAMSs and NAVCOMSTAs and are controlled by technical control operators. VHF and UHF transmitters and receivers are located at the NCTAMS and NAVCOMSTA.

Satellite transmitter and receiver operations. An NCTAMS and NAVCOMSTA is also responsible for operations of SHF and UHF satellite communications transmitters and receivers [NA-1; NA-3]. In a typical NCTAMS, SATCOM equipment installations are located in three places: the NCTAMS building (technical control

---

[3]The figure shown here has been adapted to reflect the current locations of the NCTAMSs and NAVCOMSTAs.

[4]This will be shown in detail in Chapters 13 through 15.

TABLE 3.1 MAJOR FUNCTIONS OF NAVAL COMMUNICATIONS STATIONS

Function	Capabilities for Supporting Operations
Radio transmitter and receiver operations	• Terrestrial radio (MF, HF, VHF, UHF) transmitter and receiver • VLF/LF transmitter • Antenna site
Satellite transmitter and receiver operations	• SHF SATCOM terminals • UHF SATCOM terminals • EHF SATCOM terminals • Commercial SATCOM terminals (INMARSAT, INTELSAT, etc.)
COMSEC operations	• Cryptographic equipment • Key management
Network connection	*Ashore* • DDN gateway (LDMX) • DSN gateway • PSTN gateway • NCTAMS LAN • Pier LAN • Defense secure voice gateway • DSCS gateway • FLTSATCOM gateway *Afloat* • Broadcast (NAVCOMPARS) • HF radios • Shipboard LAN
Network management	• Tactical CUDIXS • Special intelligence and general service segregation • Special subscriber expansion • Shipboard net control
Message processing	• Message preparation and delivery • NATO and allied interoperability
Technical control	• Switching • Patching • Testing • Cryptographic operations and maintenance

and UHF and SHF baseband equipment), the satellite communications facility (SHF transmitter and receiver), and the naval communications center.[5]

COMSEC Operations.   An NCTAMS and NAVCOMSTA also performs COMSEC operations. COMSEC operations are carried out in a specially secured area or vault. Encryption and decryption operations are carried out as follows. Incoming en-

---

[5]Details are given in Chapter 13 (UHF Satellite Communications) and Chapter 14 (SHF Satellite Communications).

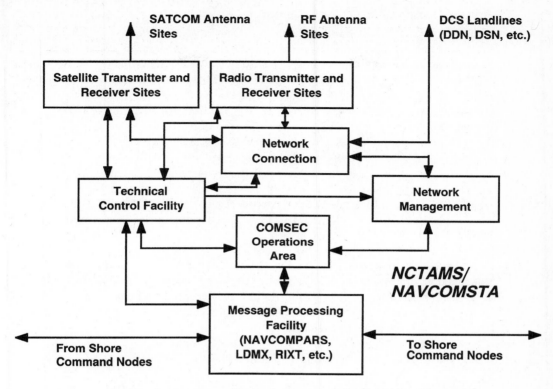

Figure 3.5   Functional Block Diagram of an NCTAMS or NAVCOMSTA

crypted signals (referred to as BLACK signals) from RF equipment, SATCOM equipment, and landlines are terminated at the main distribution frame (MDF). At this point signals are in the baseband. These encrypted signals from MDFs are routed to multiplexers, modems, and other baseband equipment, and each individual digital signal is reconstructed. The encrypted signals then are routed to the COMSEC area for decryption and distribution to the end users.

Outgoing unencrypted signals (referred to as RED signals) are encrypted by the COMSEC equipment in the vault. The encrypted signals (BLACK signals) are routed to multiplexers, modems, and other baseband equipment. Encrypted signals are routed to the MDF and sent to RF equipment, SATCOM equipment, and landlines for transmission.

Network connections.   Another function of an NCTAMS and NAVCOMSTA is maintaining network connections, which is accomplished by switching and patching. Signals are switched at NCTAMSs and NAVCOMSTAs by user request. For that purpose, an NCTAMS and NAVCOMSTA has equipment to perform circuit, message, and packet switching.

Circuit switching establishes a dedicated communications path between two locations. The circuit-switching equipment includes electromechanical or electronic switches, which are generally applied in establishing voice and real-time circuits.

Message switching is a means of transferring information by routing a message based on an address contained in the header of a message. The difference between message and circuit switching is that message switching does not need a dedicated communications path. Examples of messages are telegrams, record messages, electronic mail, and computer files, which are handled by a Local Digital Message Exchange (LDMX) and transmitted via DCS AUTODIN.

Packet switching is a system by which messages are broken down into smaller units called packets, which are then individually addressed and routed through the network. A packet is typically a few thousand bits long. Packet switching uses equipment that separates the messages into packets that are transmitted to their destination; each packet may take a different path. At the destination the packets are reassembled into the original message. Each NCTAMS and NAVCOMSTA has an entry point to the DDN packet-switched network.[6]

Technical control.   An NCTAMS and NAVCOMSTA performs technical control functions for communications circuit testing and patching. Testing of circuits and equipment is necessary for efficient operation of the station. Patching is a means for establishing a path via existing routes and allows rapid recovery of circuits when a malfunction is detected by swapping circuits or equipment.

3.4.3 Traffic Volume at a NCTAMS and a NAVCOMMSTA

The volume of record messages and other traffic handled by NCTAMSs and NAVCOMSTAs ranges from 2000 to 12,000 per day in normal peacetime conditions and in support of shore gateways for the fleet. In crisis and wartime conditions, the message volume at an NCTAMS and NAVCOMSTA could increase as much as five times the normal, peacetime traffic volume. A typical major combatant ship handles 5000 messages per day. Table 3.2 shows a typical traffic volume per day that is handled by major and smaller naval communications stations and afloat units [NA-4].

TABLE 3.2   TYPICAL TRAFFIC VOLUME PER DAY AT
VARIOUS NAVAL TELECOMMUNICATIONS NODES

Communications nodes	Messages per day
Major NCTAMSs	12,000
Small NAVCOMMSTAs	2800
Major combatant ships	5000
Small ships	1200

[6]Details of technical control are discussed in Chapter 18.

REFERENCES

[BE–1] Bell, C. and R. E. Conley, "Navy Communications Overview," *IEEE Transactions on Communications,* Vol. COM-28, No. 9, September 1980, pp. 1573–1579.

[NA–1] *Navy SHF Satellite Communications System Description,* NSHFC 301, NOSC (now NCCOSC RDT&E Div), San Diego, CA 95152-5185, August 1, 1984.

[NA–2] *Naval Communications Station Design,* NAVELEX Shore Electronics Criteria, 0101,102, Naval Electronic Systems Command (now Space and Naval Warfare Systems Command), Washington, D.C. 20363-5100, June 1970.

[NA–3] *Navy UHF Satellite Communications System Description,* FSCS-2000-83-1, NOSC (now NCCOSC RDT&E Div), San Diego, CA 95152-5185, April 15, 1991.

[NA–4] *Communications Automation Follow-on Effort (CAFE) ,* Technical Note, Space and Naval Warfare Systems Command, Washington, D.C. 20363-5100, August 1985.

[NA–5] *The Copernicus Architecture—Phase I: Requirements Definition,* Copernicus Project Office, Department of the Navy, Washington, D.C. 20350-2000, August 1991.

[RI–1] Ricci, F. J. and D. Schutzer, *U. S. Military Communications—A C3I Force Multiplier,* Computer Science Press, Rockville, MD 20580, 1986, p. 73.

4

Naval Warfare and Its Communications Connectivities

War at sea requires that each U.S. Navy battle group mutually supports its elements (air, surface, and subsurface) following the principle of defense in depth, that is, different weapons are used to engage targets at various distances from the battle group [PH-1].

4.1 NAVAL WARFARE

Naval battle group operations are divided into distinct naval warfare functions, that is, Anti-Air Warfare (AAW), Anti-Surface Warfare (ASuW), force Anti-Submarine Warfare (ASW), theater ASW, strategic submarine warfare, Strike Warfare (STW), Amphibious Warfare (AMW), and Space and Electronic Warfare (SEW). In a battle group each warfare function has a commanding officer who is responsible for carrying out the warfare function in support of the battle group. Figure 4.1[1] shows the complex ensembles of ships, aircraft, and submarines employed to support a naval battle group.

[1]This figure is reproduced, with permission, from Chester C. Phillips, "Battle Group Operations: War At Sea," *Johns Hopkins APL Technical Digest,* Vol. 2, No. 4. (1981), p. 300. Copyright 1981 by The Johns Hopkins University Applied Physics Laboratory.

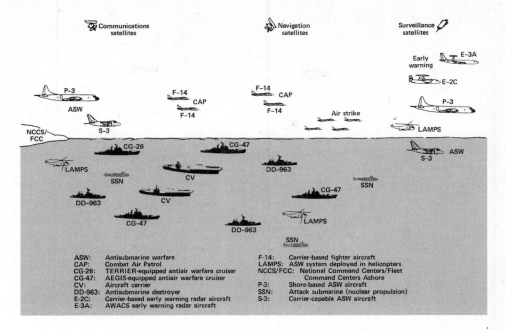

Figure 4.1 Typical Composition of a Naval Battle Group

4.1.1 Anti-Air Warfare (AAW)

The rise of air power in the 1930s added a new dimension to naval warfare by extending the offensive battle space. The task of AAW functions is the defense of a battle group against attack aircraft, bombers, and air-launched and sea-launched cruise missiles [KI-1; p. 9]. Today, the carrier-based aircraft represent one of the most important means of naval power projection, whether exercised in friendly or contested waters. But as unfriendly offensive air power grows, the task of defending a battle group also becomes more difficult. The carrier is a platform from which strike aircraft, Airborne Early Warning (AEW) aircraft, and combat aircraft patrols (CAPs) are launched. Other afloat platforms of the battle group AAW assets include screen and surface action units and picket ships. In the early days of AAW, the threat to a battle group was limited to unfriendly carrier-based aircraft with ranges limited to 100 km to 150 km (50 to 80 nmi). Today, however, threats to a battle group include not only enemy combat aircraft but also sea-launched and air-launched missiles, as well as long-range shore-based bomber forces. As a result, the range of the AAW threat has been extended to 400 km (220 nmi) and beyond.

The AAW concept is based on three defensive zones: the outer defense zone, the area defense zone, and the self-defense zone [WI-1]. In the outer defense zone, manned aircraft intercept the attackers at long range. Attackers surviving the outer defense zone are countered in the area defense zone by surface-to-air missiles (SAMs). Remaining sur-

TABLE 4.1   THE CONCEPT OF BATTLE GROUP AAW

	Defense Zones		
	Outer Defense	Area Defense	Self-Defense
Threat	• Engage enemy at long range (prior to missile launch) • Counter enemy coordination • Jam enemy radar	• Engage large, difficult threat	• Engage residual threat
Sensor	• Long-range aircraft • Surveillance satellite	• Medium-range surveillance radar	• Short-range detection radar
Threat counteraction	• Interceptor aircraft	• SAM	• Short-range weapons

vivors are then engaged by short-range weapons in the self-defense zone. This is summarized in Table 4.1 [FA-1].

Command, control, and communications provide coordination within and between the defensive zones. The connectivity of various AAW elements is illustrated in Figure 4.2, where the composition of a typical carrier battle group is shown. Assignment of the AAW command responsibility is usually given to the most capable surface action unit or group, one that can provide necessary command attention, operational control space, and required facilities. Such an assignment avoids the need for the AAW commander to compete for limited space in a carrier. Drawbacks to locating the AAW commander in a carrier include the recurring need for restrictive emission control (EMCON)[2], and the limitation of the numbers of operator consoles and other facilities in the carrier.

Due to the nature of the threats (i. e., aircraft, land-based bombers, submarine-launched cruise missiles (SLCMs), etc.), the reaction time for AAW must be on the order of a few seconds to tens of seconds. This is a very short reaction time; other warfare functions have longer reaction times, as described in the sections that follow.

Various tactical circuits required for AAW operations are listed in the table inserted in Figure 4.2. These tactical circuits are required to coordinate the interactions among defense zones, either positive for passing information between zones, or negative for friendly aircraft operating or transiting within the surface-to-air missile defense zones and in the overlap between the area and self-defense zones. Coordination of defense in all zones has substantial benefits, but the real-time coordination of area defense weapons is mandatory because of the limited time available to deal with penetrating attackers, the distance between battle group ships, and the overlap of defensive weapon coverage.

---

[2]EMCON is the control of electromagnetic emission from a naval unit afloat.

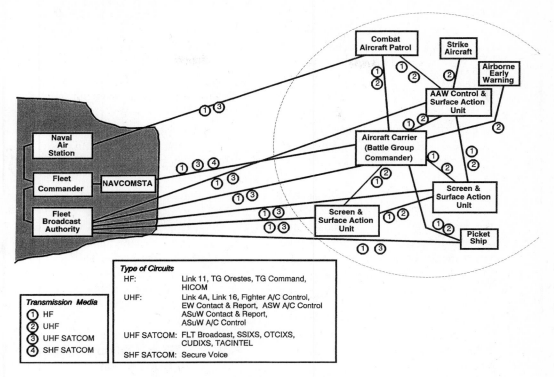

Figure 4.2   AAW Connectivity

4.1.2 Anti-Surface Warfare (ASuW)

The task of Anti-Surface Warfare (ASuW) is the defense of a battle group from surface combatant ships. The ASuW function is performed from a surface situation plot in the carrier combat direction center where there is an access to Naval Tactical Data System (NTDS) information. The ASuW function controls various strike assets, such as battle group aircraft and ship-launched missiles (e. g., Tomahawk and Harpoon), or attack submarines, assigned to support the battle group directly. Exterior communications systems aboard the carriers receive advanced threat warning information from intelligence sources for ASuW use. As organic sensors[3] gain contact with surface threats, they are displayed on the surface situation plot NTDS console. Access to data concerning the status and location of carrier-based aircraft is provided to ASuW via exterior communications systems (tactical data links, UHF SATCOM information exchanges subsystems (IXS), SHF SATCOM circuits, and others).

---

[3]An organic sensor is a radar, sonar, or other surveillance system which is part of a ship or a battle group.

Figure 4.3 shows the connectivity of various elements in the ASuW function. The connectivity is very similar to that of AAW since the enemy's combined air and surface ship attacks are generally launched against a battle group. In ASuW, there are no postulated defensive zones as in the case of AAW. The battle group deals with one large defense zone. However, primary threats are over-the-horizon weapons such as surface SLCMs. The reaction time in ASuW is on order of tens of minutes, allowing the defender to detect, identify, and track the surface ships and their attack weapons. ASuW operations are primarily defensive in nature, but they can also be offensive operations to neutralize the surface ship attackers by gunfire, or, depending on distance, by cruise missile. A list of tactical circuits required for ASuW mission support is shown in Figure 4.3.

4.1.3 Force Anti-Submarine Warfare

The function of force Anti-Submarine Warfare (ASW) is the defense of a battle group from enemy submarines. Threats to the battle group by an attack submarine are one of the most serious problems of naval warfare. The defense for a battle group becomes more difficult as the attacker's submarine becomes quieter.

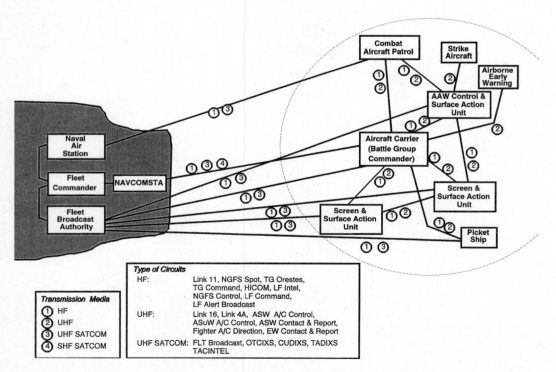

Figure 4.3   ASuW Connectivity

Force ASW is an extension of overall ASW operations in a theater. The task of area ASW is the total ASW operations in an ocean. When a battle group is in contested waters, force ASW must counter the threats within a range of 900 km (500 nmi). The force ASW concept is based upon three defensive zones. In the outer defense zone, the shore-based area ASW provides information to intercept the submarine attack at long range. Submarine attackers surviving the outer zone are countered in the middle defense zone by destroyers and their ASW weapons systems and helicopters. The surviving submarines are then engaged by inner defense weapons in the inner zone.

Command, control, and communications provide coordination within and between the ASW defense zones. The connectivity of various elements in the ASW function is shown in Figure 4.4. A typical carrier battle group's composition is shown in the figure. The force ASW function can be headed by a designated destroyer commander in a carrier battle group or by an ASW commander aboard a carrier. In the first case, the designated destroyer commander performs the force ASW function from a situation plot which provides a display of hostile and friendly submarine information. In the second case, the ASW commander aboard a carrier performs the force ASW function from a combat information center (CIC) which provides a similar display as in the destroyer.

The reaction time for ASW is on the order of tens of minutes to several hours,

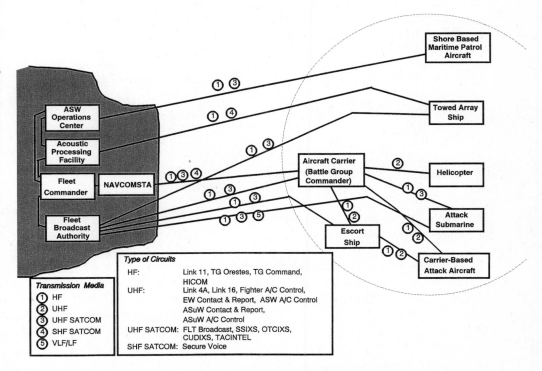

Figure 4.4   ASW Connectivity

mostly due to the speed of attacking submarines. This is in marked contrast to the reaction time requirements in AAW or ASuW as discussed in sections 4.1.1 and 4.1.2.

Tactical circuits required for ASW operations are listed in the table inserted in Figure 4.4. These tactical circuits are required to coordinate the interactions among units of the battle group. These tactical circuits include exterior communications (EXCOMM) systems (such as tactical data links, UHF SATCOM IXS, and SHF SATCOM circuits), which are used to coordinate, control, and exchange tactical situation information among battle group ASW assets.

The battle group receives overall theater situation information from the shore-based Fleet Commander (FLTCINC) via the Tactical Data Information Exchange Subsystem (TADIXS) [ME-1]. Shore-based maritime patrol aircraft (VP) in support of a battle group establish two connectivities: the first is a circuit to the ASW operations center (ASWOC)[4] for indirect communications, and the second is a circuit to the battle group for direct communications.

Surveillance Towed Array Sensor System (SURTASS) ships are used for the outer defense. Their connectivity to the battle group is via SHF SATCOM to the shore-based data processing center, which in turn sends relevant information to the battle group for ASW operations. Direct linking between the SURTASS platforms and battle group can be accomplished by HF radios and other existing EXCOMM systems. Attack submarines (SSN) can be used for outer defense against enemy submarine threats. This mission for submarines is called the submarine direct support. Submarine connectivity to the battle group and other units within the battle group such as Maritime Patrol Aircraft (MPA) of the VP class, and search and attack aircraft of the VS class,[5] is supported via the FLTSATCOM Submarine Satellite Information Exchange Subsystem (SSIXS), HF surface wave radios, acoustic communications, and blue-green laser communications [NO-1].

Light Airborne Multi-Purpose System (LAMPS) helicopters are also used in ASW missions. LAMPS helicopters can deploy both active and passive sonars to detect, classify, and track enemy submarines. They communicate with the battle group via VHF LOS circuits and Link 11 [SC-1].

Carrier-based search and attack aircraft (VS) assets are used in ASW missions. Their connectivity to the battle group is accomplished by Link 11, VHF communications, Link 4A, and other voice communications links.

Other escort ships such as destroyers and cruisers can also take part in ASW missions. Their connectivity to the battle group is via Link 11 and other existing EXCOMM systems.

4.1.4 Strike Warfare (STW)

The Strike Warfare (STW) function is carried out through the offensive operations of a battle group against enemy afloat forces. Air wings and a ship's battle staff primarily support the STW function from an aircraft carrier tactical flag command center (TFCC). The STW system

---

[4]The ASW operations center is also known as the tactical support center.

[5]VS designates the carrier-based submarine search aircraft; VP designates the land-based surveillance and patrol aircraft.

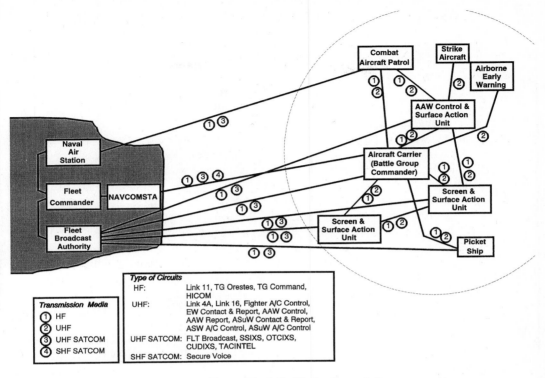

Figure 4.5   Strike Warfare Connectivity

focuses on accepting information from various sources and distributing this information to other watch stations. Information exchanged includes battle group launch plot data, order of battle and targeting data, Electronic Intelligence (ELINT) data, and Tomahawk planning data.

The connectivity of various elements in STW is shown in Figure 4.5. The STW segment composition is very similar to that of ASuW. The STW functions that support the STW commander are: (a) Over-The-Horizon (OTH) data correlation and fusion, (b) ground force Order of Battle (OOB), (c) land-based ELINT information, (d) tactical air mission planning, (e) airborne Electronic Counter Measure (ECM) mission planning,[6] and (f) Tomahawk land attack missile planning. A list of tactical circuits required for STW is included in Figure 4.5. The individual circuits are described in the previous paragraph.

4.1.5 Amphibious Warfare

Amphibious warfare as a primary means of naval warfare was developed prior to World War II and has been continually refined [EV-1]. Amphibious Warfare (AMW) is defined as an attack launched from the sea by naval and landing forces embarked in ships or craft

---

[6]The ECM is coordinated with the SEW commander.

and landing on a hostile shore. It normally requires extensive aircraft participation and is characterized by closely integrated efforts of forces trained, organized, and equipped for different combat functions both ashore and at sea.

There are two command, control, and communications (C3) functions involved in typical amphibious operations: (a) the Amphibious Task Force (ATF) and (b) the Landing Force (LF). A typical ATF consists of: (a) a primary control ship, (b) a secondary control ship, (c) landing craft, (d) transport ships, (e) a screen and surface action group, (f) mine sweepers, and (g) AAW/fire support/ASW ships. During the transit phase from sea to the beach, the overall C3 is the responsibility of the ATF commander. The LF is an embarked force and does not become active until it lands in an amphibious objective area. When the LF is on the beach, C3 becomes the responsibility of the Commander, Landing Force (CLF).

The connectivity of various elements in AMW is shown in Figure 4.6. Illustrated in the connectivity diagram is the landing force when it has landed on the beach. This connectivity must support: (a) CLF, (b) the beach master, (c) the Tactical Air Control Center (TACC), (e) aircraft for close air support, and (f) the gunfire support center.

Various tactical circuits required for AMW operations are listed in the table inserted in Figure 4.6. These tactical circuits are required to coordinate the Commander, Amphibi-

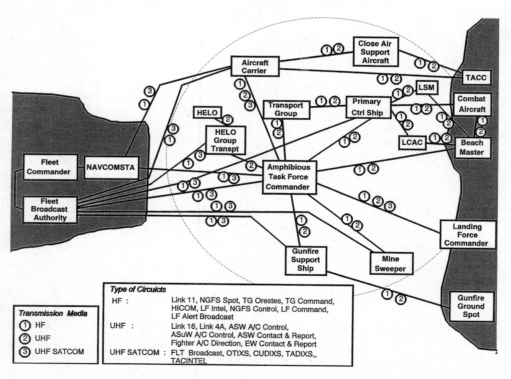

Figure 4.6 Amphibious Warfare Connectivity

ous Task Force (CATF) and the CLF. The number of circuits required to support an amphibious operation exceeds that required for any other type of naval warfare on the order of one hundred or more [PI-1].

4.1.6 Space and Electronic Warfare

Another warfare mission area in the U.S. Navy is the Space and Electronic Warfare (SEW). SEW contributes to Navy missions to gain control of space and the electromagnetic spectra, and to deny or control the enemy's use of them, and to project power by conducting offensive warfare in the space and electromagnetic spectrum dimensions. SEW is an outgrowth of several elements of naval warfare mission areas. Ever since RF communications and radars were introduced into naval operations, the enemy's electronic communications and radar signals provided opportunities for intelligence gathering by exploiting them, and the concealment and deception of one's own RF signals became the prime ingredient of surprise. These elements are communications and radar electronic warfare, and space-based and terrestrial surveillance systems.

SEW includes tactical disciplines that support other warfare mission areas such as AAW, ASuW, ASW, etc., and warfare functions that are the SEW mission itself. Warfare support disciplines are: (a) operational security, (b) surveillance, (c) command, control, communications, and intelligence (C3I), and (d) signal management. Warfare disciplines of SEW are: (a) operational deception, (b) counter-surveillance, (c) counter-C3I, and (d) electronic combat. The strategic objective of SEW is to separate the enemy leader from his forces, to render him remote from his subordinates (in effect, to take command of his forces), and to control his use of the electromagnetic spectra [NA-1]. Table 4.2 summarizes the characteristics of these eight SEW functions.

The SEW target set consists of those systems which, when destroyed, yield the strategic objective of the separation of enemy command and the denial of using the electromagnetic spectra. The target set includes the electronic capabilities of the enemy leadership at all levels including the battlefield level, its communications systems, surveillance and targeting systems, information processing, and decision and display systems, electronic combat systems, and weapons guidance systems.

The communications connectivity to support SEW and existing warfare mission areas (AAW, ASuW, ASW, STW, and AMW) are the Global Information Exchange Systems (GLOBIXSs) and the Tactical Data Information Exchange Systems (TADIXSs). The GLOBIXSs are ashore networks, which consist of the Defense Communications Systems (DCS), and terrestrial and satellite sensor nets. The TADIXSs are afloat networks, which consist of HF, UHF LOS, UHF SATCOM, SHF SATCOM, and EHF SATCOM circuits [NA-2].

4.2 Strategic Submarine Connectivity

The strategic submarine connectivity provides the one-way communications of information to the Fleet Ballistic Missile (FBM) submarines. Fleet ballistic missiles are one element of the U.S. strategic force triad, the other two being Intercontinental Ballistic Mis-

TABLE 4.2   CHARACTERISTICS OF SEW AND SEW SUPPORT FUNCTIONS

SEW		Operational Task
Warfare Support Functions	Operational security	• Measure taken to minimize hostile knowledge of ongoing and planned military operations: - Physical security - Counterespionage - Personnel security
	Surveillance	• Tactical management of all technical surveillance as a force system across the entire battle space. • Surveillance assets (national, theater and organic): - Radars - Sonars - Space-based sensors
	C3I	• Technological, organization, and doctrinal system that provides three functions: - Doctrinal delegation of forces - Information management - Intelligence dissemination
	Signal management	Measures to protect forces signals: - Frequency management - Signal security - Communications security - Computer security - Transmission security - Emission control management
Warfare Functions	Operational deception	Influence enemy plans, execute a stratagem, induce reaction over a short period, and apply pressure to act
	Counter surveillance	Deceive, degrade, evade, and attack sensors and sensor platforms - Jamming and saturation - Evasion - Diversion - Destruction
	Counter C3I	Deceive, delay, degrade, or destroy elements of hostile C4I - Link from sensors to weapon carrier - Command and control links and node
	Electronic combat	Coordination of all offensive and defensive measures across the force to provide counter-weapon protection to the forces: - Counter targeting - Counter platform - Counter weapon and terminal phase - Jamming - Destruction

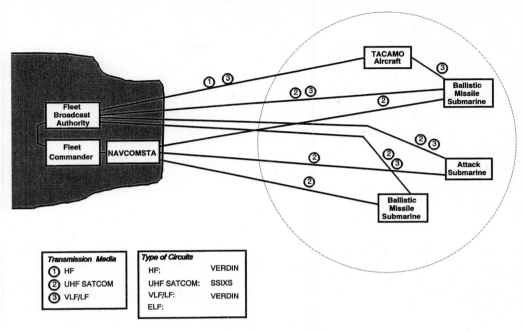

Figure 4.7   Strategic Submarine Connectivity

siles (ICBMs) and manned bomber forces. The National Command Authority (NCA) sends one-way Emergency Action Messages (EAMs), which contain strategic firing orders, to strategic ballistic missile forces [RI-1].

The connectivity of various elements in strategic submarine warfare is shown in Figure 4.7. EAMs are sent from the NCA to the nuclear ballistic missile submarine (SSBN) via various redundant transmission media: (a) LF/VLF/ELF [BU-1], (b) VLF/LF via airborne TACAMO aircraft, (c) UHF/SHF/EHF satellite communications, and (d) HF [WE-1].

Normally, VLF/LF circuits carry EAMs for submerged SSBN in depths of up to 5 m (16 ft.). The Submarine Broadcast Control Authority relays EAMs from the NCA to VLF/LF transmitting sites. As alternate means, the Submarine Broadcast Control Authority can use an airborne TACAMO aircraft, which relays EAMs to the submerged SSBN [BA-1]. In certain circumstances, UHF/SHF/EHF satellite communications can transmit EAMs to the SSBN [BO-1]. SSBNs are also equipped with HF receivers for copying[7] broadcast EAMs.

Although long-wave (VLF and LF) and ELF radio communications and UHF/SHF/EHF satellite communications may improve present submarine radio communications, they cannot provide the capabilities of high data-rate covert transmission and reception at any depth and speed. New methods of optical communications are being de-

[7]Copying is a naval communicator's expression for receiving clearly.

veloped providing better capabilities than VLF/LF and ELF radio and satellite communications do. The principal advantage of optical communications is the ability of blue-green light to penetrate seawater to a depth of about 700 m (2100 ft) [CA-1]. The principle of the optical method is to send blue-green laser light from a transmitter or relay satellite. The light passes through an outer filter and is received by the submarine via a hull-mounted sensor.

4.3 NAVAL SHIPBOARD EXCOMM CONNECTIVITY

Naval communications connectivity between the following four types of entities is described below: (a) shore facilities, (b) ships, (c) aircraft, and (d) submarines. Table 4.3 provides examples for each connectivity.

4.3.1 Shore-to-Shore Connectivity

The shore-to-shore connectivity for the Naval Telecommunications System is not a part of shipboard communications and is outside the scope of this book. However, shore-to-shore connectivity is essential in support of a battle group. Therefore, we briefly discuss here how shipboard communications are extended on shore. Most of the shore-to-shore connectivities in support of a battle group are point-to-point, long-haul communications. In the Copernicus Architecture these shore-to-shore connectivities are called the GLOBIXS.

An example of shore-to-shore connectivity in support of a battle group is the information exchange between two or more shore stations in the North Atlantic in support of

TABLE 4.3   EXAMPLES OF NAVAL SHIPBOARD EXCOMM CONNECTIVITY

	Shore	Ship	Aircraft	Submarine
Shore	• SDS • GLOBIXS • IUSS Communications	• FLT BROADCAST • OTCIXS • TADIXS	• ASWOC • HF Secure Voice	• Verdin • SSIXS
Ship	• Administration • Logistics • TG Orestes • Secure Voice	• PRITAC • Link 11 • Link 16 • Link 4A • Bridge-to-Bridge	• ATDS • Approach/ Departure Control • Aircraft Emergency	• STDL • SSIXS
Aircraft	• ASW Acoustic Data • CAS	• Sensor data • C3I message • LAMPS data link	• Intercept Control • Strike Control • Aircraft Emergency	• Tactical Coordination
Submarine	• Report-back	• Coordination in Direct Support	• Submarine Tactical Data Link (STDL)	• Tactical Coordination • Call-up

force-level ASW. For example, shore-based nodes are the Integrated Undersea Surveillance Systems (IUSS) facilities located in Norfolk, VA, and the United Kingdom. The correlation of collected information by the IUSS greatly enhances the detection probability of submarines, which in turn assists the force-level ASW. Another example is the shore-based nodes in the North Pacific in support of AAW. Location information and information on the composition of an opposing surface action group and aircraft are processed at these nodes. This information is vital for the defense of a battle group during its approach phase when it moves into a forward, contested ocean area. Data are collected at shore sites and then transferred to a fleet intelligence fusion center. The fused intelligence data are then transmitted from the fleet intelligence fusion center to a battle group where they are used for AAW threat analysis.

4.3.2 Shore-to-Ship Connectivity

The shore-to-ship connectivity provides primarily information exchange from a NCTAMS or NAVCOMSTA to a ship. Examples include the FLTBROADCAST, OTCIXS, and direct dissemination circuits from C3I nodes to ships. Most of the shore-to-ship connectivity is via long-haul circuits operating in broadcast, multicast, and point-to-point communications modes. In the Copernicus Architecture (see Chapter 20), the ship-to-shore connectivity is provided by the TADIXS circuits.

4.3.3 Shore-to-Aircraft Connectivity

Shore-to-aircraft connectivity is used for information exchange from shore-based ASWOC to their shore-based patrol aircraft (VP), such as the P-3C Orion. The most frequently used communications service is secure voice. Most of the shore-to-aircraft connectivities are either short-haul or long-haul point-to-point communications.

4.3.4 Shore-to-Submarine Connectivity

Shore-to-submarine connectivity is used for information exchange from a shore broadcast station to either submerged attack submarines (SSNs) or ballistic missile submarines (SSBNs). The connectivity is provided either by ELF/VLF/LF or FLTSAT SSIXS circuits. The ELF/VLF/LF circuits are used for broadcasting messages to submarines via the Verdin system; the SSIXS circuits are used for point-to-point communications. Both types of circuits are long-haul circuits.

4.3.5 Ship-to-Shore Connectivity

The ship-to-shore connectivity is used for two primary naval warfare applications. The first is most common, and provides the communications from navy ships to shore stations in support of naval warfare as well as administrative and logistics activities and general tactical communications via TG Orestes. The other provides the communications for the landing forces that have come ashore by amphibious ships in amphibious operations.

In the first case, the shore facilities are NCTAMSs or NAVCOMSTAs that guard a designated frequency on a continual (24-hour per day) basis to monitor any tactical message call-ups from ships. This connectivity is provided by long-haul, point-to-point, as well as multicast communications. In the Copernicus Architecture (see Chapter 20), it is a part of the TADIXS.

In amphibious operations, the shore-to-ship connectivity consists of short-haul communications, but the distance could be as much as 370 km (200 nmi) in over-the-horizon amphibious assault operations.

4.3.6 Ship-to-Ship Connectivity

Ship-to-ship connectivity supports the information exchange between ships in a battle group while in transit and also in combat. Communications uses either LOS or BLOS short-haul circuits. The ship-to-ship circuits are point-to-point, multicast, or broadcast circuits in support of all naval warfare such as AAW, ASW, ASuW, STW, and AMW. Information is in the form of voice, record messages, data, video, and TV. Examples of the ship-to-ship connectivity are Primary Tactical (PRITAC), Link 11, Link 16, Link 4A, secure voice, and bridge-to-bridge marine radio telephones. These circuits are also called TADIXS in the Copernicus Architecture.

4.3.7 Ship-to-Aircraft Connectivity

Ship-to-aircraft connectivity supports the information exchange from aircraft carriers or surface ships to carrier-based (VS) or land-based aircraft (VP) and Airborne Warning and Control System, (AWACS). This information exchange is used for the Aircraft Control System that controls and vectors in carrier-launched air wing aircraft, Air Tactical Data System (ATDS),[8] approach/departure control and aircraft emergencies. Examples of carrier-based aircraft (VS) are F/A-18 Hornet, F-14 Tomcat, E-2C Hawkeye, EA-6 Intruder, S-3B Viking, and SH-60B Seahawk Light ASW Multi-Purpose System (LAMPS) helicopters. Ship-to-aircraft communications use mostly point-to-point short-haul circuits on the VHF/UHF bands.

4.3.8 Ship-to-Submarine Connectivity

Ship-to-submarine connectivity supports the information exchange from a ship in a battle group to an attack submarine in direct support of a battle group. This type of communications submarine tactical data link (STDL) is used for the exchange of ASW related C3I information, mostly over short-haul circuits. Since submarine detection is a critical problem, low probability of intercept (LPI) communications are essential for this type of connectivity. Various HF/VHF/UHF radios and FLTSATCOM SSIXS circuits can be used for ship-to-submarine communications.

---

[8]The Air Tactical Data System (ATDS) is a part of the NTDS that is installed in aircraft platforms.

4.3.9 Aircraft-to-Shore Connectivity

Aircraft-to-shore connectivity provides communications for the information exchange from shore-based patrol aircraft (VP) such as the P-3C Orion to Naval Air Stations (NASs) or from Combat Air Patrols (CAPs) to landing forces ashore. The VP-to-NAS communications are used for the exchange of ASW-related C3I information and acoustic data, and the CAPs-to-landing force communications are used for the exchange of close air support coordination. Most of the aircraft-to-shore communications uses long-haul point-to-point circuits.

4.3.10 Aircraft-to-Ship Connectivity

Aircraft-to-ship connectivity provides for the information exchange from carrier-based aircraft or land-based AWACS or LAMPS helicopters to aircraft carriers or surface ships in all naval warfare. The aircraft or helicopters send sensor and C3I data to the Combat Information Center (CIC) and fusion centers on aircraft carriers. Point-to-point or broadcast short-haul circuits on the VHF/UHF bands are used for this purpose.

4.3.11 Aircraft-to-Aircraft Connectivity

Aircraft-to-aircraft connectivity supports the information exchange among CAPs, and between CAPs and AWACS. Examples of information exchanges are intercept control, strike control, aircraft emergency, and search and rescue (SAR). This connectivity is mostly short-haul.

4.3.12 Aircraft-to-Submarine Connectivity

Aircraft-to-submarine connectivity provides for the exchange of tactical coordination information from carrier-based aircraft to submarines in direct support of a battle group. LOS short-haul circuits are used for this type of communications. In order to protect the submarine from detection, an LPI circuit is used.

4.3.13 Submarine-to-Shore Connectivity

Submarine-to-shore connectivity supports the information exchange from attack or ballistic submarines at sea to a shore command node (SUBOPAUTH) for the purpose of exchanging report-backs. Long-haul circuits from ballistic missile submarines communicate either acknowledgment of received messages or the execution of orders. Submarine report-back is seldom exercised except in the case of an extreme emergency.

4.3.14 Submarine-to-Ship Connectivity

Submarine-to-ship connectivity provides for the information exchange from attack sub-marines in direct support of a battle group to ships in the battle group, via short-haul cir-cuits. Since submarine detection is a critical problem, low probability of intercept (LPI) communications[9] are essential. These connectivities use various HF/VHF/UHF radios and FLTSATCOM SSIXS circuits.

4.3.15 Submarine-to-Aircraft Connectivity

Submarine-to-aircraft connectivity provides for the information exchange from sub-marines in direct support of a battle group to carrier-based aircraft CAPs and surveillance aircraft. This type of communications is similar to the submarine-to-ship connectivity de-scribed in section 4.3.14. Short-haul, LPI circuits in the HF/VHF/UHF bands are used for this type of communication.

4.3.16 Submarine-to-Submarine Connectivity

Submarine-to-submarine connectivity provides for the information exchange between two submarines in direct support of battle group operations or two ballistic missile submarines via VHF/UHF radios, SSIXS, and acoustic telephones.

REFERENCES

[BA–1] Black, K. M., and A. G. Lindstrom, "TACAMO: A Manned Communication Relay Link to the Strategic Forces," *Signal,* September 1978.

[BO–1] Boyes, J. L., "A Navy Satellite System," *Signal,* May 1976.

[BU–1] Burrows, M. L., "ELF Communications in the Ocean," Conference Record 1977, *International Conference on Communications,* IEEE Pub. 77CH1209-56CSCB, June 12, 1977.

[CA–1] Callaham, M. B., "Submarine Communications," *IEEE Communications Magazine,* November 1981, pp. 16–25.

[EV–1] Evans, M.H.H., *Amphibious Warfare,* Brassey's Sea Power Series, Vol. IV, Pergammon-Brassey's International Defense Publishers, McLean, VA 22102, 1987.

[FA–1] Farris, R. S. and R. J. Hunt, "Battle Group Air Defense Analysis," *Johns Hopkins University Technical Digest,* Laurel, MD 20707, October–December 1981.

[KI–1] Kiely, D. G., *Naval Surface Weapons,* Brassey's Sea Power Series, Vol. VI, Pergammon-Brassey's International Defense Publishers, McLean, VA 22102, 1987.

[ME–1] Melancon, P. S. and R. D. Smith, "Fleet Satellite Communications (FLTSATCOM) Pro-

---

[9]LPI is the probability of a radio transmission to be detected by an enemy when operated in a hostile en-vironment. A typical situation might consist of a submarine or surface ship attempting to communicate with a satellite via a short message, and an enemy search aircraft (interceptor) whose mission is to detect the presence of RF energy of this communicated short message.

gram," *Conference Record AIAA 8th Communications Satellite System Conference,* April 20, 1980, p. 516.

[NA–1]   *Space and Electronic Warfare,* Copernicus Project Office, Department of the Navy Washington, D. C. 20350-2000, June 1992.

[NA–2]   *Copernicus Architecture—1 Phase I: Requirements Definition,* Copernicus Project Office, Department of the Navy, Washington, D. C. 20350-2000, August 1991.

[NO–1]   Nooney, J. A., "UHF Demand Assigned Multiple Access (UHF DAMA) System for Tactical Satellite Communications," *Conference Record 1977 International Conference on Communications,* IEEE Pub. 77CH1209-6, June 12, 1977, pp. 195–200.

[PH–1]   Phillips, Chester C., "Battle Group Operations: War At Sea," *Johns Hopkins University Technical Digest,* Laurel, MD 20707, October–December 1981, Vol. 2, No. 4, pp. 299–301.

[PI–1]   Terry C. Pierce, "Amphibious Maneuver Warfare—Who's in Charge," *Naval Institute Proceedings,* Annapolis, MD, August 1991, pp. 32–37.

[RI–1]   Ricci, F. J. and D. Schutzer, *U. S. Military Communications—A C3I Force Multiplier,* Computer Science Press, Rockville, MD 20850, 1986, p. 73.

[SC–1]   Schoppe, W. J., "The Navy's Use of Digital Radio," *IEEE Transaction on Communications,* Vol. COM-127, December 1979.

[WE–1]   Weiner, T. F. and S. Karp, "The Role of Blue/Green Laser Systems in Strategic Submarine Communications," *IEEE Transactions on Communications,* Vol. COM-28, No. 9, September 1980, pp. 1602–1607.

[WI–1]   Winnefield, J. A., "Winning the Outer Air Battle," *Proceedings of U.S. Naval Institute,* Annapolis, MD, August 1989.

5

Naval Telecommunications System (NTS) Circuits

The Naval Telecommunications System (NTS) provides the connectivity for Navy Command, Control, Communications, and Intelligence systems, which consists of shore command and support systems and afloat platforms. There are three types of nodes[1] in the Naval Telecommunications System. The first type consists of various afloat units aggregated around a battle group, the second type is for shore commands or shore facilities, and the third consists of strategic submarine forces. Figure 5.1 shows these NTS nodes, relative locations, and how they interrelate [BE-1]. Communications between NTS nodes are different because of varying distances and operating platforms and are divided into the following three categories:

- Naval Telecommunications in support of various afloat units aggregated around a battle group. It serves tactical needs and consists of short-haul communications between members of the battle group. Members of the battle group are separated by a distance of no more than four or five hundred km.
- Long-haul communications between shore command nodes and the Battle Group. It supports tactical or strategic long-distance communications for distances from 740 km to 11,000 km (400–6000 nmi) depending on the type of operation that is supported.

[1]A node is a facility in a communications network.

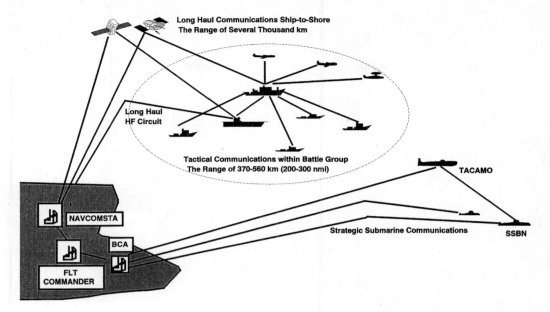

Figure 5.1   Naval Telecommunications—Three Categories of Communications Circuits/Links/Nets

- Strategic Submarine Communications between the submarine communications authority and strategic submarine fleets. Although strategic submarine communications are similar to long-haul communications, they are different since they support communications with submerged platforms.

5.1 TACTICAL CIRCUITS/LINKS EMPLOYED IN NAVAL WARFARE

Tactical communications provide the means for afloat units to: (a) perform command and control functions, (b) support the dissemination and distribution of sensor data, and (c) control combat and weapon systems. The range of these command and control communications is contained within a circular area with a radius of 400 km (220 nmi).

Most tactical communications are provided by HF, UHF, and satellite links. They are capable of supporting both analog and digital communications. Tactical communications consist of record traffic, voice, data, and imagery, which are generally transmitted in encrypted form. Most tactical communications are also resistant to electronic jamming.

Short-haul tactical communications are supported predominantly in the UHF band. Without relay aircraft or unmanned aerial vehicles (UAVs), ships separated by 50 km (27 nmi) or more rely on HF communications. Another means of communications available for ships dispersed beyond the line-of-sight is satellite communications.

The following is a brief description of various tactical links and circuits employed in naval warfare and discussed in a previous chapter. Table 5.1 lists tactical links and circuits and their characteristics.

TABLE 5.1 DESCRIPTION OF TACTICAL LINKS, CIRCUITS, AND NETS

Links and Circuits	Frequency band	Usage
Tactical Group (TG) communications		
• Fleet tactical	UHF	• Forces meeting at sea; guarded by all ships underway
• TG command	HF/UHF	• Tactical group commander's circuit
• Tactical group ORESTES	HF/UHF	• General tactical communications
• TG broadcast	HF/UHF	• Intra-tactical group communications for operations/administration
• TG PRITAC	UHF	• Primary tactical communications for maneuver signals and urgent warning
• TG SI	HF/UHF	• Special intelligence and operations communications
ASW communications		
• Screen tactical	HF/UHF	• Maneuver signals and urgent messages for screen ships
• ASW control	HF/UHF	• ASW action messages for surface action units
• ASW air coordination	HF/UHF	• Ship-to-air and air-to-air exchange of ASW information
• ASW helo coordination	HF/UHF	• Ship-to-helicopter and helicopter-to-air exchange of ASW information
ASuW communications		
• ASuW coordination	HF/UHF	• Ship-to-ship exchange of tactical information
• ASuW C&R	UHF	• Ship-to-ship exchange of air, surface, and sonar information
AAW communications		
• Air coordination	HF	• Dissemination of information concerning aircraft control
• Air reporting	HF	• Air raid reporting between ships with air defense capability
• AAW weapons coordination	HF/UHF	• Coordination of gunfire, missiles, and fighter intervention
EW communications		
• EW coordination	HF/UHF	• Control of jamming, search, and DF
Air operations communications		
• MPA C&R	HF	• Communications between MPAs and their controlling station
• Helicopter control	UHF	• Control and coordination of helicopter transit between ships
• Carrier coordination	HF/UHF	• Coordination of carrier air operations
• Air homing control	UHF	• Exchange net for aircraft homing to carriers
• Land/launch control	UHF	• Net to control launch and recovery of aircraft
Tactical air communications		
• Fighter control	UHF	• CAP aircraft communications before ship allocation
• AEW control	UHF	• Control and reporting between ship and AEW aircraft
• Air strike control	HF/UHF	• Air strike aircraft control
• Fighter air direction	UHF	• Controlling aircraft in the conduct of intercepts
• Tactical air traffic control	UHF	• Positive control of all aircraft entering/leaving objective area
• Tactical air command	HF	• TACC to direct aircraft for close air support

TABLE 5.1 (*Cont.*)

Links and Circuits	Frequency band	Usage
Amphibious communications		
• Beachmaster coordination	VHF	• Communications between beachmaster and offshore units
• Control ship coordination	HF/UHF	• Control of amphibious ships and LF in ship-to-shore movement
• LST command	HF/VHF/ UHF	• Communications between LST unit and its commander
• LF command	HF/VHF/ UHF	• Net for LF command and control
• LF reconnaissance	HF/VHF	• For coordination of reconnaissance effort with LF
• LF intelligence	HF/UHF	• Intelligence dissemination between LF and its combat elements
• LF logistics	HF	• Coordination of LF logistics support
Naval gunfire support communications		
• Naval gunfire control	HF	• For control and coordination of naval gunfire support
• Naval gunfire ground spot	HF	• For ground spotter and gunfire support ships
• Naval gunfire air spot	HF/VHF	• For air spotter and naval gunfire support ships
• LF gunfire support	HF/VHF	• For LF to request naval gunfire support
Submarine communications		
• Submarine coordination	HF/UHF	• For coordination among submarines and between submarines and ships
• Submarine operations and distress	HF/UHF	• For exchange of operations signals and safety between ships and submarines
Data Link		
• Link 11	HF/UHF	• For exchange of naval tactical data for NTDS ships
• Link 14	HF/UHF	• For exchange of naval tactical data for non-NTDS ships
• Link 16	UHF	• For exchange of joint tactical data
• Link 4A	UHF	• For exchange of tactical data for control of fighter aircraft
Distress communications		
• Search and rescue	HF/VHF/ UHF	• For SAR operations and emergency
Mine countermeasure communications		
• MCM tactical group command	UHF	• Command net between tactical group and MCM unit command
• MCM tactical group operations	UHF	• Exchange of maneuver signals and urgent MCM messages
• MCM tactical group reporting	UHF	• For reporting of sonar contact and intelligence report on mines
Harbor communications		
• Bridge-to-bridge	UHF	• Marine radiophone for harbor communications

5.1.1 Tactical Group (TG) Communications

Fleet Tactical (FLETAC) Circuit.   This circuit is used for communicating with those afloat units that are not yet a member of a battle group but are ordered to meet the battle group at sea. These units are usually supporting forces such as shore-based aircraft or additional surface forces. These units do not hold the supported force communications plan, which delineates the operational purpose and associated frequency of communications circuits for each user. Initial communications between the battle group and tactical group and its supporting forces are established on the FLETAC circuit and then shifted as directed by the Officer-in-Tactical-Command (OTC). The circuit is guarded[2] by all U.S. ships underway individually and at least one ship of a battle group and tactical group. The circuit operates primarily in the UHF band.

TG command net.   TG Command Net is used for exchange of information among command elements of the battle group using UHF satellite, HF, and UHF bands.

TG ORESTES net.   TG ORESTES (TGO) is a TTY network which provides: (a) the general intercommunications and exchange of operational and administrative messages between ships in a task organization and (b) transfer of messages to the ship-to-shore relay ships. This net uses either the HF or UHF band. The net is on-line Radio TTY (RATT).

TG broadcast.   This net is broadcast keyed by the OTC for operational and administrative traffic to ships under his command. The net uses either the HF or UHF band.

TG tactical (PRITAC) circuit.   The PRITAC circuit is a primary tactical circuit which is used for tactical maneuvering signals and urgent short tactical messages. It is guarded at all times. This circuit may also be used for urgent warning, tactical alerts, or initial contact reports of vital information concerning immediate safety of ships in company. This circuit is controlled by the OTC and guarded by all ships of the battle group/TG. The net uses either the HF or UHF band.

TG SI (ORESTES) circuit.   This circuit is used for passing special intelligence, operational, and coordination information for suitably equipped ships in a tactical group. The circuit is carried in the HF or UHF band.

5.1.2 Anti-Submarine Warfare (ASW) Communications

Screen tactical net.   This is a network for use by screen ships for maneuvering signals and urgent short tactical messages. Signals are transmitted in the HF or UHF band.

ASW control net.   This network is guarded by ships of the surface action unit for traffic associated with an ASW action. This circuit uses either the HF or UHF band.

[2]The term "guarded" means that a ship is constantly tuned to a communications network.

ASW air coordination circuit.   This circuit is used for ship-to-air and air-to-air exchange of ASW information in the HF or UHF band.

ASW helicopter coordination circuit.   This circuit is for communications with and between ASW helicopters on screen duty or on ASW patrol. Either the HF or UHF band is used for this circuit.

5.1.3 Anti-Surface Warfare (ASuW) Communications

ASuW coordination circuit.   This circuit is used for ship-to-ship exchange of tactical information in the HF or UHF band.

ASuW contact and reporting circuit.   This circuit is used by surface action units for exchange of air, surface, and sonar information. The UHF band is used for this circuit.

5.1.4 Anti-Air Warfare (AAW) Communications

Air coordination circuit.   This circuit is used for the dissemination of information for aircraft control. It is an HF circuit. The information transmitted includes: (a) stationing and relieving Combat Air Patrol (CAP), (b) assignment of CAP to air control ships, (c) interception instructions, (d) progress of interception activities, (e) CAP position reports, (f) CAP missile coordination, (g) "tally-ho"[3] reports, (h) report of flight deck conditions and status of airborne aircraft, (i) coordination and direction of air phase of search and rescue (SAR), (j) coordination of homing or downed aircraft, and (k) jamming and electronic support measures (ESM)[4] [KI-1; p. 17].

Air reporting net.   The Air Reporting Net is an HF network used for air raid reporting between ships with air defense capabilities.

AAW weapon coordination circuit.   This circuit is used for the coordination of gunfire, missiles, and fighter interceptions in defense of the battle group formation. It may also be used for passing simplified air situation status to small ships. This circuit is in the HF or UHF band.

5.1.5 Electronic Warfare (EW) Communications

EW coordination circuit.   This circuit is used for the control of jamming, search, and direction finding, for the alerting and steering of search and direction finding (DF) activity, and for the exchange of information. This circuit uses either the HF or UHF band [KI-1; p. 5 and p. 93].

---

[3]Tally-ho is derived from hunting terminology, in this case, meaning a call of an aircraft sighting an enemy aircraft.

[4]Electronic support measures (ESM) are one of the three principal disciplines of electronic warfare (EW). ESM is concerned with interception of electronic information using sensor and surveillance systems. The other two disciplines are electronic countermeasures (ECM) and electronic counter-countermeasures (ECCM).

5.1.6 Air Operations Communications

Maritime patrol aircraft (MPA) reporting and control circuit.   This HF circuit is used for communications between MPA and their controlling maritime head-quarters. It may also be used between maritime patrol aircraft and cooperating ships.

Helicopter control circuit.   This UHF circuit is used for coordinating and con-trolling helicopter transit between ships.

Carrier coordination circuit.   This circuit is for the coordination of carrier air operations. It uses the HF or UHF band.

Air Homing control net.   This UHF net is allocated exclusively for aircraft homing on carriers.

Land/Launch control net.   This UHF net is allocated exclusively to aircraft carriers and ships equipped with helicopters for control of launch and recovery of aircraft.

5.1.7 Tactical Air Communications

Fighter control circuit.   This UHF circuit is for communications with CAP air-craft before they are allocated to individual ships to control.

AEW control circuit.   This UHF circuit is used for control and reporting be-tween ships and Airborne Early Warning (AEW) aircraft.

Air strike control net.   This net is used for controlling strike aircraft and oper-ates in the UHF or HF band.

Fighter air direction circuit.   This UHF circuit is used by the air control ele-ments of the tactical group to control aircraft during intercept.

Tactical air traffic control circuit.   This UHF circuit is used to exercise posi-tive control of all aircraft entering and leaving the operating area.

Tactical air command circuit.   This HF circuit is used by the Tactical Air Control Center (TACC) to direct aircraft groups and to provide aircraft control for close air support and AAW missions.

5.1.8 Amphibious Communications

Beachmaster coordination net.   This UHF net is used for providing commu-nications between the beachmaster and offshore units to control boat traffic in the imme-diate vicinity of the beaches and communications to adjacent beaches.

Control ship coordination net.   This net is used for controlling amphibious ships and landing force elements during the ship-to-shore movement and follow-on phases. This net operates in the HF or UHF band.

LST command net.   This is a command net operating from the LST[5] unit commander to the subordinate LST element commander whenever a separate LST control organization is employed. This net uses the HF, VHF, or UHF band.

Landing force (LF) command net.   This command net is used by the Commander, Landing Force (CLF) to exercise command and control over the major LF components. This net is established during amphibious operation in the HF, VHF, or UHF band.

Landing force (LF) reconnaissance net.   This net is used for coordination of the reconnaissance effort within the landing force. Information collected by reconnaissance units is transmitted directly to the CLF via this net, which uses the HF or VHF band.

Landing force (LF) intelligence net.   This net is used for rapid collection and dissemination of intelligence information between the CLF and the major combat elements of the landing force. This net uses the HF or UHF band.

Landing force (LF) logistics net.   This HF net is used for the coordination of logistic support with the landing force.

5.1.9 Naval Gunfire Communications

Naval gunfire control circuit.   This HF circuit is used for communications between naval gunfire support ships and other authorities for control and coordination, including aircraft safety [KI-2; p. 16].

Naval gunfire ground spot net.   This HF net provides communications between the naval gunfire spot, the fire support ships, and the battalion naval gunfire liaison office to conduct fire missions against targets. Its primary use is to call a naval gunfire ground spotter and adjust gunfire; its secondary use is to exchange vital information between stations on the net, such as front line positions.

Naval gunfire air spot net.   This net is used for air spotting of naval gunfire. It is used to call for and adjust fire by the air spotter in the same manner as the ground spotting net. This net uses the HF or VHF band.

Landing force (LF) gunfire support net.   This net provides a means for requesting naval gunfire support and coordinating the employment of gunfire ships in general support of the landing force. It uses the HF or UHF band.

[5]Landing Ship, Tank (LST) is an amphibious ship class.

5.1.10 Submarine Communications

Submarine coordination circuit. This circuit is used for coordination between submarines and, when required, between submarines and ships during the direct support of a battle group. This circuit uses the HF and UHF band.

Submarine operations and distress net. This net is used for exchange of operational signals between submarine operating authorities (SUBOPAUTH), submarines, and ships assigned to submarine operations and safety. This net uses the HF or UHF band.

5.1.11 Data Links

As shipboard sensors, command and control, and weapon systems become increasingly sophisticated, it is necessary to transfer digital signals between two or more shipboard, airborne, or shore nodes. Modern naval ships therefore employ data links with sophisticated protocols and network management concepts. Table 5.2 lists tactical data links for shipboard use.

TABLE 5.2   LIST OF TACTICAL DATA LINKS

NATO Data Link Designation	U.S. Data Link Designation and Format	Data Rate	Frequency Band	Use
Link 11	TADIL A	2.25 kbps	HF 2–30 MHz	NTDS
Link 16	TADIL J	28.8 kbps, 57.6 kbps or 115.2 kbps	UHF 950–1150 MHz	JTIDS
Link 4A	TADIL C	5 kbps	UHF	Aircraft Data System
CHBDL	—	100 kbps (ship-to-aircraft); 274 Mbps (aircraft-to-ship)	X-band (9.7–10.5 GHz); Ku-band (14.4–15.5 GHz)	Sensor Data by AN/USQ-123 (V) Radio
LAMPS Data Link	—	25 Mbps	G-band (4–6 GHz)	Sensor Data by AN/SRQ-4 (V) Radio
HAVE QUICK II	—	16 kbps	UHF 225–400 MHz	Command & Control
Link 1	TADIL B	2.4 kbps	Landline	Air Defense[1]
Link 14	TADIL A	75 bps	HF/UHF	NTDS

[1]Link 1, using the TADIL B format, normally interconnects two shore nodes by a dedicated, point-to-point, full duplex digital data link [DD-2].

Link 11.   Link 11 is a data link that is used for the exchange of tactical data, such as contact reports, and for coordinating platforms in the area of operations. Link 11 uses the TADIL A data format, described in MIL-STD-188–203–1 [DD-1]. Link 11 normally interconnects participating tactical units, such as afloat units, aircraft, and shore nodes. Primarily operating in the HF band, this link is also supported at UHF for platforms within the line-of-sight [SC-1]. Only platforms capable of handling and displaying status and target information have Link 11 equipment.

Link 11 supports the transmission of the Naval Tactical Data System (NTDS) among battle group elements, using HF radios at a rate of 2275 bps. NTDS is a Navy shipboard tactical command and control system that supports various warfare commanders. Link 11 uses a polling technique to provide for communications among the various units and exchange target information.

Link 16.   Link 16 supports an integrated communications, navigation, and Identification of Friend-or-Foe (IFF) capability among battle group elements. Link 16 uses TADIL J as data format, described in NRaDWarm-92-TRG-001 [NR-1]. The TADIL J system normally interconnects participating tactical units such as afloat units, aircraft, and shore nodes. Link 16 is used to exchange joint tactical data using a combination of VHF/UHF radios with an applique for anti-jam communications.

Link 4A.   Link 4A is a UHF data link used for controlling fighter aircraft from ships and aircraft with Link 4A capability. This link uses TADIL C as data format, described in MIL-STD-188-203–3 [DD-3]. Link 4A normally interconnects tactical and support aircraft to the aircraft control units. TADIL C is a 5-kbps, half duplex digital data transmission format used to transfer aircraft control and target information between a control station and a controlled aircraft. It uses a TDMA technique to communicate with various units and exchange target information.

The Link 4A was originally designed to support an aircraft automatic landing system. Its use was extended into a means to coordinate an airborne early warning E-2C Hawkeye aircraft and F-14A Tomcat fighter aircraft by exchanging status and target data. All carrier-based aircraft are equipped with Link 4A data link. This link provides one-way air intercept control communications and is used by all the Navy's carrier-based aircraft.

Common high bandwidth data link (CHBDL).   The CHBDL is a data link that is used for imagery data communications from airborne platforms, such as reconnaissance aircraft, to ships [PS-1; NA-1]. It provides automated communications between aircraft carriers for acquiring sensor signals from airborne reconnaissance aircraft and other data link-equipped aircraft. The operating frequencies of the CHBDL are the X-band (9.7 GHz to 10.6 GHz) and the Ku-band (14.4 GHz to 15.56 GHz). The CHBDL aircraft-to-ship link transmits sensor data at a rate of 10.71 Mbps to 274 Mbps, and the CHBDL ship-to-aircraft link transmits command and control data at a rate of 200 kbps.

Light airborne multi-purpose system (LAMPS) data link.   The Light Airborne Multi-Purpose System (LAMPS) Data Link is a tactical microwave digital data link between a ship and SH-60B Seahawk LAMPS helicopters [BL-1; LA-1; NA-2]. It provides an automated exchange of radar and sonar data acquired in airborne sensors systems to the LAMPS parent ship.

The operating frequency of the LAMPS data link is the G-band[6] (4 GHz to 6 GHz). The LAMPS data link transmits sensor data at a rate of 25 Mbps.

HAVE QUICK II.   HAVE QUICK II is a tactical UHF radio that is used for exchange of tactical digital data between ships and various nodes such as other ships, aircraft, and ashore units. It provides an ECCM capability to existing UHF radios such as AN/ARC-182 and AN/WSC-3. The operating frequency of the HAVE QUICK II radio is in the VHF and lower UHF band (225 MHz to 400 MHz). The HAVE QUICK II radio transmits digital data at the rate of 16 kbps.

Link 1.   Link 1 is a data link using landlines. It provides automatic exchange of air defense data. Link 1 automatically reformats data for the exchange of air defense information between two data links by using a data buffer device. Link 1 transmits a formatted message at a rate of 2.4 kbps [DD-2].

Link 14.   Link 14 is a data system operating at both HF and UHF. It provides a computer-controlled broadcast of the tactical data by designated ships equipped with Link 11 and other platforms. Link 14 transmits a standard 100 words/min teletype to minimize tactical data processing equipment costs of those force elements not directly involved in area defense or offense roles supported by the tactical data exchange.

5.1.12 Distress Communications

Search and rescue (SAR) net.   The SAR net provides communications for available personnel and facilities to render aid in distress situations. SAR missions require a continuous communications interface between the participating rescue command, shore activities, and surface ships and aircraft. SAR operations use the following frequencies: (a) internationally designated CW and voice circuits in various UHF and VHF bands, (b) U.S. Navy emergency sonobuoy communications and homing signals in the VHF band monitored by ASW aircraft, and (c) the NATO voice (AM) SAR circuit in the VHF band.

5.1.13 Mine Countermeasure (MCM) Communications

MCM TG command net.   This command net provides for the exchange of mine countermeasure command and control messages between the tactical group command and the minesweeper unit commander. This net operates in the VHF or UHF band.

MCM TG tactical net.   This net is used for the exchange of maneuvering signals and urgent short tactical messages between the tactical group commander and the MCM unit commander. This net uses the HF and UHF bands.

[6]The frequency region and band designation are given in Figure 11.6.

MCM TG reporting net.   This net is used for reporting of sonar contacts, intelligence, and situation reports (SITREPS). It is also used for the exchange of radar and plot information on surface, subsurface, and air contacts. This net is guarded continuously by all components of a tactical group when underway regardless of the tactical situation. It cannot be used for passing command orders or administrative traffic.

5.1.14 Harbor Communications

Bridge-to-bridge circuit.   When operating on navigable waters of the United States or in inland waters, U.S. law requires that all ships maintain a continuous radiotelephone guard on the VHF (FM) band for safety of navigation. This radiotelephone is for the exclusive use of the commanding officer and officer-in-charge of a ship. It is not a military circuit and operates in the UHF band from 156 MHz to 174 MHz.

5.2 LONG-HAUL COMMUNICATIONS IN NAVAL WARFARE

Long-haul communications provide Beyond-Line-of-Sight (BLOS) communications between fleet elements and shore commanders or between independent fleet elements. The range of long-haul communications typically exceeds 185 km (100 nmi). Typical long-haul circuits support: (a) transmission of intelligence data, (b) dissemination and distribution of sensor data, (c) force direction, (d) control over the combat and weapon systems, (e) administrative and logistics support, and (f) in crises, detailed exchange of information pertaining to the situation by the on-scene commander. Messages are in the form of record traffic, voice, data, and imagery; they can be encrypted or in plain text. Many circuits are also resistant to electronic jamming.

There are four categories of long-haul communications: (a) wideband SHF SATCOM, (b) fleet satellite broadcast system in UHF SATCOM, (c) general-purpose FLTSAT information exchange channels, and (d) HF communications.

Long-haul communications supported by HF media and satellite communications complement each other. Satellite circuits have advantages over the HF circuits, but they are vulnerable to certain electronic threats. HF circuits augment satellite circuits in support of tactical operations. In addition, they provide connectivity to small ships that do not carry SATCOM terminals and are also used for interoperation with allied navies. The following provides a brief description of various long-haul circuits. Table 5.3 provides an overview of long-haul circuits and their characteristics.

5.2.1 Wideband SHF Satellite Communications

The Navy uses SHF SATCOM terminals on key command ships with small antennas (1.2 m) [NC-1]. SHF SATCOM supports jam-resistant communications to large shore terminals with large antennas (18 m) via the SHF DSCS satellites.

TABLE 5.3   DESCRIPTION OF LONG-HAUL CIRCUITS

Circuits/Links	Frequency band	Usage
Wideband SHF SATCOM • JRSC • SURTASS data relay	 SHF SHF	 • For exchange of operational information between shore command and ships • For exchange of acoustic data between SURTASS ships and shore facility
FLTSATCOM Broadcast • Fleet broadcast • Submarine broadcast	 SHF (uplink) UHF (downlink) VLF, HF, UHF SATCOM	 • Broadcast of operational, weather, intelligence, and logistics messages to ships • Broadcast of operational messages to submarines
UHF FLTSATCOM • OTCIXS • TACINTEL • CUDIXS • SSIXS • TADIXS • FLTSAT secure voice	 UHF UHF UHF UHF UHF UHF	 • For exchange of C2 and targeting information between OTC and ship/shore • For exchange of intelligence messages • For exchange of ship-to-shore messages • For exchange of submarine messages between submarines and SUBOPAUTH • For exchange of tactical data from shore node to ships • Two voice communications between shore sites and ships
Long-haul HF Communications • Commander's ship-to-shore • Fleet primary ship-to-shore • Primary ship-to-shore (NATO) • HICOM	 HF HF HF HF	 • For exchange of C2 information between shore and afloat commander • "On-call" ship-to-shore circuit, continuously guarded by NCTAMS/NAVCOMSTA • Long-haul primary ship-to-shore circuit for allied maritime forces • For voice communications circuit between shore and afloat commanders

Jam-resistant secure communications (JRSC).   This circuit is used for the exchange of operational and intelligence information between shore commands and major combatant ships at sea. This circuit uses the DSCS SHF satellite and requires special anti-jam modems operating at low data rates.

SURTASS acoustic data relay circuit.   This circuit is used for the exchange of acoustic data and operational information between SURTASS ships and shore acoustic processing facilities [MU-1]. This circuit uses the DSCS SHF satellite; data are transmitted at 32 kbps.

5.2.2 UHF Fleet Satellite Broadcast

The need for shore-to-ship jam-resistant communications at a cost considerably lower than the one dictated by the use of SHF terminals fostered the hybrid SHF/UHF Fleet Broadcast system. The Fleet Broadcast system provides long haul satellite communciations from shore to ship [NC-2]. A large shore terminal transmits a spread-spectrum signal to the satellite, providing substantial resistance to electronic jamming on the potential vulnerable uplink. This signal is processed onboard the satellite and retransmitted at UHF, where it can be received with small UHF omni-directional antennas and inexpensive receivers onboard the ships at sea.

The shipboard equipment for broadcast repeaters are AN/SSR-1 (UHF broadcast receiver) and AN/WSC-3 (UHF transmitter/receiver) terminals. Nearly all ships and many naval aircraft are equipped to receive UHF Fleet Broadcast.

The Fleet Broadcast is a primary one-way circuit which is broadcast from shore to all ships in specified areas. This circuit supports several channels, including operational, weather, intelligence, administrative, and logistics traffic. All ships except submarines receive the fleet broadcast either directly or through a guard ship arrangement.[7]

The Submarine Broadcast net provides primary one-way broadcast to submarines in a specified area. The content of broadcast messages is similar to those of fleet broadcast. The primary frequency band for broadcast to submarines is the VLF band. The secondary submarine broadcast channels are provided by UHF satellites and HF radios.

5.2.3 UHF Fleet Satellite Communications

Another long-haul satellite communications capability is provided by the general-purpose FLTSAT Information Exchange Subsystem (IXS) channels [NC-2; SC-1]. The AN/WSC-3 shipboard terminal is used for these FLTSATCOM IXS channels. Most major combatant ships have both fleet broadcast and IXS subsystems, and multiple AN/WSC-3 (two-way) installations. These satellite channels provide the umbilical cord from naval platforms to the shore command. The demand for FLTSATCOM has grown and exceeds the capacity of channels presently available in the UHF satellite. In order to make more efficient use of the available channels, a demand assigned multiple access (DAMA) concept has been implemented. It is used to increase the data flow through each of the UHF repeater channels and to exploit the low-duty factor of most users by allowing channel sharing on demand [NO-1].

FLTSATCOM includes the following six circuits: (a) Officer-in-Tactical Command Information Exchange Subsystem (OTCIXS), (b) Tactical Intelligence Information Exchange Subsystem (TACINTEL), (c) Command User Data Information Exchange Subsystem (CUDIXS), (d) Submarine Satellite Information Exchange Subsystem (SSIXS), (e) Tactical Data Information Exchange Subsystem (TADIXS), and (f) FLTSATCOM Se-

---

[7]A guard ship is a 24-hour per day designated station in a tactical group.

cure Voice. These circuits and their equipment are described in detail in Chapter 13 (UHF Satellite Communications).

5.2.4 Long-Haul HF Communications

HF communications are not as reliable as SATCOM on a routine day-to-day basis. They are subject to ionospheric propagation conditions. However, not all naval ships can afford the more complex satellite terminals. Consequently, HF continues to be an important means of long-haul communications. In order to overcome ionospheric propagation problems and noise, sophisticated HF sounders are used to optimize the frequency assignment. HF automatic link establishment (ALE)[8] is an emerging technology to improve operations by tuning of receivers and transmitters to the best available frequencies.

In order to overcome vulnerability to jamming and interception, improvements in anti-jamming technology are incorporated in HF radios. Modern HF radios are also designed to minimize interference from harmonic and intermodulation products[9] between various HF signals and other shipboard emanations that produce a noise background for onboard receivers.

Commander's ship-to-shore circuit.   This circuit is used for command and control between afloat commander and shore stations in the US Navy and allied maritime forces. This circuit normally requires a full-period ship-shore-ship termination.[10]

Fleet primary ship-to-shore circuit.   This is an "on-call" ship-to-shore circuit for the U.S. Navy. It is continuously guarded by a Naval Computer and Telecommunications Area Master Station or a Naval Communications Station (NCTAMSs/NAVCOMSTAs) to provide unscheduled access for fleet units for the purpose of delivery of ship-to-shore messages whenever required. NCTAMSs or NAVCOMSTAs deliver traffic to ships on this type of circuit only in emergencies. Unless otherwise directed, ships are not required to maintain a continuous guard on this circuit.

Primary ship-to-shore (NATO) circuit.   This is a long-haul primary ship-to-shore circuit for allied maritime forces. This circuit includes multichannel and single-channel terminations with NCTAMSs and NAVCOMSTAs. The multichannel termination handles a set of multiplexed TTY channels in a single transmission, which may include up to 16 75-bps TTY circuits in an HF circuit or UHF satellite communications at a rate of 2400 bps. The single-channel termination is a single TTY circuit operating at a rate of 2400 bps in an HF circuit or UHF satellite communications circuit.

---

[8]See section 10.6 (Automatic Link Establishment Techniques).

[9]See section 17.4 (Shipboard Interference Analysis).

[10]Full-period termination is a method of ship-to-shore or shore-to-ship transmission. During this transmission, a circuit is completely dedicated on a 24-hour-per-day basis.

HICOM net.   The HICOM Net is a clear (nonencrypted) voice net, used for exchange of tactical data between the Fleet Commander (FLTCINC) and the battle group commander.

5.3 STRATEGIC CIRCUITS/LINKS EMPLOYED IN SUBMARINE WARFARE

The Navy's ballistic missile submarines and attack submarines share similar communications needs which present special challenges to Navy communications systems. Both types of submarines are built to move freely throughout the world's oceans to avoid detection and to either attack or pursue enemy submarines which are seeking to avoid detection. Communications requirements to receive information from the shore and to send information back must be met without compromise or constraint.

The following provides a brief description of various strategic submarine circuits employed in strategic submarine warfare. Table 5.4 lists strategic submarine circuits and characteristics.

Verdin circuit.   The Verdin VLF submarine communications system is a broadcast system which transmits submarine messages from large VLF radio transmitter facilities located around the world. These stations radiate very high power signals (1 MW to 2 MW) from large antennas. Submarines are able to receive the VLF signals worldwide with trailing wire or buoy-mounted antennas. Backup transmitters operating at LF or HF provide shorter range communications.

Since all of the foregoing systems depend on fixed, land-based facilities which are subject to preemptive attack, the Navy operates a survivable airborne VLF communications aircraft known as TACAMO [BA-1]. This large transport type aircraft (EC-130) carries a fully equipped radio room to receive and acknowledge messages via a number of receivers and transmitters, as well as a high-power VLF amplifier and long (approximately 5 km) trailing wire antenna. One or more of these aircraft is constantly on patrol and ready to relay messages to submarines if the fixed land-based facilities are destroyed.

TABLE 5.4   DESCRIPTION OF STRATEGIC SUBMARINE COMMUNICATIONS

Circuits/Links	Frequency band	Usage
• VERDIN	VLF/LF	• Broadcast of submarine operational messages from shore to submarines
• TACAMO	VLF/LF	• Relay of submarine operational messages from TACAMO aircraft to submarines
• SSIXS	UHF SATCOM	• For exchange of submarine operational messages via a FLTSATCOM channel
• ELF (Sanguine)	ELF	• Broadcast of submarine operational messages from shore to submarines
• MILSTAR SSIXS	EHF SATCOM	• For exchange of submarine operational messages via a MILSTAR channel

UHF FLTSATCOM.   A UHF FLTSATCOM circuit is also used for strategic submarine communications. The FLTSATCOM circuit is the Submarine System Information Exchange System (SSIXS) [NC-2]. The SSIXS allows submarines to receive information globally except for in the high northern and southern latitudes that are not covered by the FLTSAT. During brief periods of surfacing, the submarine is subjected to detection by radars since SSIXS terminals require a mast-mounted antenna.

ELF submarine communications system (Sanguine).   The ELF submarine communication system (Sanguine) is a very low data rate, submarine broadcast system that augments the Verdin VLF submarine communications system [BU-1]. The ELF submarine communications system can transmit a signal to a submarine with a trailing wire antenna at up to 100 m (up to 330 feet) below the ocean's surface because of the extremely low attenuation rate in seawater at ELF (0.03 dB/m). With current VLF systems on submarines, the Navy must depend on a special towed buoy to carry antennas and receiving equipment.

MILSTAR EHF SATCOM.   An EHF-band SSIXS channel is provided in the MILSTAR satellite, which can be employed using small periscope- or mast-mounted antennas. The combination of wideband signal processing available in this frequency band and small high-gain transmitting antennas provide substantial resistance to both jamming and undesired transmission interception. Downlink frequency selection and signal design support direct reception by the submarines and TACAMO with moderate satellite power requirements.

REFERENCES

[BA–1] Black, K. M. and A. G. Lindstrom, "TACAMO: A Manned Communication Relay Link to the Strategic Forces," *Signal,* September 1978.

[BE–1] Bell, C. and R. E. Conley, "Navy Communications Overview," *IEEE Transactions on Communications,* Vol. COM-28, No. 9, September 1990, pp. 1573–1579.

[BL–1] Blake, B., ed., "LAMPS MK III," *Jane's Underwater Warfare Systems 1989–1990,* Jane's Information Group, Inc., Alexandria, VA 22314-1651, 124.

[BU–1] Burrows, M. L., "ELF Communications in the Ocean," *Conference Record 1977 International Conference on Communications,* IEEE Pub. 77CH1209-56CSCB, June 12, 1977.

[DD–1] *Subsystem Design and Engineering Standards for Tactical Digital Information Link (TADIL) A,* MIL-STD-188-201-1, Department of Defense, Washington, D. C. 20350-2000, September 10, 1982.

[DD–2] *Subsystem Design and Engineering Standards for Tactical Digital Information Link (TADIL) B,* MIL-STD-188-201-2, Department of Defense, Washington, D. C. 20350-2000, March 23, 1984.

[DD–3] *Subsystem Design and Engineering Standard for Tactical Digital Information Link (TADIL) C,* MIL-STD-188-201-3, Department of Defense, Washington, D. C. 20350-2000, October 5, 1983.

[KI–1]   Kiely, D. G., *Naval Electronic Warfare,* Pergammon-Brassey's International Defense Publishers, McLean, Va. 22102, 1988.

[KI–2]   Kiely, D. G., *Naval Surface Weapons,* Pergammon-Brassey's International Defense Publishers, McLean, Va. 22102, 1988.

[LA–1]   Law, P., "AN/SRQ-4 (LAMPS MK III)," *Shipboard Antennas,* Artech House Inc., Dedham, MA 02026, 1983, pp. 502–504.

[MU–1]   Muehldorf, E. I., P. Hildre, and R. C. Hobart, "Global Ship-Shore Satellite Data Transmission System for Connecting Digital Data Processors," *AIAA 10th Communications Satellite System Conference,* March 19–22, 1984, Orlando, FL, pp. 380–386.

[NA–1]   *Tactical Data Link Assessment,* Copernicus Project Office, Croesus Data Link Architecture Team Department of the Navy, Washington, D. C. 20350-2000, June 15, 1992.

[NA–2]   *AN/SRQ-4 Radio Terminal Set,* Technical Manual, EE185-AA-OMI-0120, NAVELEX (now Space and Naval Warfare Systems Command) Washington, D. C. 20363-5100, September 1, 1984.

[NC–1]   *Navy SHF Satellite Communications System Description,* FSCS-2000-83-1, NOSC (now NCCOSC RDT&E Div.), San Diego, CA 95152-5158, November 1983.

[NC–2]   *Navy UHF Satellite Communications System Description,* NSHFC 301, NOSC (now NCCOSC RDT&E Div.), San Diego, CA 95152-5158, December 31, 1991.

[NR–1]   *Link 16 Communications Planning—Quick Reference Guide,* NRaDwarm-92-TRG-001, NCCOSC RDT&E Division, Warminster, PA, August 1, 1992.

[NO–1]   Nooney, J. A., "UHF Demand Assigned Multiple Access (UHF DAMA) System for Tactical Satellite Communications," *Conference Record 1977 International Conference on Communications,* IEEE Pub. 77CH1209-6, June 12, 1977, pp. 195–200.

[PA–1]   Pakenham, W. T. T., *Naval Command and Control,* Pergammon-Brassey's International Defence Publishers, McLean, VA 22102, 1989, p. 106.

[PS–1]   *Paramax Data Sheet on Common High Bandwidth Data Link-Ship Terminal AN/USQ–123(V) ,* Paramax Systems Corporation, Salt Lake City, UT, 84116, 1993.

[SC–1]   Schoppe, W., "The Navy's Use of Digital Radio," *IEEE Transaction on Communications,* December 1979, pp. 1938–1945.

6

Shipboard Exterior Communication (EXCOMM) Requirements

Naval shipboard exterior communications (EXCOMM) requirements are defined in terms of five types of communications services: (a) voice, (b) record message, (c) data, (d) facsimile, and (e) video imagery [NA-1].

Figure 6.1 shows the bandwidths needed to transmit the five shipboard communications services identified above. A bandwidth of 3 kHz to 1.544 MHz is required to transmit computer data. A wide bandwidth is required to transmit a real-time 30-frame per second TV signal; it is approximately 4.5 MHz. Voice and facsimile services require a voice-grade circuit as a minimum; some voice transmissions need wider RF circuits of up to 25 kHz.[1] The naval shipboard record message service bandwidth can be as low as 75 Hz[2] as is the case when using teletypewriters (TTY).

Figure 6.2 shows the available RF media used for the naval shipboard communications service. Voice circuits can be established over RF transmission media (HF, VHF, and UHF) and all satellite links (UHF, SHF, and EHF). Record messages are transmitted via all RF and satellite circuits. Naval tactical computer data and facsimiles are sent over

---

[1]Encrypted voice requires digital links of 2.4 kbps or 4.8 kbps for narrowband digitized voice generated by equipment such as the ANDVT or the CV-3333 (see Chapter 13); higher data rates are required for the wideband secure voice equipment.

[2]Usually, several TTY or TTY and data channels are multiplexed on a 3-kHz circuit.

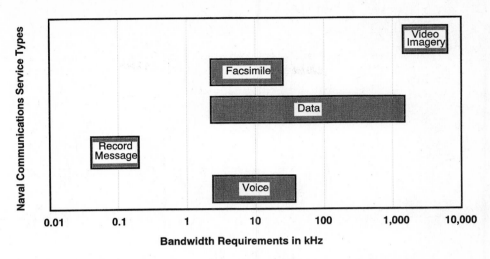

Figure 6.1   Bandwidth Requirements of Various Naval Communications Services

all RF media (HF, VHF, and UHF) and satellite circuits (UHF, SHF, and EHF). The video imagery service is provided by special wideband Ku-band links as well as by SHF wideband circuits.

Table 6.1 shows the characteristics of the naval shipboard services in terms of bandwidth, data rate, terminal processing, and mode of transmission.

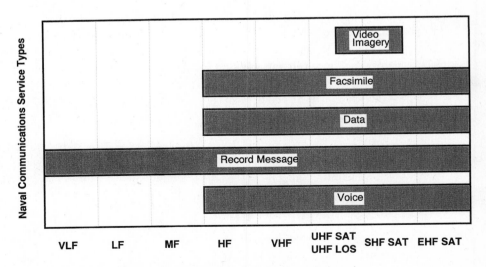

Figure 6.2   RF Media Used in Naval EXCOMM Systems

TABLE 6.1   CHARACTERISTICS OF NAVAL SHIPBOARD COMMUNICATIONS SERVICES

Service Type	Bandwidth	Data Rate	Terminal Processing	Mode of Transmission
Voice	• 3 kHz	• Digital (2.4 kbps, 4.8 kbps, 16 kbps, 50 kbps)	• Analog (Channel vocoder) • Digital (LPC-10, CVSD, CELP)	• Half Duplex • Full Duplex
Record Message (TTY)	• 150 Hz • 3 kHz	• 75 bps • 2.4 kbps	• Message Processing	• Half Duplex • Full Duplex
Data	• 3 kHz • 50 kHz • 1 MHz	• 2.4 kbps • 16 kbps • 64 kbps	• Communications Processing	• Half Duplex • Full Duplex
Facsimile	• 3 kHz	• 2.4 kbps • 4.8 kbps • 16 kbps	• Data Compression (typical compression ratios 5:1 and 3:1)	• Half Duplex
Video Imagery	• 4.5 MHz	• 128 kbps	• Video Processing	• Half Duplex

6.1 VOICE

Voice service is the most frequently used service in Navy shipboard communications. With few exceptions, the voice signal is sent by radio equipment. From the standpoint of bandwidth occupancy, the voice signal can be either narrowband (3 kHz) or wideband (25 kHz or 50 kHz). When digitized narrowband and wideband voice signals are defined in terms of the data rate, the narrowband data rates are 2400 bps and 4800 bps, and the wideband data rates are 16 kbps, 32 kbps, or 50 kbps.

Narrowband voice circuits are used in naval applications for:

- Tactical Force (TF) battle management
- High Command (HICOM) voice
- Maneuvering
- Bridge-to-bridge maritime communications
- Naval Gunfire Support (NGFS) air spot
- NGFS ground spot
- NGFS control
- Search and Rescue (SAR)
- Medical Evacuation (MEDVAC)

Wideband voice circuit applications are:

- Primary Tactical (PRITAC)
- Anti-Air Warfare (AAW) control
- AAW report
- Anti-Surface Warfare (ASuW) Contact and Report (C&R)
- Anti-Submarine Warfare (ASW) Aircraft Control
- Amphibious Control

Voice processing comprises all operations required to transform the analog voice signal into a form appropriate for transmission. Naval shipboard voice transmission, like most military voice transmission, is performed in a secure mode whenever required. The process of encoding the voice signal generates a digital signal that can be readily encrypted and provides jamming protection by bandwidth spreading.

Several frequently used voice digitization techniques are listed in Table 6.2. The earliest voice digitizer used in naval voice service was a channel VOCODER (HY-2) that operated at a rate of 2.4 kbps over a voice-grade circuit. The channel VOCODER is a device where the voice is first separated into several narrow frequency bands (typically 300 Hz); the average amplitude in these bands represents the voice signal. In addition, pitch (fundamental frequency of the speaker's voice) information is extracted and added to the signals of the narrow bands. The composite of pitch and amplitude variation requires significantly less bandwidth to transmit than the original voice signal; it can be used to reconstruct the voice signal with reasonable fidelity. (The achieved fidelity is a function of the compression—more compression results in less fidelity.) The fidelity required for military communications needs to be sufficient to recognize the speaker in order to authenticate the communication. When the narrowband composite channel VOCODER signal is digitized, a 2400 bps data rate is sufficient to transmit the voice with acceptable fidelity.

A more advanced voice digitizer is the Advanced Narrowband Digital Voice Terminal (ANDVT) that uses linear predictive coding (LPC) and operates at a rate of 2.4 kbps (or occasionally 4.8 kbps) over a voice-grade circuit. LPC employs a one-level or multi-

TABLE 6.2   VARIOUS NAVAL DIGITAL VOICE SYSTEMS

Voice Digitization Technique	Data Rate	Equipment
Channel VOCODER	2.4 kbps	HY-2
Linear Predictive Coding (LPC-10)	2.4 kbps	ANDVT [SU-1]
Code-Excited Linear Predictive (CELP) Coding	4.8 kbps	STU-III
Continuously Variable Slope Delta Modulation (CVSD)	16 kbps	KY-57 (Vinson)

level sampling system in which the value of the signal at each sample time is predicted to be a particular linear function of the past values of the quantized signal [TR-1; FS-1].

The Code-Excited Linear Predictive (CELP) voice digitization technique is related to LPC in that both use adaptive prediction of a set of coefficients $\{a_i\}$ [FS-2; AT-1; SC-1]. These coefficients $\{a_i\}$ are associated with the expansion of voice signal $s(t)$ by a set of orthogonal signals $\{f_i(t)\}$ as shown below:

$$s(t) = \sum a_i f_i(t).$$

The CELP voice digitizer is a speech encoder using two time-varying linear recursive filters excited by white noise; CELP has a much better speech synthesis performance than LPC. However, the CELP voice digitizer operates at a minimum data rate of 4800 bps since it requires an additional 2400 bps for the code excitation. The CELP performance deteriorates rapidly below 4800 bps. The CELP technique is applied in the STU-III secure voice terminal.

Another high-quality system uses the continuously variable slope delta (CVSD) modulation. The secure voice terminal implemented with the CVSD modulation is the Vinson terminal (KY-57). This technique generates a digital signal at a rate of 16 kbps that may be transmitted over RF circuits [KA-1].

6.2 RECORD MESSAGE TRANSMISSION

The transmission of record messages is a very important communications service in the Navy's shipboard communications since it provides a narrative or formatted TTY message that provides a record and requires little bandwidth. It can be stored at a node and then forwarded to the final destination. Examples of record message systems are: Fleet Broadcast (FLTBROADCAST), Naval Tactical Data System (NTDS), Tactical Group Orestes (TGO), ASW Warning, and Tactical Force Administration. A typical naval message is formatted according to the AUTODIN formats JANAP 128(I) [JC-1], and ACP 121 [NT-1; NT-2], when used in shore activities, or semiautomatic tape relay ACP 127 [NT-3; NT-4] when used in shipboard EXCOMM shore-to-ship communications.[3]

Associated with naval messages is a precedence level. The precedence level indicates the required speed of delivery to the addressee, the order of handling and delivery by communications personnel, and the order in which the addressee should process the messages. The originator determines the level of precedence for each message by selecting one of the following levels: ROUTINE, PRIORITY, IMMEDIATE, and FLASH [JC-1].

An important part of a naval message is the address. Two kinds of addresses are used in the naval message: the first is the plain language address (PLA) (e. g., NCTAMS MED Naples), and the second is the routing indicator (RI) (e. g., RUSZD12). The second is used to identify communications facilities, commands, activities, and units with a code

---

[3]See section 13.2.1 (Fleet Satellite Broadcast Subsystem).

TABLE 6.3   SAMPLE NAVAL MESSAGE

== SAMPLE ==

UUUUUUUUUUUUUUUUUUUUUUUUUUUUUU
U   UNCLASSIFIED   U
UUUUUUUUUUUUUUUUUUUUUUUUUUUUUUUU

ROUTINE

R 182322Z DEC 93
FM NEEACT PAC PEARL HARBOR HI

TO COMSPAWARSYSCOM WASHINGTON DC

INFO COMNAVCOMTELCOM WASHINGTON DC   NAVOCEANPROFAC FORD ISLAND HI
NTCC FORD ISLAND HI   NAVCAMS EASTPAC HONOLULU HI

UNCLAS //N02300//

SUBJ: NAV MSG REQUIREMENTS

A.   NAV MSG REQUIREMENTS WORKING GROUP MEETING OF 24   - 26 OCT 90

1.   THIS MSG OF INTEREST TO xxxxxx xxx xx xxxx xx xxx xxxxxx x xxxxx xx
      xxxxx xxx xxxxx xx xxx xxxxxxx xxxxxxxxxx xxxxx xx xxxx xxxxxxxxxxx
      xxxxxx xxx xxxxxxxx xx xxxx xxxxxx xxxxx xxx xxxxxxxx xxxxx xxxx xxxxx
      xxxxxxxx xxx xxxxxxxx xxx xxxxxxx xxxx xxxxxxxxx xxxxx xxxxx xxxx
      xxxxxxxxx xxx xxxxxxxxx xxxxx xxxx xxxx xxxxx xxxxx xxxxxxxx xxx
      xxxxxxxxx xxxxxxxxx xxx xxxxxxx xxxx xxxxxxxxx xxxxx xxxxxx xxxx xxxxxxx.

2.   HOWEVER, DURING THE INTERIM xx xxxxxx xxxx xxx xxxxxxx xxxxxxx xxx
      xxxxxxxxx xxxxx xxxx xxxx xxxxx xxxxx xxxxxxxx xxx xxxxxxxxx x xxxxx
      xx xxxxx xxx xxx xxxxx xx xxx xxxxxx xxxxxxxxxx xxxxxxx xx xxxx
      xxxxxxxxxxx xxxxxx xxx xxxxxxxx xx xxxx xxxxxx xxx.

3.   THIS REQUIREMENT HAS CREATED xxxxxxxxx xx xxxxxxxxx xxxxxxxxx xxx
      xxxxxxxxx xxxxx xxxx xxxx xxxx xxxxx xxxxx xxxxxxx xxx xxxxxxxx xxxxxxxx
      xxx xxxxxxx xxxx xxxxxxxxx xxxxx xxxxx xxxx xxxxxxxxxx x xxxxx xx
      xxxxx xxx xxxxx xx xxx xxxxxxx xxxxxxxxxx xxxxx x xxxxx xx xxxxx
      xxx xxx xxxxx xx xxx xxxxxxx xxxxxxxxxx xxxxxxx xx xxxx xxxxxxxxxx
      xxxxxx xxx xxxxxxx xx xxxx xxxxxx xxx.

UUUUUUUUUUUUUUUUUUUUUUUUUUUUUU
U   UNCLASSIFIED   U
UUUUUUUUUUUUUUUUUUUUUUUUUUUUUUUU

== SAMPLE ==

of seven alpha-numeric characters [JC-1]. Naval record messages can be classified in 5 levels, that is, unclassified, confidential, secret, top secret, and special intelligence.

A sample naval message is shown in Table 6.3. The top portion of the message shows the classification. The first line designates the precedence level of the message. The date-time-group (DTG) is the second line. The DTG indicates Zulu time, day, month, and year.[4] The DTG is not only used as the date but also as an identifier of naval record messages. The next three lines are the originator, the destination, and the distribution of the message in PLA. The main body of the message follows after that.

Message processing consists of: (a) the reformatting of the message received from the user terminal to a text recognizable by the end user, (b) the storage of messages, (c) the formation of message traffic queues based on the priority/preemption, (d) the incorporation of appropriate preamble data on outgoing messages, (e) the recognition of these preambles on received messages, and (f) the coordination with radio rooms and NAVCOMSTAs. The Naval Communications Processing and Reporting System (NAVCOMPARS) and Local Digital Message Exchange (LDMX) System are two shore-based message processing systems for handling fleet broadcast messages and AUTODIN messages. The Naval Modular Automated Communications System (NAVMACS) is the corresponding shipboard message processing system.

Record messages in shipboard communications are also sent via an extension of naval shore-based communications as well as parts of the Defense Communications System (DCS), such as the AUTODIN, the Defense Data Network (DDN), the Defense Secure Network 1 (DSNET1), DSNET2, DSNET3.

6.3 DATA

The expanding use of computers for shipboard combat data processing has created a demand for communications systems that: (a) can transmit data in a form suitable for automatic introduction into shipboard and shore computer systems, and (b) after processing, can send data to other ships and/or shore facilities.

Data are processed by a communication processor that carries out the following functions: (a) formatting of computer data as dictated by the shipboard tactical system, (b) synchronization of bit streams, and (c) error detection and correction including automatic repeat request (ARQ). Communications processors for shipboard data service do not include the end processor that operates on the information content of data signals; these processors are considered to be Command and Control (C2) rather than communications processors.

Examples of Navy shipboard data systems are the NTDS, the Joint Tactical Information Distribution System (JTIDS), and the Aircraft Control System. The NTDS is a tactical data system that processes contact reports by a target data processor, and it exchanges and coordinates the data by Link 11. The JTIDS is an integrated tactical data system that processes and exchanges and distributes tactical data by Link 16. The Aircraft

---

[4]Zulu time is equivalent to Greenwich mean time (GMT).

Control System is another tactical data system that coordinates airborne early warning aircraft, fighter aircraft, and their aircraft carrier by exchanging status and target data with a data processing system, and it exchanges the data by Link 4A.

6.4 FACSIMILE

Facsimile is a method of transmitting hard copy that is transformed to an analog waveform by optical scanning. Facsimile differs from video imagery in two ways: (a) facsimile requires less bandwidth than video imagery, and (b) facsimile is presented to the user as hard copy. Facsimile is transmitted over a telephone circuit, that is, the bandwidth is typically 3 kHz, whereas the video bandwidth is in the neighborhood of 4.5 MHz.

Facsimile equipment type definition: Facsimile equipment which provides for the transmission and reception of an image can be of:

(a) Type I—with black and white information only, or
(b) Type II—in shades of gray, as well as black and white.

Facsimile equipment uses two modes of transmission, that is, a basic mode and a handshake mode. In the basic mode, the facsimile equipment sends an image in the simplex and broadcast mode of operation. In the handshake mode the facsimile equipment sends an image in the duplex mode of operation. In the basic mode, the transmitter does not pause after calling the receiving unit to wait for an acknowledgment before sending an image. In the handshake mode the transmitter pauses after calling the receiving unit to wait for an acknowledgment before sending an image. In the handshake mode, communications protocol is used between a facsimile transmitter and receiver. The code words and singaling sequences used in Types I and II facsimile are defined in Table 6.4.

TABLE 6.4   SYNCHRONIZATION CODE WORDS AND SIGNALING SEQUENCE

Name	Definition
• Beginning of Intermediate Line Pair (BILP)	(00000000000000001)
• Beginning of Line Pair (BOLP)	(00000000000000001)
• End of Line (EOL)	(0000000000001)
• End of Message (EOM)	16 consecutive S_1 code words
• Not End of Message (NEOM)	16 consecutive inverted S_1 code words
• Return of Control (RTC)	EOL EOL EOL EOL EOL EOL
• Start of Message (SOM)	S_0 S_1 (N clock periods) S_1 S_0 (where N is the number of clock periods between the pairs of code words)
• S_0	(111100010011010)
• S_1	(111101011001000)
• Fill	Variable length string of 0's
• Stuffing	Variable length string of 1's
• Preamble	Variable length string of all 1's or all 0's

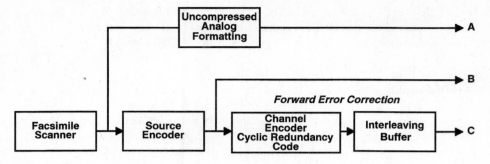

Figure 6.3   Facsimile Transmitter Block Diagram

Figure 6.3 shows a block diagram of the structure of a facsimile transmitter. There are three outputs in the facsimile transmitter: A for the uncompressed mode, B for the compressed mode, and C for the compressed with forward error correction mode. Figure 6.4 shows the general structure of a facsimile receiver. As the facsimile transmitter has three types of outputs, the receiver has three inputs: A for the uncompressed mode, B for the compressed mode, and C for the compressed mode with forward error correction mode.

Table 6.5 shows the technical characteristics of a typical naval shipboard facsimile service [DD-2]. Facsimile transmission is one-way. Naval shipboard facsimile is usually transmitted by either RF radios or SATCOMs.

Examples of facsimile applications in naval shipboard communications are transmission of text, tactical maps, and meteorological maps.

6.5 VIDEO TRANSMISSION (TV)

Video (television) is a communications system in which a picture is converted into an analog waveform, transmitted, and converted back into a picture. The video image is scanned by an electro-optical device. TV employs interlaced scanning at a high frame rate (30 frames per second) to eliminate flicker. (On the other hand, each facsimile is required to send a stationary image faithfully.) A single frame is associated with the integration of

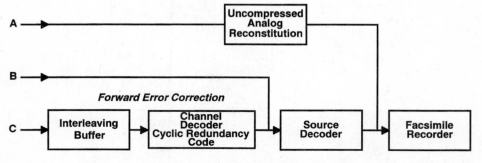

Figure 6.4   Facsimile Receiver Block Diagram

TABLE 6.5   TECHNICAL CHARACTERISTICS OF A TYPICAL NAVAL FACSIMILE SERVICE

Features	Characteristics
Operations	Full- and half-duplex modes
Shades of gray	• Black/white 4, 8, 16 • Selectable gray shades
Resolution	• Fine: 8×7.7 lines/min • Standard: 8×3.85 lines/min • Fast: 4×3.85 lines/min
Compression ratio	• Average 5:1 for black/white • 3:1 for gray scale
Data interface	• MIL-STD-188-161 [DD-2] • MIL-STD-188-114 [DD-3] • RS-232C
Transmission time	• Copy dependent • Typically under 25 s per page of text at 9600 bps
Channel rate	From 1.2 kbps to 32 kbps
Encryption equipment	TSEC/KG-13, KG-30 family, KG-84C, and KY-57 Vinson cryptos, TRI-TAC DVST and DLED

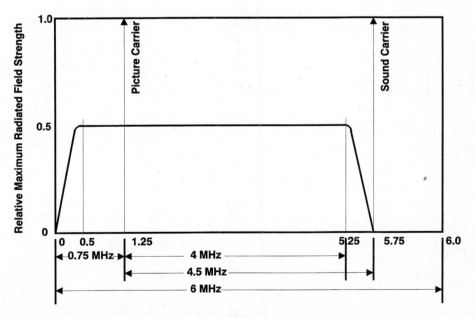

Figure 6.5   A Typical Spectrum of a TV Signal of NTSC

TABLE 6.6   TECHNICAL CHARACTERISTICS OF VIDEO SERVICE
(ANALOG VIDEO FORMAT OF NTSC)

Features	Characteristics
Picture frame dimension	525 lines; 60 Hz; 2:1 interlace scanning ratio
Aspect ratio	4:3
Interlacing	Odd-line
Bandwidth	• Total: 6MHz • Video: 4.5 MHz

the individual image by the human eye. TV can transmit image motion and also other media such as text, graphics, still images, and audio signals (as in commercial television circuits). Table 6.6 shows the technical characteristics of a typical analog video signal.

A distinctive characteristic of video imagery is its large signal baseband requirement. 500 scanning lines each with 600 picture elements result in 300,000 total elements in each frame of the picture. To permit smooth merging of a flicker-free picture through persistence of vision, 30 complete pictures are transmitted per second. If each picture element is reproducible by a half cycle of a signal wave, then a bandwidth of 4.5 MHz is required. This bandwidth is used for a standard video broadcast, which is three orders of magnitude larger than the bandwidth required for voice communications or facsimile. Consequently, video imagery is used only for important tactical purposes. Figure 6.5 shows the spectrum characteristics of a typical analog video signal according to the National Television Standards Committee (NTSC).

Examples of Navy shipboard TV transmission for tactical applications requiring picture detail and object motion include tactical sensor imagery, low-light television (LLTV) imagery, and infrared imagery. For these applications the video information is often collected by surveillance aircraft or other platforms and then transmitted to the receiving ship.

REFERENCES:

[AT-1] Atal, Bishnu and Manfred Schroeder, "Stochastic Coding of Speech at Very Low Data Rate," *Proceedings of IEEE 1984 International Conference on Acoustics, Speech and Signal Processing,* pp. 1610-1613.

[DD-1] *Common Long-Haul and Tactical Communications System Technical Standards,* MIL-STD-188-100, Department of Defense, Washington, D. C. 20350-2000, November 17, 1976.

[DD-2] *Interoperability and Performance Standards for Digital Facsimile Equipment,* MIL-STD-188-161A, Department of Defense, Washington, D. C. 20350-2000, July 4, 1988.

[DD-3] *Electrical Characteristics of Digital Interface Circuits,* MIL-STD-188-114A, Department of Defense, Washington, D. C. 20350-2000, December 13, 1991.

[DD-4] *Interoperability and Performance Standards for Tactical Digital Information Link (TADIL) A,* MIL-STD-188-201A, Department of Defense, Washington, D. C. 20350-2000, March 1987.

[FS-1] *Analog to Digital Conversion by 2400 bps Linear Predictive Coding,* Federal Standard 1015, General Service Administration, Washington, D. C. 20407, November 28, 1984.

[FS-2] *Analog to Digital Conversion of Radio Voice by 4800 bps Code-Excited Linear Predictive Coding (CELP) ,* Federal Standard 1016, General Service Administration, Washington, D. C. 20407, February 14, 1991.

[JC-1] *Automatic Digital Network (AUTODIN) Operating Procedures,* JANAP-128(I), JCS Publication, Washington, D. C. 20350-2000, March 1983.

[KA-1] Kang, G. and D. A. Heide, "Canned Speech for Tactical Voice Message System," *Proceedings of the IEEE Tactical Communications Conference,* Ft. Wayne, IN, April 29-30, 1992, pp. 47–56.

[NA-1] *Copernicus Architecture—Phase I Requirements Definition,* Copernicus Project Office, Department of the Navy, Washington, D. C. 20350-2000, August 1991.

[NC-1] *Link 16 Communications Planning—Quick Reference Guide,* NRaDWarm-92-TRG-001, NCCOSC, RDT&E Division, Warminster, PA, August 1, 1992.

[NT-1] *Communications Instructions—General,* ACP-121, NATO Communications Procedures, National Communications System Office (NTS-TS), Arlington, VA 22204-2198.

[NT-2] *Instructions—General (U.S. Supplement No. 1) ,* ACP-121 US SUPP-1, NATO Communications Procedures, National Communications System Office (NTS-TS), Arlington, VA 22204-2198.

[NT-3] *Communications Instructions—Tape Relay,* ACP-127, NATO Communications Procedures, National Communications System Office (NTS-TS), Arlington, VA 22204-2198.

[NT-4] *Communications Instructions—Tape Relay, (U.S. Supplement No. 1) ,* ACP-127 US SUPP-1, NATO Communications Procedures, National Communications System Office (NTS-TS), Arlington, VA 22204-2198.

[SC-1] Schroeder, Manfred R. and Bishnu S. Atal, "Code-Excited Linear Prediction (CELP): High-Quality Speech at Very Low Bit Rates," *Proceedings of IEEE 1985 International Conference on Acoustics, Speech and Signal Processing,* Tokyo, Japan, 1985, pp. 937–940.

[SU-1] Sudhoff, R. I., "The ANDVT Program of Equipment," *Proceedings of Military Speech Technology 87,* November 17-19, 1987, pp. 54–63.

[TR-1] Tremain, T., "The Government Standard Linear Predictive Coding Algorithm: LPC-10," *Speech Technology,* April 1982, pp. 40–49.

7

Performance Measures

Every shipboard communications circuit is used with one objective in mind: to get the message from the sender to the intended recipient. Messages can have various characteristics: they can be long or short, and can consist of alphanumeric symbols, graphic pictures, or voice. Messages can be real-time (voice) or could be non-real-time (teletypewriter messages, facsimile). Circuits can be line-of-sight (ship-to-aircraft) or long-haul (as much as halfway around the globe). For all the different conditions encountered in naval communications, operationally acceptable circuits must deliver messages to the recipient as reasonable replicas of what has been transmitted, that is, the circuits must perform appropriately. The performance of circuits can be defined and measured in terms of the following parameters:

- Coverage
- Link quality
- Link capacity (data rate, bandwidth)
- Reaction time
- Availability.

These parameters are interrelated. The coverage can be increased if the link capacity is reduced; the reaction time can be shortened (down to real time) if the link capacity is increased; and the link capacity can be increased if the quality is decreased. These interrelationships are complex; they can be expressed in terms of link design equations. In this chapter, the key parameters determining the performance of a communications link are discussed.

7.1 COVERAGE

The coverage describes the distance[1] that can be bridged by a communications circuit. The coverage is a function of:

- The antenna gain at the transmitter and receiver
- The transmitter power
- The noise (background, receiver input)
- The signal-to-noise ratio
- The receiver sensitivity
- The received signal quality
- The link capacity (bandwidth and data rates).

7.1.1 Antenna Gain and Directivity

An omnidirectional antenna radiates equal power into all directions, or receives equally well from any direction. An omnidirectional antenna is an idealized antenna, and while it is useful as reference and for theoretical considerations, is not practical. All practical antennas have a directional pattern; they radiate more power into some directions than into others. The ratio of the power radiated into a direction by an antenna to what an omnidirectional antenna would radiate is the gain. Generally, the antenna gain is expressed in dB with respect to the gain of an omni-directional antenna. Typical gains are 20 dB for specially shaped antennas (e. g., log-periodic antennas) and 40 dB or more for reflector antennas.

Antennas have an effective area, which is the area that receives the power density $[Wm^{-2}]$ and transforms it into a signal voltage that appears at the antenna output port and the receiver input port. The effective antenna area is a function of the antenna type (e. g., dipole, reflector, or array) and the signal carrier wavelength. Table 7.1 summarizes the gain and effective area of some antennas (λ is the wavelength, A the physical area of the antenna).

When highly directional antennas are used, the solid angle covered by the effectively radiated power in the main beam is small. Hence, highly directional antennas

---

[1]The distance between transmitter and receiver, in particular for line-of-sight transmission, is also referred to as range.

TABLE 7.1   GAIN AND EFFECTIVE AREA OF ANTENNAS

Type of Antenna	Maximum Gain	Effective Area
Omnidirectional	1 (0 dB)	$\lambda/4\pi$
Elementary Dipole	1.5 (1.76 dB)	$1.5\lambda^2/4\pi$
Halfwave Dipole	1.64 (2.1 dB)	$1.64\lambda^2/4\pi$
Parabolic Reflector	6 to 7.5 A/λ^2	0.5 to 0.6 A
Array (ideal, broadside)	$4\pi A^2/\lambda^2$	A

need to be steerable so that they can be pointed into the direction of the receiver (or transmitter). This requires that steerable antennas be placed on a ship in a manner such that, over the entire range of a ship's motion, the antenna's directional beam can be kept accurately oriented, and will not become obscured by the ship's structure. As a result, the higher part of the ship's structure is very much in demand for the placement of directional antennas.[2]

7.1.2 Transmitter Power

The power of a transmitter is also an important factor for determining the distance that can be covered by a communications circuit. Generally, the higher the power, the larger the distance that can be covered.

There are modifications to this rule. For HF circuits, very long distances can be covered with little power (10 W–100 W) by exploiting ionospheric reflection and concentrating (launching) the signal into the right direction. The latter is accomplished by using directional antennas with an appropriate pattern; in HF communications, the launching direction of the signal is very important. On the other hand, transmissions from very high power HF stations (100 kW and more) may sometimes not reach the destination because of HF propagation constraints.

Generally, the effective transmitted power is expressed as effective isotropic radiated power (EIRP); the EIRP (into a specific direction) is the product of the transmitter power and the antenna gain (into that direction). The EIRP is often used when determining the performance of communication links.

The received power is given by the range equation. The received power P_R is a function of the transmitted power P_T, the transmitter and receiver antenna gains G_T and G_R, the path loss L,[3] and a design margin M. The general expression is:

$$P_R\,[\text{dBW}] = P_T\,[\text{dBW}] + G_T\,[\text{dB}] + G_R\,[\text{dB}] - L\,[\text{dB}] - M\,[\text{dB}] \qquad (7.1)$$

---

[2]This is illustrated in a later chapter. Also, the closeness of antennas on a ship leads to electromagnetic interference; this subject, too, will be discussed in a later chapter.

[3]The path loss L is a function of the range. For satellite links, the expression applicable for L is discussed in detail in Chapter 12.

7.1.3 Noise

Any communications receiver generates noise, even without a signal applied to it. This noise is a natural phenomenon that must be accepted; signals must be reproduced sufficiently accurately despite the noise. Generally, the signal must be more powerful than the noise to be discernible.[4]

There are several sources of receiver noise; they are a result of the receiver physics and the quantitative nature of the flow of electricity, that is, the movement of electrons and their collision. The receiver input noise can be modeled by a resistor matched to the receiver input impedance operating at a certain temperature; the higher the resistor's temperature, the faster the electrons move, the more often they collide, and the more noise power they produce. This type of noise is called thermal noise. From this notion, the *equivalent noise temperature* T_R has been adopted as a measure for receiver noise.

In addition to the receiver noise, there is background noise, that is, the noise that exists in the signal path and that enters the antenna jointly with the signal. There are various sources of background noise, ranging from atmospherics (impacting mainly HF signals); man-made noise from generators, electric lighting, and electric machinery (impacting VHF and UHF signals); galactic and solar noise (affecting SATCOM signals). Figure 7.1, combining information from various sources, shows a typical plot of the representative background noise sources. In Figure 7.1, the background noise is given in terms of the equivalent antenna noise temperatures. Other presentations, such as given in the USAF Electromagnetic Compatibility Handbook DH 1–4 [AF-1], show this quantity in μV/m of electric field; however, such presentations are normalized for a certain bandwidth (10 kHz) and the use of a half-wave dipole antenna.

For the design of VLF, LF, and HF links it is often useful to know the background noise in terms of an electric field strength. The electric field strength E_n of the noise can be related to the radio bandwidth B_{TR} of the signal, the effective antenna areas (and hence the frequency f of the carrier) as follows [JA-1]:[5]

$$E_n \text{ [dB}\mu\text{Vm}^{-1}] = T_{eq} \text{ [dBK]} + B_{TR} \text{ [dBMHz]} + 2f\text{[dBMHz]} - 121.3 \qquad (7.2)$$

---

[4]This statement applies to the total signal energy. By spreading the signal spectrum over a wide bandwidth, the signal *power* in a given bandwidth slice may be smaller than the noise, but after processing, the signal *energy* will be concentrated, and the level of the processed signal can be raised above the noise. Such spread-spectrum systems are used to protect against jamming and for LPI communications.

[5]The noise temperature shown in Figure 7.1 will generate in an omnidirectional antenna the noise power $N = kT_{eq}B_{TR}$. Dividing this noise power by the effective area of an omnidirectional antenna $A = \lambda/4\pi = c^2/4\pi f^2$ produces the background noise power density n, i.e., $n = kT_{eq}B_{TR}4\pi f^2/c^2$. This power density n is related to the rms value of the electric field E_n of the background noise by $E_n^2 = 120 \ \pi n$ [JA-1]; the factor 120π is also referred to as the radiation impedance of free space. When inserting the background noise power density n one obtains:

$$E_n^2 = kT_{eq}B_{TR}480\pi^2 f^2/c^2.$$

When the field is converted from [Vm^{-1}] to [μVm^{-1}] and the carrier frequency f from [Hz] to [MHz], then

$$E_n^2 = 10^{24}kT_{eq}B_{TR}480\pi^2 f^2/c^2.$$

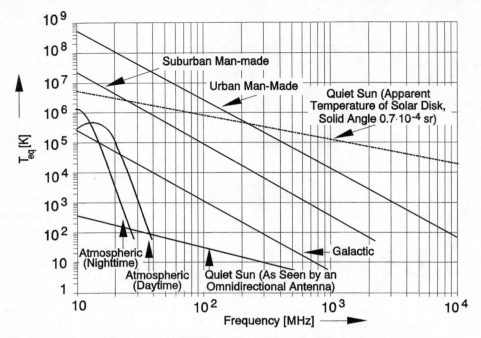

Note: The noise temperatures are given as they would be seen by an omnidirectional antenna
(For the Quiet Sun the disk temperature is also shown as seen by a directional antenna with a gain of 46 dB or more)

Figure 7.1   Noise from Various Sources, Expressed as Equivalent Operating Temperature

As it can be seen in Figure 7.1, the background noise contribution is different in each part of the frequency band. The background noise together with the receiver noise will determine the overall system performance.[6]

An additional signal interference phenomenon is multipath, that is, the same signal is received via different paths.[7] When the path lengths differ by a full wavelength or a multiple

In logarithmic form this becomes

$$E_n[\text{dB}\mu\text{Vm}^{-1}] = T_{eq}[\text{dBK}] + B_{TR}[\text{dBHz}] + 2f[\text{dBMHz}] - 121.3.$$

[6]Note that in Figure 7.1 the noise temperatures from various sources are given as they would be seen by an omnidirectional antenna near the Earth's surface. Antennas with a gain larger than 1 will receive less than the total background noise, hence the noise temperature would be reduced by the gain. The Sun is a point source and the noise from the Sun can be a major contribution in SATCOM links using highly directional antennas that may be pointing at the Sun (the solar disk occupies a solid angle of $0.7 \cdot 10^{-4}$ sr). Hence, two lines are shown in Figure 7.1, one for the quiet Sun as seen by an omnidirectional antenna, and one as seen by a highly directional antenna. Incidentally, the noise from the active or disturbed Sun is significantly more powerful than from a quiet Sun.

[7]This is particularly noted in HF transmission, where it is caused by ionospheric reflections; however, it is just as destructive in VHF and UHF communications between ships and aircraft, where it can be caused by reflections from a ship's structure or from the surface of the sea.

thereof, the signals are added; however, if the path lengths differ by one half, three halves, etc., wavelengths, the signals cancel each other. This causes fading. Multipath voice signals can superpose to create a hollow sound, an effect that is also known as barreling.[8]

Noise is a random variable, that is, a noise voltage has a random amplitude and phase. Each individual noise voltage sample, generated by a source, such as a resistor or a front-end amplifier, considered over a certain time interval, is different. There are, however, common characteristics attributable to noise voltage samples. The sum-total of all noise samples representing the thermal noise of a source can be characterized as a gaussian random variable and described by two parameters: a mean value and a variance. Hence, receiver noise is referred to as gaussian noise. Impulse noise and HF fading are described in a different manner; their descriptions are more complex.

Communications systems use various methods for overcoming noise. Redundancy coding can be applied to guard against signal elements lost to noise. Various codes can be used for the different types of noise; for example, Hamming codes are effective against gaussian noise, and Fire codes protect against certain forms of impulse noise [PE-1]. Fading can also be overcome by redundancy coding; however, signal interleaving to bridge the periods of fades will be more effective to provide reliable communications. Fading periods can vary from seconds to minutes, and long-term fading is very difficult to overcome by coding; however, frequency diversity reception, that is, reception of the same signal transmitted at different frequencies, or space diversity, that is, reception of the same signal transmitted at a single frequency and received by physically separated antennas, can be used to mitigate signal deterioration from fading.

Another type of noise is quantization noise. In naval communications, signals are generally not transmitted in an analog format but digitally in order to make use of encryption. The required analog-to-digital (A/D) conversion produces a quantization error. This error is a function of the quantum steps used in the A/D conversion. The quantization error is considered to be a continuous random variable with a uniform amplitude distribution over the quantization step α having the limits $+\alpha/2$ and $-\alpha/2$. Then, the quantization noise N_q can be expressed as:

$$N_q = \frac{\alpha^2}{12} = \frac{A^2}{12\,(L-1)^2}, \tag{7.3}$$

where L is the number of quantum steps covering the full range of the quantization, A is the peak-to-peak signal amplitude, and N_q is the average noise power resulting from the quantization in a 1-ohm resistor.

Quantization noise and system noise will both affect the quality of the received signal. They will combine in their effect; a more detailed discussion is given in [FI-1].

Two limiting cases are of interest:

1. The system noise is predominant, (i.e., e_B is 10^{-5} or more) and the quantization noise is negligible (i. e., the system uses many quantization levels, generally 128 or more). Then the $(S/N)_{out}$ is:

---

[8]This term originates from the sound of voice in a large barrel, where acoustic reflections create a multipath effect.

$$(S/N)_{out} \approx 10 \log e_B - 6, \tag{7.4}$$

where e_B is the bit error rate.

2. The system noise is negligible (i.e., e_B is 10^{-6} or less) and the quantization noise is predominant. Then the $(S/N)_{out}$ is:

$$(S/N)_{out} \approx \left[\frac{S_{in}}{N_q} \right] = 20 \log L \approx 6n, \tag{7.5}$$

where n is the number of bits in the code.

It needs to be understood that in a quantized digital system a single bit error may have a varying effect: if the most significant bit of a code word is in error, the error is more severe than if the least significant bit would be affected.

7.1.4 The Signal-to-Noise Ratio

In a communications link signal and noise are combined in the receiver. The signal-to-noise ratio (SNR) is the primary parameter that characterizes system performance; it will be used in the more detailed discussions in other parts of this book.

The SNR is the ratio of the power of the useful signal (S) to the total noise power (N); it is denoted by (S/N). It is a dimensionless number and is customarily given as a logarithmic value in dB. The value of (S/N) at a receiver/demodulator input (in the transmission bandwidth B_{TR}) is the input SNR, $(S/N)_i$; at a receiver/demodulator output (in the baseband B_{out}) it is $(S/N)_{out}$. In general, the noise power N in a bandwidth B is given by $N = kT_{op}B = N_0B$, where $N_0 = kT_{op}$ is the noise power density, k is Boltzmann's constant, and T_{op} the operating temperature. Thus, with the above assumptions, the noise power in the radio band is $N_{in} = kT_{op}B_{TR}$, and in the baseband $N_{out} = kT_{op}B_{out}$. From that follows that $(S/N)_i = S/kT_{op}B_{TR}$ and $(S/N)_{out} = S/kT_{op}B_{out}$.

The relationship of signal and noise power at a receiver-demodulator is illustrated in Figure 7.2. At the input of a receiver-demodulator the power is S, the noise power density N_0. The signal power S' at the output of the receiver-demodulator is, for most practical purposes, equal to S, the signal power at the input. For many communications links the noise at the input is gaussian with a constant density N_0 over the transmission bandwidth B_{TR}. At the demodulator output the noise power N'_0 is a function of the demodulation process [DA-1].[9] For a coherent detector, as it is often used in digital systems, the signals are shifted from the radio band to the baseband, and then filtered over the width of the baseband B_{out}, the noise power density N'_0 in the baseband is, for all practical purposes, equal to N_0, the noise power density in the radio band. The SNR at the output $(S/N)_{out}$ can be larger than at the input $(S/N)_i$; this is accomplished by signal processing; for example, in spread-spectrum communications for anti-jamming (AJ) protection.

In the design of VLF, LF, and HF links it is often customary to use the field strength

---

[9]For example, a square-law detector device will transform gaussian noise into chi-square distributed noise [DA-1].

Front-End Signal In:
Signal Power = S
Bandwidth = B_{TR}
Noise Power Density = N_0
Noise Power = $N_0 B_{TR}$
SNR = $(S/N)_i$ = $S/N_0 B_{TR}$

Demodulator Signal Out:
Signal Power = S'
Bandwidth = B_{out}
Noise Power Density = N'_0
Noise Power = $N'_0 B_{out}$
SNR = $(S/N)_{out}$= $S'/N'_0 B_{out}$

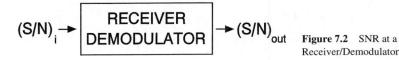

$(S/N)_i \rightarrow$ **RECEIVER DEMODULATOR** $\rightarrow (S/N)_{out}$ **Figure 7.2**   SNR at a Receiver/Demodulator

of the received signal, E_R, and the field strength of the background noise. When it is observed that both fields are received by the same antenna, that is, both signals are seen by the same effective antenna area, and are terminated in the same port impedance, then the SNR at the receiver input is

$$(S/N)_i = (E_R/E_n)^2. \tag{7.6}$$

This equation links the noise field strength as given in eq. (7.2) to the field strength of the received signal.

For some considerations, the *carrier-to-noise* C/N_0 is more useful than the SNR. If one visualizes an unmodulated carrier signal of power C and considers the SNR in a 1-Hertz band, one obtains $\{S/N\}_{(1Hz)} = C/N_0$ with the logarithmic measure [dBHz]. When the carrier is modulated and its power spreads over the bandwidth B_{TR}, the carrier power C becomes the signal power S, the noise power in the transmission band becomes $N_0 B_{TR}$, and $(S/N)_i = (C/N_0)(1/B_{TR})$.

For communications systems, the performance is generally given as a function of the SNR, for example, for an analog video link an SNR of 30 dB will produce acceptable picture quality. Hence, for an analog communications link, the SNR $(S/N)_{out}$ is usually specified. The SNR $(S/N)_{out}$ is related to $(S/N)_i$; when considering that the receiver demodulator will produce the signal with a low-pass bandwidth B_{out}, and designating the bandwidth ratio $b_r = B_{TR}/B_{out}$, then one has

$$(S/N)_i = \frac{S}{kT_{op}B_{out}} \frac{B_{out}}{B_{TR}} = \frac{(S/N)_{out}}{b_r}, \tag{7.7}$$

and

$$(S/N)_{out} = (S/N)_i b_r. \tag{7.8}$$

For example, when a broadband FM signal is demodulated, the output bandwidth B_{out} will be a fraction of the transmission bandwidth B_{TR}, and $(S/N)_{out} > (S/N)_i$.

For digital signals, however, where signal processing is performed, the duration τ_{sy} of the transmission symbol is important. One form of signal processing consists of integration of the signal and noise mixture within the band B_{TR} in the detector over the duration τ_{sy}. The performance then depends on $S\tau_{sy}$, the symbol energy. This leads to:

$$(S/N)_{out} = \frac{S\tau_{sy}}{N_0 B_{out} \tau_{sy}} = \frac{E_{sy}}{N_0 B_{out} \tau_{sy}}, \tag{7.9}$$

or

$$\frac{E_{sy}}{N_0} = (S/N)_{out} B_{out} \tau_{sy}. \tag{7.10}$$

The parameter E_{sy}/N_0, that is, the ratio of symbol energy to the noise power density,[10] is also known as the signal's energy ratio and denoted by ε_r. Furthermore, considering that $(S/N)_i = (S/N)_{out} (B_{out}/B_{TR})$, one obtains from eq. (7.10)

$$\varepsilon_r = \frac{E_{sy}}{N_0} = (S/N)_i B_{TR} \tau_{sy}. \tag{7.11}$$

The expression $B_{TR}\tau_{sy}$ is the time-bandwidth product of the signal, (also named the WT product); it expresses the signal processing gain; for a small $(S/N)_i$ one can obtain a sizable ε_R. The error rate performance of digital communications systems is a function of ε_r (or E_{sy}/N_0). Hence, for digital communications systems, where the error rate is generally specified, system design strives to accomplish a high performance by choosing a modulation method that offers a high processing gain.

When the received power P_R[11] from the range equation (see relationship 7.1) is combined with the noise power at the receiver input, that is, $N_0 = kT_{op}B_{TR}$, one obtains the SNR $(S/N)_i$; in logarithmic form it is:

$$(S/N)_i[dB] = P_T[dBW] + G_T[dB] + G_R[dB] - L[dB] - M[dB] - kT_{op}B_{TR}[dBW]. \tag{7.12}$$

This equation is a fundamental link design equation.

7.1.5 Receiver Sensitivity

The receiver sensitivity is a measure of the ability of a receiver to amplify and detect a signal so that it will be distinguishable from the environmental noise. A receiver always adds noise to a received signal; from a system viewpoint the background (external) noise, entering together with the signal via the antennas and the receiver noise, are both factors affecting the system performance. The input signal level that may be satisfactorily processed by a receiver to produce an acceptable performance is called receiver sensitivity [RH-1].

There are several definitions of receiver sensitivity; the measure will depend on whether it is for an analog receiver, that is, AM or FM, or for a digital receiver, that is, FSK, PSK, etc. The definition may also differ, whether it is for a receiver in an LOS or long-haul link.

---

[10]Note that the noise power density has the dimension of energy, i.e., [WHz^{-1}]; this is a spectral density and must not be confused with the power density.

[11]P_R corresponds to S in Figure 7.2.

AM sensitivity.   The sensitivity of an AM receiver is expressed in terms of an input voltage[12] [BE-1] that will produce a specified SNR. The customary definition uses as input signal a carrier modulated at a specified percentage (usually 30%), modulated by a tone (usually 1000 Hz), and a specified input bandwidth (usually 6 kHz). The sensitivity is the voltage (in μV) that produces a specified output SNR (usually 10 dB).[13] An alternative is specifying the input power required to produce this voltage across the receiver input impedance (usually 50 Ω).[14] A sensitivity of 1 μV would translate into -107 [dBm] (dB with respect to 1 mW) when the receiver impedance is 50 Ω.

FM sensitivity.   For FM receivers a similar definition is used. Here the input signal is a carrier modulated by a tone (usually 1000 Hz) such that it produces a specified deviation (e.g., 3 kHz) and specifying the receiver baseband width (e.g., 3 kHz). The sensitivity is the voltage (in μV) that produces a specified SNR at the output (e.g., 10 dB).[15]

Noise figure.   For some applications, the noise figure F is a useful measure of the noise contribution of the receiver front-end amplifier. It is a measure that compares how a receiver performs with respect to an ideal noiseless receiver. The noise figure (in dB) is given by the expression:

$$F[\text{dB}] = 10\log\left[1 + \frac{T_R}{290}\right]$$ (7.13)

Operating temperature.   This approach is used for LOS and SATCOM links. The receiver front-end noise is expressed as an equivalent receiver noise temperature T_R, that is, the temperature of a matched input resistor to produce the front-end noise. From the receiver noise temperature and the effective receiver bandwidth B_{TR}, the receiver input noise N_R can be expressed as $N_R = kT_RB_{TR}$, where k is Boltzmann's constant. External noise (produced by the environment) is added in terms of an equivalent environmental input noise temperature T_{eq}, so that the total noise is then given by $N_T = k(T_{eq} + T_R)B_{TR}$ $= kT_{op}B_{TR}$ where T_{op} is the operating temperature of the communications link. Examples on how to determine the operating temperature for a SATCOM link are given in Chapter 12.

---

[12]Note that the receiver input voltage is a function of the electrical field strength E_R at the receiving antenna and the so-called effective height h of the antenna. The voltage V at the antenna port (its terminals) is $V = hE_R$; the antenna height h is the distance along the antenna over which the field acts. For example, for a dipole of length l that is situated parallel to the E-field the effective height is l; if the dipole is inclined at an angle θ, the effective height is $l \cdot \sin\theta$ [BE-1].

[13]Actually, the 10 dB are a ratio of $(S+N)/N$ (i.e., signal plus noise to noise) at the receiver output, since the sensitivity is an empirical measure that cannot be verified in the absence of noise.

[14]The input power (received by an omnidirectional antenna) can also be obtained when the signal power density $p[\text{dBWm}^{-2}]$ from the transmitter is known at the receiver. The input power is the p/A, where A is the effective antenna area. The power density p and the field strength E_R of the received signal are related by $p = E_R/120\pi$ [JA-1].

[15]For FM the 10 dB are a ratio of $(S+N+D)/N$ (i.e., signal plus noise plus distortion to noise) at the receiver output. This ratio is also known as SINAD.

Table 7.2 summarizes the sensitivity for various receivers as given in data sheets. It is intended to provide an overview of current practices how the sensitivity for a receiver is given and not to be an exhaustive listing of the sensitivity of receivers in use for naval communications.

7.1.6 Link Quality

Link quality is a measure of the integrity of the data arriving at the destination. In the transmission of information a message is sent; the message can be characterized as a random process that is embedded in other random processes (noise). Fundamentally, the message will always have errors since the noise is added to it. The design of a communications link must then *reduce the error rate* (message elements received in error versus message elements received correctly) *to a level meeting an operational requirement for a link.*

For digital communications systems, the link quality is measured by the bit error rate (BER) e_B. The bit error rate depends on the received signal power to noise ratio and the demodulation method used; the latter usually involves integration over the transmission symbol duration τ_{sy}. Much fundamental work has been done to characterize the BER in digital transmission systems as a function of the detection method; good references on

TABLE 7.2 A SAMPLE OF RECEIVER SENSITIVITY VALUES

Receiver	Operating Frequency	Modulation	Sensitivity for 10 dB $(S + N)/N$ or 10 dB SINAD
AN/URR-69	0.5–30 MHz	AM FM	1.0 μV (30% modulation, 1 kHz tone, 4 kHz bandwidth) 0.8 μV ($\pm$ 8 kHz deviation, 1 kHz tone, 25 kHz bandwidth)
AN/URC-94	1.5–80 MHz	SSB AM	0.5 μV 0.6 μV
AN/URC-103(V)	2.0–30 MHz	AM	0.5 μV (or -113 dBm)
AN/ARC-182	30–88 MHz 108–156 MHz 225–400 MHz 156–174 MHz 225–400 MHz	FM AM AM FM FM	-112 dBm (2.4 kHz deviation, 1 kHz tone) -103 dBm (30% modulation, 1 kHz tone) -103 dBm (30% modulation, 1 kHz tone) -110 dBm (2.4 kHz deviation, 1 kHz tone) -110 dBm (2.4 kHz deviation, 1 kHz tone)
AN/ARC-201 (SINCGARS)	30–88 MHz	FSK	-109 dBm (25 kHz bandwidth)
JTIDS	969–1008 MHz 1053–1065 MHz 1113–1206 MHz	MSK MSK MSK	-95 dBm (3 MHz bandwidth) -95 dBm (3 MHz bandwidth) -95 dBm (3 MHz bandwidth)
AN/WSC-3	225–400 MHz	AM FM	3.5 μV 3.0 μV (25 kHz bandwidth)

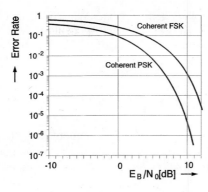

Figure 7.3   Error Rate as a Function of the Bit Energy to Noise Power Density Ratio (E_b/N_0)

signal detection are [VI-1; SC-1]. With demodulation using integration over the duration of a transmission symbol, the error rate depends on the transmission symbol energy. In many systems the integration is carried out over each bit, and thus the error rate is generally given as a function of the bit energy to noise power density ratio E_B/N_o (or the energy ratio ε_r). Figure 7.3 shows the error curves for digital systems using synchronous coherent detection. The bit error rate $e_{B[PSK]}$ for coherent PSK detection is given by $e_{B[PSK]} =$ (1/2) erfc ($\sqrt{\varepsilon_r}$), and the error rate $e_{B[FSK]}$ for coherent FSK is given by $e_{B[FSK]} =$ (1/2) erfc ($\sqrt{2\varepsilon_r}$), where the error function complement erfc(x) is defined by the expression[16] erfc $(x) = 1 - (1/2\sqrt{\pi}) \int_x^\infty \exp(-z^2)dz$. As shown in Figure 7.3, PSK performs by 3 dB better than FSK, hence the error curve for FSK is shifted by 3 dB to the right of the errorcurve for PSK. An error rate of 10^{-5} is obtained for $E_B/N_0 = 9.6$ dB for PSK and at $E_B/N_0 = 12.6$ dB for FSK.

Additional protection against errors, that is, reduction of the error rate, can be accomplished by coding. The classical work of Peterson [PE-1] discusses the coding theory; practical applications are found in the book by Viterbi [VI-1].

Typical acceptable error rates for practical digital systems are:

- Computer-to-computer data links 10^{-7}
- Voice 10^{-4}
- Record messages (TTY) 10^{-4}
- Facsimile 10^{-3}
- Video imagery 10^{-4}.

The quality of voice signals is often measured by intelligibility. The intelligibility of voice is based on the articulation index (AI), a measure of the percentage of single and multi-syllable words that are understood when transmitted. The word should be chosen from phonetically balanced (PB) word lists. The theory and practice of using PB word lists and performing AI measurements has been developed by the telephone companies [BE-2]. Generally, an 80% understandability of single-syllable PB words will result in

[16]See reference [SC-1].

better than 95% intelligibility of spoken language. When voice is transmitted in digitized and compressed form (compressed by a VOCODER) with an error rate of 10^{-4} or less, the intelligibility (barring nonlinear distortions from analog amplification and voice reproduction equipment) generally exceeds 95%.

7.2 LINK CAPACITY, BANDWIDTH, AND DATA RATE

The link capacity is given as an operational requirement. It is the data rate that can be transmitted with a specified error rate.[17] The link design parameters, such as transmitter power, transmission bandwidth, and link quality, need to be chosen such that the requirement can be met.

The link capacity, its data rate, and its bandwidth are connected by a complex relationship. This relationship is defined by the interdependence of the various parameter defining a digital transmission system. A generalized block diagram of a digital communications link is shown in Chapter 2, Figure 2.2. For a digital transmission system the modulation process transforms the information bits of duration τ_b, arriving at the rate $R_b = 1/\tau_b$, into transmission symbols. These symbols (or waveforms) are often sinusoids of duration τ_{sy}, where the information is contained in the amplitude, frequency, or phase.[18] In general, each symbol can transfer m bits (by using multiple amplitudes, frequencies, or phases).[19] The symbol rate R_{sy} resulting from the modulation process is the inverse of the symbol duration τ_{sy}, that is, $R_{sy} = 1/\tau_{sy}$. Because there are m bits per symbol, the bit rate R_b and the symbol rate R_{sy} are related by $R_b = mR_{sy} = m/\tau_{sy}$.

The stream of transmission symbols occupy the one-sided bandwidth B_{out}, that is, the low-pass bandwidth of the signal to be transmitted. The ratio R_b/B_{out} (in bit s^{-1} Hz^{-1}) is a bandwidth effectiveness ratio [SK-1];[20] here the bandwidth effectiveness is denoted by η_B, that is, $\eta_B = R_b/B_{out}$, or $\eta_B = 1/(\tau_b B_{out})$, where τ_b is the effective duration of each bit. The symbol stream is generally filtered, in order to contain B_{out} to the smallest value possible.

For transmission, the stream of symbols generally needs to be translated into the radio band; this translation can be accomplished by a second modulation step that puts the signal onto a radio carrier. The resulting signal occupies a radio channel with the bandwidth B_{TR}. The translation process can be a frequency shift, as it is accomplished by SSBAM, where $B_{TR} = B_{out}$, or by some other method that may spread the spectrum, such as DSBAM or FM, where $B_{TR} > B_{out}$. The bandwidth B_{TR} will also be contained so that there is no interference with other radio signals that occupy adjacent radio channels.

[17]Or alternatively, the bandwidth required to transmit an analog signal with a specified quality.

[18]Typical examples are amplitude shift keying (ASK), frequency shift keying (FSK), and phase shift keying (PSK).

[19]For example, 8-ary FSK, implemented in the MILSTAR uplink, uses a set of eight symbols of different frequency, where each symbol transmits three bits (see also Figure 15.11).

[20]This ratio is also called bandwidth efficiency by Sklar [SK-1]; however, the values of an efficiency range between 0 and 1, while the value of R_b/B_{out} can assume values larger than 1. Therefore, the designation "bandwidth effectiveness" is more appealing.

Given an error rate and a modulation method (that determines m, that is, the number of bits per transmission symbol) one obtains an energy ratio ε_r (or E_b/N_0) at which the link needs to operate.[21] Then, since $\varepsilon_r = (S/N)_i B_{TR}\ \tau_{sy}$, and since $N_i = B_{TR}\ \tau_{sy}$, one obtains $\varepsilon_r = (S/N_0)\ \tau_{sy}$. From the other link design parameters (power, range, receiver noise) the value of S/N_0 (which is equivalent to C/N_0) is computed, and τ_{sy} follows. The last expression for ε_r describes the relationship between the bit error rate and the symbol rate $R_{sy} = 1/\tau_{sy}$, and from this also the bit rate $R_b = m/\tau_{sy}$ that the link can support.

In the complex process of communications link design various choices need to be made, and there is a lot of room for trading among the link parameters, that is, data rate, bandwidth, power, and receiver sensitivity.

Thus, when a long communications range needs to be covered, the available power is low, and a high gain receiver antenna cannot be used, then a link with reduced data rate could be used. The system would then operate below its level of requirement, but it would still provide some minimum communications services.

Low data rates are typically 75 bps and 2400 bps. 75 bps is generally the speed for teletypewriter (TTY) communications, and 2400 bps is used for high-speed TTY; 2400 bps is also the minimum rate for encoded voice using compression.[22]

Generally, in order to accomplish a mission, circuits for different services are required. These services include:[23]

- Computer-to-computer data transmission
- Voice
- Record message TTY
- Facsimile
- Video imagery.

A shipboard communications system design must be balanced to meet mission requirements. To meet all shipboard communications requirements, several links are generally needed consisting of a mixture of media; VHF and UHF LOS provide short-haul communications, and HF and SATCOM provide for long-haul communications. Prudent design practices dictate that certain essential mission capabilities be retained under the most adverse conditions, such as record message TTY circuits; and that diversity, redundancy, and alternate routing be provided for contingencies.

[21]Figure 7.3 shows two examples of system performance curves that relate the ε_r to the bit error rate. These curves are for binary systems, i.e., the bit and the transmission symbol are one and the same, or, $m = 1$.

[22]Using a low data rate is not necessarily a panacea for increasing range, unless special receivers are used. Phase-locked loop receivers require a certain bandwidth and power to stay in lock; the break-even point corresponds to about 300 bps. This is to say that a phase-locked loop receiver will require the power of approximately 300 bps at an error rate of 10^{-5} to maintain lock. Special receiver designs, however, can be used for very low data rate links (1 bps or less for ELF, 50 bps or less for VLF) where the available bandwidth is very low and signal loss due to penetration into seawater is high. The relationship between data rate, bandwidth, and SNR for VLF is discussed in Chapter 9.

[23]Table 6.1 summarizes the bandwidth and data rate requirements of various naval communications services.

7.3 REACTION TIME

The reaction time of a link is the time required to deliver a message from the time it was sent. This time is a function of many variables, most of them nontechnical, that depend on the operational procedures established for a particular link. These variables include message routing, message priority, link availability (not in the sense of an operational availability A_o, but access to a link to send the message, which in turn depends on priorities). Such factors as equipment capabilities, message handling delays by personnel, and equipment configuration enter into the consideration.

Reaction time is a matter of mission requirements and matching a communications link to these requirements. Examples of response times to accomplish a mission successfully are:

- Missile warning—2 to 10 minutes
- Aircraft and submarine threat warning—10 to 15 minutes
- Priority messages—15 to 30 minutes
- Routine messages—10 to 120 minutes.

The general rule for determining a reaction time is:

- Threat detection messages arrive in time for a decision-making officer to take action to avoid damage of a task force.
- Command and control messages need to arrive in time for the decision-making officer to coordinate his efforts to positively support the task force.
- Other messages (weather) need to arrive in time to be useful (e.g., storm warnings are needed before the storm arrives).
- Routine messages need to arrive while they are still imparting valid news.

7.4 AVAILABILITY

Link availability is the probability of a link being operational when the service is required. The design goal of an operational availability for circuits should be 95% or more, depending on the criticality of the link for accomplishing a mission.

Generally, the operational availability, A_o, is given by the relationship (NA-1):

$$A_o = \frac{\text{Uptime}}{\text{Total Time}} = \frac{\text{Uptime}}{\text{Uptime} + \text{Downtime}}. \tag{7.14}$$

Uptime is the time during which the system performs its required function. Downtime is the time during which the system is not performing its required function, either because of failure, maintenance actions, or logistic delays.

The relationship given above defined A_o in principle, but it does not show how to control the uptime and downtime through system design factors. To do this the uptime and downtime must be defined in terms of reliability, maintainability, and supportability, which are measurable factors. They are defined as follows:

- *Reliability* is the probability that an item[24] performs its function for a specified time under specified conditions. It is controllable by design. Its measure is the mean time between failures (MTBF). The MTBF is a measure of failure-free performance, it is a parameter in the expression of an item's operation as a function of time [BA-1; NA-1].[25] The MTBF is an item's uptime.
- *Maintainability* is the capability of an item to be restored to its specified operating condition when maintenance and repair is performed by skilled and trained personnel. Its measure is the mean time to repair (MTTR). The MTTR is a measurement of the inherent maintainability designed into the item. The MTTR is a component of the item's downtime.
- *Supportability* is the ability of satisfying all conditions needed to restore a failed item to its operational condition. Its measure is the mean logistic delay time (MLDT). The MLDT is a measure of delays not attributable to hands-on maintenance; it includes providing parts, repair at other levels, and other logistic delays. The MLDT is a component of the item's downtime.

With the given definitions A_o becomes:

$$A_o = \frac{\text{MTBF}}{\text{MTBF} + \text{MTTR} + \text{MLDT}}. \tag{7.15}$$

This expression is the design equation for determining a system's A_o. It is valid for systems that are in continuous use. If a system is not in continuous use, this relationship is modified as follows:

$$A_o = \frac{k\text{MTBF}}{k\text{MTBF} + \text{MTTR} + \text{MLDT}}. \tag{7.16}$$

The factor k is the ratio of total calendar time (Cal) less the downtime divided by the uptime, that is,

$$k = \frac{\text{Cal} - \text{MTTR} - \text{MLDT}}{\text{MTBF}}. \tag{7.17}$$

---

[24]The term *item* is a general term; an item can be a replaceable unit, a subsystem, or a system.

[25]The reliability is a probability and a function of time. For items that are not subject to wear-out (e.g., light bulbs or electric motors), the reliability is given by $R(t) = exp\,(-\frac{t}{\text{MTBF}})$ (exponential reliability). The MRBF is then given by $\text{MTBF} = \int_0^\infty R(t)\,dt$; this relationship needs to be used if the MTBF of a system with a complex reliability model is computed [BA-1].

7.4.1 Availability Modeling

When designing a system, the A_o can be calculated from the A_{oj} of its components, usually referred to as line replaceable units (LRUs). Since the availability is a probability, the rules for modeling a system from its components are as follows:

- For systems without redundancy, where the failure of a single one of the n LRUs with the availability A_{oj} would cause the system to fail, each of the LRUs is considered to be in series in the A_o model. Each LRU is a single point of failure in the system. In this case, the composite A_o is the product of the A_{oj} (serial A_o).
- For systems with redundancy, where n LRUs with an A_{oj} are in parallel, all LRUs would have to fail to result in a system failure, and the LRUs are considered to be in parallel for the A_o model. In this case, the composite unavailability U_{oj} ($U_{oj} = 1 - A_{oj}$) is the product of the U_{oj} and the composite A_o can be computed from it (parallel A_o).
- For systems where some LRUs are redundant (in parallel in the A_o model) and some nonredundant (in series in the A_o model), a composite A_o must be constructed from the serial and parallel groups.

From these rules, the mathematical expressions for the system A_o can be derived. In the model, each LRU is represented by a block. The serial A_o for J LRUs (or subsystems), each with an A_{oj}, is:

$$A_{o[\text{ser}]} = \prod_{j=1}^{J} A_{oj} \qquad (7.18)$$

The parallel A_o is obtained by first computing the unavailability U_o from the U_{oj} for the blocks

$$U_{o[\text{par}]} = \prod_{j=1}^{J} U_{oj} = \prod_{j=1}^{J} (1 - A_{oj}) \qquad (7.19)$$

and with $A_o = 1 - U_o$ one arrives at

$$A_{o[\text{par}]} = 1 - \prod_{j=1}^{J} (1 - A_{oj}) \qquad (7.20)$$

This is illustrated in Figure 7.4, where a serial, a parallel, and a combination of serial and parallel A_o are shown.[26]

The effect of A_o improvement using redundancy is shown in Figure 7.5. It shows the A_o for two items in series and two items in parallel.

7.4.2 Improving the Availability of Communications Circuits

In order to improve the A_o of a communications system, one must raise the system's MTBF and reduce the MTTR and MLDT. The MTBF can be improved by providing redundancy. Figure 7.5 indicates that the A_o of a system can be significantly improved by

---

[26]The same model structure can be used for system reliability models. If every term A_{oj} in Figure 7.4 is replaced by R_j, one would obtain the reliability of a system.

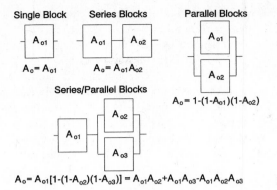

$A_o = A_{o1}[1-(1-A_{o2})(1-A_{o3})] = A_{o1}A_{o2}+A_{o1}A_{o3}-A_{o1}A_{o2}A_{o3}$ **Figure 7.4** A_o Models

providing redundancy. For communications systems, that translates into providing redundant equipment (such as hot spares) and alternate circuit routing. If a circuit is lost, it can be established by rapidly patching in a standby unit or establishing an alternate route via another medium.

When interruptions of a link are critically impacting the accomplishment of a mission, then the MTTR and MLDT need to be limited to a maximum allowed interruption that will result in an acceptable A_o. The MTTR then needs to be interpreted as a "time to reestablish communications" rather than an MTTR in the classical sense. The MLDT can

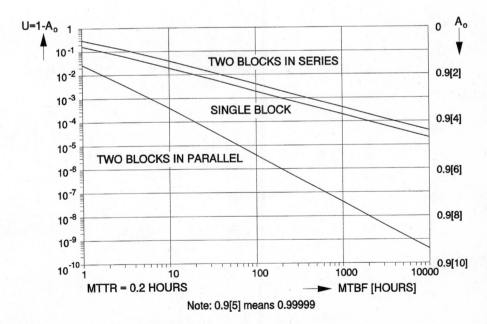

Figure 7.5   Availability of a Simple Model (Serial and Parallel)

be reduced by having a sufficiently large inventory of spare parts on hand. Maintenance and repair can then be expeditiously performed without excessive logistic delays.

In naval communications, the elements of a system are often controlled by different system commands.[27] For example, in a satellite link the shore ground station reports to one commander, the ground links are under another command, the shipboard segment is designed by another command, and it is maintained by another one. In such cases the A_o elements contributing to the overall system A_o need to be allotted to each command. This can be done by generating a decision coordinating document by the office of the Chief of Naval Operations (CNO) allotting to each command a certain A_o such that the overall system A_o will meet the needs put forth in the system's operational requirement (OR).

REFERENCES

[AF-1]   "Electromagnetic Compatibility," *AFSC Handbook DH 1-4, fourth ed.*, Doc. No. AFSC DH 1-4, 2 March 1984, HQ AF Systems Command, Andrews AFB, Washington, D.C. 20334.

[BA-1]   Bazovsky, I., *Reliability Theory and Practice*, Prentice Hall, Englewood Cliffs, N.J., 1961.

[BE-1]   Berkowitz, R. S., *Modern Radar Analysis, Evaluation and Design*, John Wiley and Sons, New York, N.Y., 1965.

[BE-2]   Beranek, L., *Acoustic Measurements*, John Wiley & Sons, New York, N.Y., 1962.

[DA-1]   Davenport, W. B. and W. L. Root, *An Introduction to the Theory of Random Signals and Noise*, McGraw-Hill, New York, N.Y., 1958.

[FI-1]   Filipowsky, R. F. and E. I. Muehldorf, *Space Communication Systems*, Prentice Hall, Englewood Cliffs, N.J., 1965.

[JA-1]   Jasik, H., *Antenna Engineering Handbook*, McGraw-Hill, New York, NY, 1961.

[NA-1]   *Operational Availability Handbook*, OPNAVINST 3000.12, Department of the Navy, Office of the Chief of Naval Operations, Washington, D.C. 20350, 29 Dec 1987.

[PE-1]   Peterson, W., and E. J. Weldon, Jr., *Error Correction Codes*, 2nd ed. MIT Press, Cambridge, Mass., 1972.

[RH-1]   Rhode, U. L. and T. T. N. Bucher, *Communications Receivers Principles and Design*, McGraw-Hill, New York, N.Y., 1990.

[SC-1]   Schwartz, M., *Information Transmission, Modulation, and Noise*, McGraw-Hill, New York, N.Y., 1970.

[SK-1]   Sklar, B., "Defining, Designing, and Evaluating Digital Communication Systems," *IEEE Communications Magazine*, Vol. 31, No. 11, Nov. 1993, pp. 92–101.

[VI-1]   Viterbi, A., *Principles of Coherent Communications*, McGraw-Hill, New York, N.Y., 1966.

[27]Examples are the Naval Sea Systems Command, the Naval Computer and Telecommunications Command and the Space and Naval Warfare Command.

8

AJ Communications

Electronic Counter-Countermeasures (ECCM) are an important part of military communications. A common countermeasure (taken by the adversary) is jamming. The counter-countermeasures are protection against the jamming, also known as anti-jamming or AJ techniques.

8.1 INTRODUCTION

AJ techniques are based on band spreading of the signal in the radio frequency (RF) band. Considering the relationships between the channel capacity, the signal bandwidth, and the signal-to-noise ratio, one can write the following expressions for the input and output of a receiver plus a demodulator (see also Figure 8.1):[1]

$$C_{TR} = B_{TR} \log_2 [1 + (S/N)_i], \qquad (8.1)$$

$$C_{out} = B_{out} \log_2 [1 + (S/N)_{out}], \qquad (8.2)$$

where C_{TR} and C_{out} are the channel capacities as defined by Shannon [SH-1], B_{TR} and B_{out} the channel bandwidths, and $(S/N)_{TR}$ and $(S/N)_{out}$ the signal-to-noise power ratios.

[1]This figure is similar to Figure 7.2 but it explicitly includes the demodulation processing gain γ_{PR} that can be realized from band spreading.

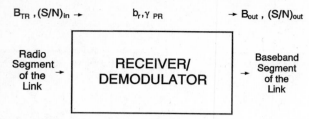

$B_{TR}, (S/N)_{in} \rightarrow$ b_r, γ_{PR} $\rightarrow B_{out}, (S/N)_{out}$

Radio Segment of the Link $\rightarrow$ **RECEIVER/ DEMODULATOR** $\rightarrow$ Baseband Segment of the Link

B_{TR} .... Transmission (RF) Bandwidth
B_{out} .... Bandwidth in the Baseband
$(S/N)_{in}$ .. Signal to Noise Power Ratio at the Receiver Input
$(S/N)_{out}$. Signal to Noise Power Ratio at the Receiver Output
b_r .... Bandwidth Ratio (B_{TR}/B_{out})
γ_{PR} .... Processing Gain $(S/N)_{out}/(S/N)_{in}$

Figure 8.1   Relationships at the Receiver/Demodulator

Equation (8.1) applies to the radio frequency (RF) segment of the channel, Eq. (8.2) to the baseband segment. Note that the channel capacity is the data rate which the channel could carry *error-free*.

Assuming that the receiver plus demodulator is ideal and does not change the channel capacity, that is, the receiver and demodulator does not degrade the signal or introduce additional noise. That is a condition generally closely approximated in most practical systems. Then Eqs. (8.1) and (8.2) can be combined [HA-1]:

$$B_{TR} \log_2 [(1 + (S/N)_i] = B_{out} \log_2 [1 + (S/N)_{out}], \qquad (8.3)$$

or, by introducing b_r for B_{TR}/B_{out},

$$b_r \log_2 [1 + (S/N)_i] = \log_2 [1 + (S/N)_{out}]. \qquad (8.3a)$$

From Eq. (8.3a) one obtains

$$(S/N)_{out} = [1 + (S/N)_i]^{b_r} - 1 \qquad (8.4)$$

by raising both sides to the power of 2.

If $b_r > 1$, that is, the modulation spreads the signal bandwidth so that the radio signal occupies a wider bandwidth, then dividing Eq. (8.4) on the right and the left side by $(S/N)_i$, and substituting $\gamma_{PR} = (S/N)_{out}/(S/N)_i$, one obtains the processing gain:

$$\gamma_{PR} = \frac{[1 + (S/N)_i]^{b_r} - 1}{(S/N)_i}. \qquad (8.5)$$

Equation (8.5) describes the processing gain γ_{PR} resulting from the bandspread b_r as a function of the SNR $(S/N)_i$ in the radio transmission segment of the channel. For $(S/N)_i \ll 1$, $\gamma_{PR} \approx b_r$; for $(S/N)_i \gg 1$, $\gamma_{PR} \approx (S/N)_i^{b_r} - 1$. A plot of the processing gain is given in Figure 8.2. In AJ communications, the desired signal is in contention with other users' signals (the clutter), noise, and the jammer. In AJ systems many users share the full bandwidth of the available RF spectrum, and generally $(S/N)_i \ll 1$, that is, the processing gain is equal to the bandspread.

To gain an insight into the operation of an AJ link, the relationships of the various signal and interference power components in a radio channel are analyzed for a SATCOM

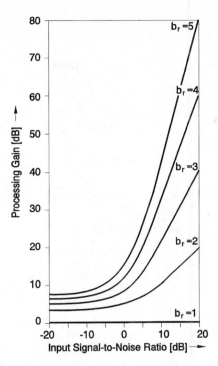

Figure 8.2   Processing Gain γ_{PR} as Function of the Input SNR $(S/N)_i$ and the Bandspread b_r

system. A system with M controlled users is assumed, each transmitting the equivalent isotropic radiated power (EIRP) P. A jammer with the EIRP J is interfering.

Figure 8.3 illustrates the various components of the signal, noise, clutter, and jamming power at the satellite repeater. The point at the repeater after the limiter and before the final amplifier is used as the reference point in this analysis. At this reference point the power of the desired signal is βP, the clutter (the sum of the signal power from all other users) is $\beta \sum_{1}^{m-1} P_m$, the jammer power is βJ, and the satellite front-end amplifier noise power is $kT_s B_{TR}$.[2]

An approach customarily taken to evaluate the system performance of an AJ link is to consider the total power, and separate it into a desired fraction D (resulting from P, the power of the signal of interest), and an interference power fraction I (resulting from the clutter, the jammer power, and the satellite front-end noise). The sum of all power components at the reference point (in a satellite transponder after the front-end amplifier and hard limiter, but before the repeater output amplifier, see Figure 8.3) can be expressed as:

$$D+I=\alpha\beta P+\alpha\beta \sum_{1}^{M-1} P_m + kT_s B_{TR} + \beta J, \qquad (8.6)$$

where

$$D=\alpha\beta P \qquad \text{and} \qquad I=\alpha\beta \sum_{1}^{M-1} P_m + kT_s B_{TR} + \beta J, \qquad (8.7)$$

[2]The factor β represents the total signal amplitude change in the uplink; it consists of a combination of the satellite receiver antenna gain G_{sr} and the loss in the uplink path L_{sup}; it is expressed as $G_{sr}L_{sup}$ (or $G_{sr} + L_{sup}$, (in dB)). Note that L_{sup} (in dB) is a negative number.

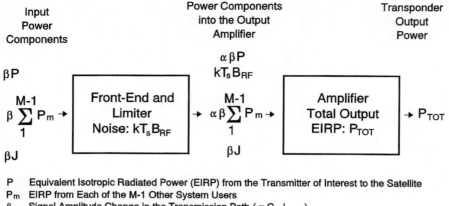

P   Equivalent Isotropic Radiated Power (EIRP) from the Transmitter of Interest to the Satellite
P_m EIRP from Each of the M-1 Other System Users
β Signal Amplitude Change in the Transmission Path ($= G_{sr} L_{sup}$)
G_{sr} Gain of the Satellite Receiver Antenna for the Signal P
L_{sup} Space Loss in the Uplink
T_s Equivalent Noise Temperature of the Satellite Repeater Front-End
k   Boltzmann's Constant
B_{TR} Bandwidth of the Transmitted Signal in the Radio Channel
α Small Signal Suppression by the Limiter
J   Jammer EIRP
P_{TOT} Power of the Satellite Transponder Radiated to the Earth Station

Figure 8.3   Signal, Noise, and Jammer Power Components at the Satellite

Note that the jammer power is considered to be large compared to P and P_m; consequently the limiter suppresses the small signals by a factor α.[3]

The total power $D + I$ available at the reference point is amplified by a factor A and sent to the ground; the power transmitted from the satellite is designated as P_{TOT}, where $P_{TOT} = A(D + I)$. The signal power S_{REC} at the ground station receiver is:

$$S_{REC} = \delta P_{TOT} + kT_R B_{TR}. \tag{8.8}$$

In the above expression δ is the total signal amplitude change[4] in the transmission path, and $kT_R B_{TR}$ is the front-end noise power of the ground receiver that is added to the undesired signal components.

At the ground receiver the total power is again separated into a desired part D' and the interference[5] I', where

$$D' = \alpha\beta\delta AP \qquad \text{and} \qquad I' = \alpha\beta\delta A \sum_{1}^{M-1} P_m + \delta A k T_s B_{TR} + \beta\delta AJ + kT_R B_{TR}. \tag{8.9}$$

---

[3]For details on small-signal suppression see section 8.2.

[4]Like the factor β used in the uplink, the factor δ describes the amplitude change in the downlink; it consists of the satellite transmitter antenna gain G_{st}, the ground station antenna gain G_R, and the downlink path loss L_{sdn}; it is expressed as $G_{st} G_R L_{sdn}$ (or $G_{st} + G_R + L_{sdn}$, in dB); L_{sdn} (in dB) is negative.

[5]The interference for the downlink is the sum of the powers of the clutter, the noise in the satellite front-end, the jammer, and the noise in the ground receiver front-end.

The ratio of the desired signal to the interference at the ground receiver is not a customary SNR but a signal-to-interference ratio, here designated by (S/J'). This ratio is:

$$(S/J') = \frac{\alpha\beta\delta AP}{\alpha\beta\delta A\sum_{1}^{M-1} P_m + \delta AkT_s B_{TR} + \beta\delta AJ + kT_R B_{TR}}. \tag{8.10}$$

As discussed in Chapter 7, the error rate in a digital communications system depends on the energy ratio ε_R of the received signal.[6] For the AJ link discussed here, (S/J') takes the place of $(S/N)_i$, and then

$$\varepsilon_R = (S/J')\,B_{TR}\tau_{sy}. \tag{8.11}$$

For many systems $B_{TR}\tau_{sy}$ equals the processing gain,[7] such that[8]

$$\varepsilon_R = (S/J')\,\gamma_{pr}. \tag{8.12}$$

In this analysis, the formulation given in Eq. (8.12) will be used, with the understanding that γ_{PR} may be replaced with $B_{TR}\eta_B\tau_{sy}$ or $B_{TR}\eta_B/R_{sy}$ where appropriate, where η_B is the bandwidth effectiveness.

When replacing the (S/J') in Eq. (8.10) with the signal energy ratio ε_R from Eq. (8.12), Eq. (8.10) becomes:

$$\varepsilon_R = \frac{\alpha\beta\delta AP\gamma_{PR}}{\alpha\beta\delta A\sum_{1}^{M-1} P_m + \delta AkT_s B_{RF} + \beta\delta AJ + kT_R B_{RF}}. \tag{8.13}$$

Although somewhat complicated, Eq. (8.13) is very useful to evaluate the performance of an AJ link.[9] In order to assess Eq. (8.13), values typical for a SATCOM link are

---

[6]See Chapter 7, Eq. (7.11).

[7]For an AJ link, where the jamming and clutter power are the prevalent interference, which are much larger than the power of the desired signal, the signal-to-interference ratio (S/J') is significantly smaller than 1. In this case, as it can be seen from Figure 8.2, the processing gain γ_{PR} is approximately equal to the bandspread B_{TR}/B_{out} or b_r.

[8]In Eq. (8.11) the term $B_{TR}\,\tau_{sy}$ is the signal time-bandwidth product. The signal duration τ_{sy} and the modulation method determine the baseband bandwidth B_{out}. The symbol rate R_{sy} equals $1/\tau_{sy}$. In general the ratio of the data rate to the bandwidth in the baseband R_{sy}/B_{out} is the bandwidth efficiency (see Chapter 9, section 9.4) that is designated here as η_B. As an example, consider coherent binary PSK without signal shaping, where $\eta_B = 1$ and $\tau_{sy} = B_{out}$; in such a system $B_{TR}\,\tau_{sy} = \gamma_{pr}$.

[9]It is of interest to note that Eq. (8.13) can also be used to determine the performance of a normal SATCOM link (without clutter and jamming). By deleting the terms for clutter and jamming, and noting that without a jammer $\alpha = 1$, Eq. (8.13) reduces to

$$\varepsilon_R = \frac{\beta\delta AP\gamma_{PR}}{\delta AkT_s B_{TR} + kT_R B_{TR}} = \frac{(\gamma_{PR}/B_{TR})}{(\frac{kT_s}{\beta P}) + (\frac{kT_R}{\beta\delta AP})}.$$

The first term $kT_s/\beta P$ in the denominator equals $1/(C/N_0)$ for the uplink and the second term, $kT_R/\beta\delta AP$, equals $1/(C/N_0)$ for the downlink. Noting that $1/(C/N_0)_{total} = 1/(C/N_0)_{uplink} + 1/(C/N_0)_{downlink}$, the right side of the expression above is recognized to be $(\gamma_{PR}/B_{TR})(C/N_0)_{total}$. In Chapter 12 the same expression for $1/(C/N_0)_{total}$ is used to compute the C/N_0 for the total satellite link for calculating SATCOM link budgets.

TABLE 8.1 TYPICAL SIGNAL AND NOISE VALUES

Parameter	Value in dB	Numerical Value
k	-228.6 [dbWHz^{-1}K^{-1}]	1.38×10^{-23} [WHz^{-1}K^{-1}]
$L_{sup} = L_{sdn}$	-202 [dB]	6.31×10^{-21}
B_{TR}	77 [dBHz]	5×10^{7} [Hz]
G_{sr}	15 [dB]	3.16×10^{1}
G_{st}	30 [dB]	10^{3}
G_{R}	50 [dB]	10^{5}
A	100 [dB]	10^{10}
P	80 [dBW]	10^{8} [W]
ΣP_{m}	90 [dBW]	10^{9} [W]
J	90 [dBW]	10^{9} [W]
T_{s}	30 [dBK]	10^{3} [K]
T_{R}	20 [dBK]	10^{2} [K]
α	-7 to 0 [dB]	0.2 to 1
β	-187 [dB]	2×10^{-19}
δ	-122 [dB]	6.31×10^{-13}
$A\delta\beta$	-209 [dB]	1.26×10^{-21}
$A\delta$	-22 [dB]	6.3×10^{-3}
$kT_{R}B_{TR}$	-131.6 [dBW]	6.92×10^{-14} [W]
$A\delta kT_{s}B_{TR}$	-143.6 [dBW]	4.36×10^{-15} [W]
$A\delta\beta J$	-119 [dBW]	1.26×10^{-12} [W]
$A\delta\beta\Sigma P_{m}$	-119 [dBW]	1.26×10^{-12} [W]

used. These values are summarized in Table 8.1. It is assumed that a total of 11 users with equal power P are participating in the system, and each user can transmit 80 dBW. One user transmits the signal of interest, the other 10 users generate the clutter. In addition, one jammer with a power of 90 dBW is assumed to generate the hostile interference.

From the values in Table 8.1 it can be seen that the noise contributions can be neglected with respect to the clutter and jammer power, and hence Eq. (8.13) can be reduced to:

$$\varepsilon_R \approx \frac{\gamma_{PR}}{\frac{1}{P}\sum_{1}^{M-1} P_m + \frac{J}{\alpha}P}. \tag{8.14}$$

The result given in Eq. (8.14) indicates that in an AJ link the power P is subject to small-signal suppression, and the signal energy ratio is a function of the clutter power and jammer power.

Equation (8.14) can be used to examine the performance of an AJ system for the case that all P_m are equal (and also equal to P, i.e., all ground stations use the same power). Then, $\sum_{1}^{M-1} P_m = (M-1)P$, and Eq. (8.14) becomes

$$\varepsilon_R \approx \frac{\gamma_{PR}}{(M-1) + \frac{J}{\alpha P}}. \tag{8.15}$$

Number of Circuits that can be Maintained at an Error Rate of 10^{-5}

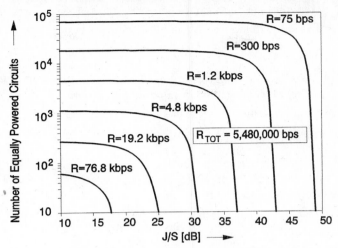

Figure 8.4   Number of Participants as Function of J/P

If it is further assumed that $M > 10$ (i. e., there are enough participants so that the clutter is significantly larger than the power from a single user), and that all participants operate with the same data rate, then 1 in the parentheses in the denominator can be neglected, that is, $M - 1 \approx M$. Equation (8.15) then develops into $\varepsilon_R \approx \dfrac{\gamma_{PR}}{M + J/\alpha P}$, and from that[10]

$$M \approx \frac{\gamma_{PR}}{\varepsilon_R} - \frac{J}{\alpha P} \qquad \text{or} \qquad M \approx \frac{B_{TR}\eta_B}{\varepsilon_R R_{sy}} - \frac{J}{\alpha P}. \tag{8.16}$$

An illustration of the performance of an AJ system based on Eq. (8.16) is given in Figure 8.4. The assumptions used to develop Figure 8.4 are summarized below:

- There are M disciplined (controlled) users emitting the same EIRP ($M > 10$).
- The bandwidth B_{TR} is 50 MHz (77 dBHz).
- The value for the small-signal suppression factor is taken to be 1.
- Coherent PSK modulation is used ($\eta_B = 1$), with an error rate of 10^{-5}, which is obtained for the signal energy ratio $\varepsilon_R = 9.6$ dB.
- A broadband noise jammer is assumed (i.e., the jamming power is spread over the whole bandwidth B_{TR}).

The number of circuits that can be maintained are shown in Figure 8.4 as function of the jammer to signal ratio (J/P) with the information transmission data rate R_{sy} as parameter. Figure 8.4 shows the following:

[10]The second approximation given in Eq. (8.16) results from a substitution of $B_{TR}\eta_B/R_{sy}$ for γ_{pr}

- For low J/P the system performance is determined by the clutter, that is, the interference in one channel from the signals of the other system users.
- The system has a maximum throughput data rate $R_{TOT} = B_{TR}\eta_B/\varepsilon_R$ (at $J/P = 0$) that is distributed among the different circuits, as long as the power in each circuit is balanced to produce the same rate.
- As soon as the jammer begins to take over (where $J/P \approx B_{TR}\eta_B/\varepsilon_R R_{sy}$) the capability to maintain circuits rapidly drops to zero.
- The performance of circuits with different data rates can be interpolated in this figure, for example, if the data rate increases by 3 dB, the curve shifts downward and inward by 3 dB.
- If the jammer power can be reduced by θ dB (through reduction of the satellite antenna gain in the direction of the jammer) the curves shift upward and to the right (outward) by θ dB.

The relationships given in Eqs. (8.14) to (8.16) can also be used to evaluate how users with different power levels would share the time-bandwidth resource of an AJ system. If there are M_1 users with the capability to transmit with an EIRP of P[W] and M_2 users with the capability to transmit with an EIRP of Q[W], the clutter will be composed of the sum of the power transmitted by the M_1 and M_2 users. If the desired signal is P, the relationship between the signal energy ratio, the clutter and the jamming, given in Eq. (8.14) expands to

$$\varepsilon_R \approx \frac{\gamma_{PR}}{\dfrac{1}{P}\displaystyle\sum_1^{M_1-1} P_{m_1} + \dfrac{1}{P}\displaystyle\sum_1^{M_2-1} Q_{m_2} + \dfrac{J}{\alpha}P}. \qquad (8.17)$$

Following the steps that led from Eq. (8.14) to (8.16), Eq. (8.17) develops into[11]

$$M_1 \approx \frac{R_{TOT}\eta_B}{R_{sy}} - \frac{Q}{P}M_2 - \frac{J}{\alpha P}. \qquad (8.18)$$

From Eq. (8.18) it is seen that at $J = 0$ the number of channels M_1, available for the user operating with the EIRP P, depends on the number of channels M_2, occupied by user operating with the EIRP Q, and the ratio Q/P. Two cases can be distinguished. In the first case $Q/P \gg 1$, and the clutter added by the M_2 (more powerful) users is significant. The users with the low operating power are disadvantaged with respect to the user with the large operating power.[12] One can readily see from Eq. (8.18) that when M_2 (Q/P) exceeds $R_{TOT}\eta_B/R_{sy}$, the clutter would become so large there is no resource left for the low power users, even if $J = 0$, and M_1 would be 0. In the second case $Q/P \ll 1$, and the added clutter, coming from the M_2 (less powerful) users has little effect.

An illustration is shown in Figure 8.5. The applicable parameter values are summa-

[11]In the steps developing Eq. (8.18) the substitutions $\gamma_{PR} = B_{TR}\eta_B/R_{sy}$ and $B_{TR}\eta_B/\varepsilon_R = R_{TOT}$ are also used.

[12]A typical disadvantaged user is an AN/WSC-6 terminal operating in a network where other users have AN/FSC-79 terminals. The disadvantage may be as much as 20 dB.

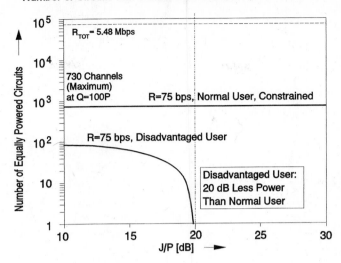

Figure 8.5   Mixture of Normal and Disadvantaged Users

rized in Table 8.2. In this case $Q \gg P$, that is, the M_2 users are more powerful. If the system had users with only a single power level, it could sustain 73098 75-bps channels without jamming. (This corresponds to the topmost curve shown in Figure 8.4.) However, because of the power imbalance of the users, only 730 high-power users can be allowed to participate in the system; otherwise the clutter would overpower the low-power users.

Figure 8.5 shows resource sharing of the high- and low-powered users. In the left part of the figure the operational characteristics of the disadvantaged users are shown when the number of high-powered users is constrained to 730. Without jamming only the clutter from the M_2 high-powered users interferes with the low-powered users. Under jamming, the number of low-powered users that can be sustained is reduced, and, in this example, at J/P of about 20 dB, the low-powered users can no longer operate.

The performance of an AJ system can also be characterized by plotting the error

TABLE 8.2   VALUES USED IN DEVELOPING
FIGURE 8.5

Parameter	Value
B_{TR}	50 MHz
$\varepsilon_R \, (\eta_B = 1)$	9.12 (9.6 dB)
$R_{TOT} \, (B_{TR}\eta_B/\varepsilon_R)$	5.48 Mbps
R_{sy}	75 bps
P	70 dBW
Q	90 dBW
Q/P	100 (20 dB)
M_2	730
M_1 (for $J=0$)	98

rate as function of J/P with M and R_{sy} as parameters. The expression $\sum_{1}^{M-1} P_m = (M-1)P$ is used to compute the error rate, the latter being a function of the signal energy ratio ε_R; for coherent PSK [SC-1]

$$P_e = \frac{1}{2}\mathrm{erfc}(\sqrt{\varepsilon_R}) \tag{8.19}$$

where erfc(z) is the complement of the error function, that is,

$$\mathrm{erfc}(z)=1-\frac{2}{\sqrt{\pi}}\int_z^\infty e^{-x^2}\,dx.$$

Figure 8.6 shows the error rate as a function of M for $R_{sy} = 75$ bps. As the figure shows, for $M = 100{,}000$ and $75{,}000$, which would require an R_{TOT} in excess of the 5.48 Mbps, a P_e of 10^{-5} cannot be realized.

As a final step, the performance of one link in an AJ system is analyzed. Let it be assumed that the total number M of channels is small, and the clutter from all participants in the system, whose power is expressed by $(1/P) \cdot \sum_{1}^{m-1} P_m$, can be neglected with respect to the jamming power J. Then, from Eq. (8.14) follows

$$\varepsilon_R \approx \frac{\gamma_{PR}}{\dfrac{J}{\alpha P}} = \frac{B_{TR}\eta_B}{R_{sy}}\frac{J}{\alpha P}. \tag{8.20}$$

Using Eq. (8.20) one can relate the symbol rate R_{sy} of the link could give support to the required signal energy ratio ε_{Rreq}

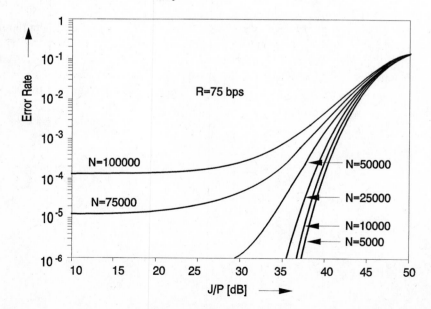

Figure 8.6   Error Rate for 75-bps Circuits

$$R_{sy} \approx \frac{1}{\varepsilon_{Rreq}} \frac{\alpha P}{J} B_{TR} \eta_B. \tag{8.21}$$

Assuming a broadband jammer, a specific modulation system (that allows to choose required ε_{Rreq} for a desired error rate performance and also determine the modulation efficiency η_B) one can convert Eq. (8.22), by taking the logarithms on the right and the left sides, into (all values in dB)

$$R_{sy} \approx -\varepsilon_{Rreq} + \alpha + P - J + B_{TR} + \eta_B. \tag{8.22}$$

If the satellite antenna pattern can be modified by controlling it through electronic beam forming and steering, a selective coverage pattern with a lower gain into the direction of the jammer can be obtained. Let the gain reduction into the direction of the jammer be designated θ, then Eq. (8.22) can be rewritten as:

$$
\begin{aligned}
R_{sy}\left[\text{dB bit s}^{-1} \right] = &-\varepsilon_{Rreq}[\text{dB}] + \alpha[\text{dB}] + P[\text{dBW}] - J[\text{dBW}] \\
&+ \theta[\text{dB}] + B_{TR}[\text{dBHz}] + \eta_B\left[\text{dB bit } s^{-1}\text{Hz}^{-1} \right].
\end{aligned} \tag{8.23}
$$

Note that $\alpha \leq 1$ and $\theta \geq 1$; hence, in dB the value for α will be zero or negative, and the value for θ positive.

As a numerical example, consider ε_{Rreq} to be 9.6[dB], $\alpha = 0$ [dB], $B_{TR} = 76$ [dBHz], and the modulation to be coherent PSK (with a bandwidth effectiveness $\eta_B = 1$ [bit s^{-1} Hz^1]), then Eq. (8.23) becomes

$$R_{sy} \, [\text{dBbit s}^{-1}] \approx 67.4[\text{dBbit s}^{-1}] + P\,[\text{dBW}] - (J - \theta)\,[\text{dBW}]. \tag{8.24}$$

Figure 8.7 is a parametrized plot of Eq. (8.24). The effective jammer power is $J - \theta$. Note that the bandwidth B_{TR} is chosen to be 77 dBHz (corresponding to 50 MHz). For

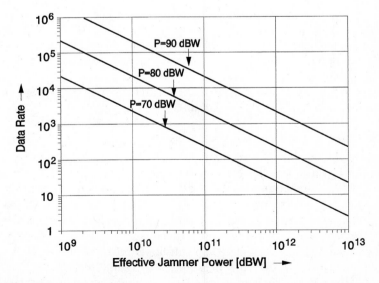

Figure 8.7 Available Data Rate as Function of Transmitter Power and Effective Jammer Power

different bandwidths, the lines for the power P would shift up (if the bandwidth exceeds 77 dBHz) or down by the number of dB by which the bandwidth differs from 77 dBHz.

8.2 SUPPRESSION OF SIGNALS IN A REPEATER BY A STRONGER SIGNAL

When a satellite repeater amplifies multiple signals, generally a power balancing of the input signals is required in order to distribute the available repeater output in the correct proportion among the signals transmitted by the satellite. This power balancing is not possible if one strong signal, the jammer, is present. While under normal circumstances the repeater can be shared among the various users and act as a linear (or near-linear) amplifier, the strong signal will overdrive the repeater amplifier into nonlinear operation.

In a nonlinear amplifier, a strong signal tends to capture the transponder, that is, the weak signals will be suppressed with respect to the strong signal. The capture results from the fact that the nonlinearity produces harmonics and intermodulation signals that will demand their share of the total available repeater output power.

This effect has been analyzed in several papers by Aein [AE-1], Berglund [BE-1], Bond and Meyer [BO-1], Cahn [CA-1], Jones [JO-1], and Sevy [SE-1]. All analyses assume that the strong signal is significantly larger than the weak signals; the analyses are complex and assume a power law device with a characteristic as shown in Figure 8.8. The characteristic is described by

$$y = ax^\nu \qquad \text{for } x \geq 0,$$
$$y = -a|-x|^\nu \qquad \text{for } x < 0,$$
$$0 \leq \nu < 1.$$

For $\nu = 0$ the power law device is a hard limiter, for $\nu \neq 0$ it is a soft limiter.

The methods of analysis in the aforementioned papers differ. Some authors, for example, Berglund [BE-1], assume constant envelope gaussian signals, while other authors, for example, Sevy [SE-1], assume multiple sinusoids and FM signals. All analyses lead to the result that a suppression of the small signal will occur. For the discussion presented here, it is assumed that there is a strong (jammer) signal J and one or more weak signals S_i.

The amount of suppression, that is, the ratio α of small-signal amplitude at the input to the small signal at the output, $\alpha = S_{in}^{(i)}/S_{out}^{(i)}$ is less than one, and there will be a power loss that can be several dB. The factor is a function of the power law device. For $\nu = 0$ (hard limiter), the results of the analyses using large unmodulated sine waves show α to be 1/4, corresponding to a power loss of approximately 6 dB. As ν increases, α decreases, and for an ideal, perfectly linear amplifier, with unlimited dynamic range, $\alpha = 1$, or the power loss is 0 dB. For a saturated traveling wave tube (TWT) amplifier, which has a characteristic corresponding to a soft limiter, the power suppression for sinusoids may range from 0 to 4.5 dB rather than 6 dB as in a hard limiter.

If the strong signal is not significantly larger than the weak signals, that is, S_J is ex-

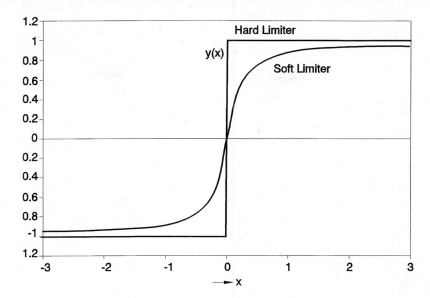

Figure 8.8   Typical Transponder Characteristics

ceeding S_i by a factor between one and ten, and a TWT power amplifier is used, the TWT is likely not to be driven fully into saturation, and the small-signal suppression factor may range from 1 to 2 dB.

Figure 8.9 shows the typical weak signal suppression. The analysis is based on two constant-envelope sinusoidal FM signals with the amplitude S_1 and S_2 plus narrowband gaussian noise. Figure 8.9[13] shows that when the amplitudes are equal no suppression occurs. There is also no significant suppression when the input SNR $(S_1/N)_i$ is very small. For a large input SNR, the weaker signal (S_2) is suppressed with respect to the strong signal (S_1). The suppression factor $[(S_2/S_1)_o/(S_2/S_1)_i]$ shown in Figure 8.9 equals the factor used throughout this chapter.

If the signals are gaussian processes, the suppression of the smaller signal will be $\pi/4$ or 1.05 dB [BE-1]. Figure 8.10[14] shows the suppression for gaussian signals and an unmodulated sinusoid. For $v = 0$ (ideal limiter), the degradation is seen to be 6 dB for sinusoids and about 1 dB for gaussian processes. For $v = 1$ (linear amplifier), the degradation is zero, as expected.

Generally, the worst case suppression will be experienced with two constant envelope signals, with the larger exceeding the smaller by a factor of four (6 dB). When the larger of the two signals can be characterized as a gaussian process the small signal is reduced by about 1 dB [SC-2]. When many constant envelope signals with nearly equal

[13]This figure is from J. J. Jones, "Hardlimiting of Two Signals in Random Noise," *IEEE Trans IT, Vol. IT9,* Copyright 1963 IEEE.

[14]This figure is from C. N. Berglund, "A Note on Power Law Devices and their Effect on Signal to Noise Ratio," *IEEE Trans Vol. IT10,* Copyright 1964 IEEE.

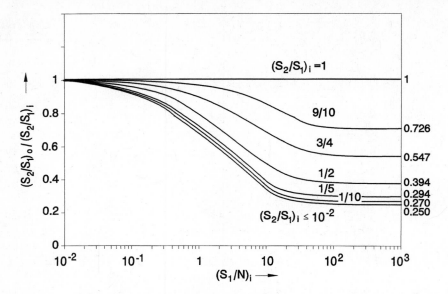

Figure 8.9   The Ratio of the Output Signal-to-Signal Ratio to the Input Signal-to-Signal Ratio as Function of the Input SNR of the Larger Signal

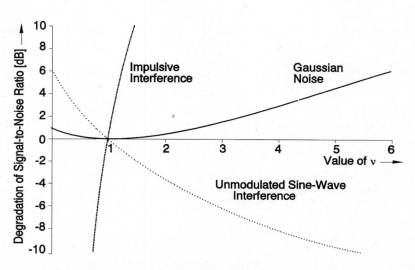

Figure 8.10   Suppression of Signals in a Limiter when $(S/N)_i$ Is Small as a Function of ν

power arrive simultaneously at the transponder they have the effect of a gaussian signal, and a constant envelope signal among them would be suppressed by about 1 dB.

8.3 SIGNALS FOR BANDSPREAD SYSTEMS

Signals can be characterized in both the time and frequency domain. In the time domain their duration is τ_s, and in the frequency domain their bandwidth is B. In a communications system, the communications resource, that is, the available transmission bandwidth B_{TR} and signal time frame T_f of a time division multiplex, must be shared among the signals of all system users. In order to distinguish and separate the signals most effectively, the signals need to be chosen for easy separation. A universally used and near optimal choice for this purpose is to use orthogonal signals $s_i(t)$.

Orthogonality is expressed in the time domain by the relationship

$$z_{ij} = \int_0^{\tau_s} s_i(t)s_j(t)dt \qquad \begin{cases} = \varepsilon_s & \text{if } i = j \\ = 0 & \text{if } i \neq j \end{cases} \qquad (8.25)$$

where τ_s is the signal duration, ε_S the signal energy.[15]

The signals $s_i(t)$ (in the time domain) and their spectra $S_i(\omega)$ (in the frequency domain) are related by a pair of Fourier transforms

$$S_i(\omega) = \int_{-\infty}^{\infty} s_i(t)e^{-j\omega t}dt$$

$$s_i(t) = \frac{1}{2\pi}\int_{-\infty}^{\infty} S_i(f)e^{-j\omega t}df \qquad \omega = 2\pi f.$$

The signal orthogonality can be expressed in the frequency domain by

$$Z_{ij} = \int_{-\infty}^{\infty} S_i(\omega)S_j(\omega)d\omega \qquad \begin{cases} = \varepsilon_s & \text{if } i = j \\ = 0 & \text{if } i \neq j \end{cases} \qquad (8.26)$$

Equation (8.25) characterizes signals for a channelization called time division multiple access (TDMA)[16] and Eq. (8.26) characterizes signals for a channelization known as frequency division multiple access (FDMA).[17] One can also use a set of signals that are

---

[15]It is assumed that the signals are normalized and all signals have an equal energy ε_s.

[16]To visualize time division multiple access (TDMA) consider a time frame T_f with M time slots (1,2,3, . . . M) and M different signals (A,B,C, . . . M) (see also Figure 8.11). The signals are sampled and the samples associated with the time slots. Samples of A go into slot 1 of a frame, but A is zero elsewhere in that frame; samples of B into slot 2, but B is zero elsewhere in the frame, and so forth. If the product of A and B is taken over a frame, this product will be zero, i.e., A and B are orthogonal over the frame. The same is true for signals A and C, B and C, and so forth.

[17]To visualize frequency division multiple access (FDMA) consider the frequency region B_{TR} with N bands (1,2,3, . . . N) and N different signals (A,B,C, . . . N) (see also Figure 8.11). The signals are shifted in frequency and each signal occupies one frequency band. Signal A is located in band 1, but A is zero in all other bands of the frequency region B_{TR}, B is in band 2, but B is zero in all other bands of the frequency region, and so forth. If the product of A and B is taken over the frequency region B_{TR}, the product will be zero, i.e., A and B are orthogonal. The same is true for signals A and C, B and C, and so forth.

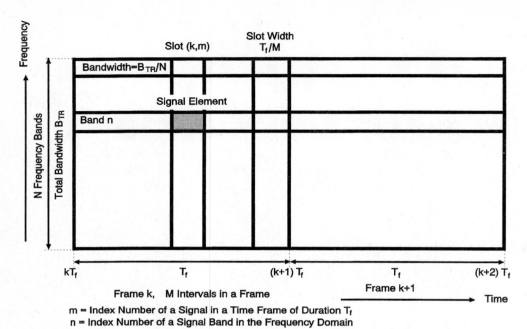

Figure 8.11   The Signal Resource Plane (Time-Bandwidth Product Plane)

orthogonal codes, and that use both the time and frequency domain. By definition, orthogonal codes meet the condition given in Eq. (8.25). Systems that use orthogonal codes are called code division multiple access (CDMA) systems.[18]

In general, signals can be portrayed in a time-bandwidth plane; this plane is then called the signal resource plane of the system. This plane is shown in Figure 8.11. The total available bandwidth B_{TR} is subdivided into N frequency bands; each band occupies the bandwidth B_{TR}/N. The time is divided into frames of duration T_f, and each frame is subdivided into M time intervals or slots of width T_f/M. The rectangle formed by the time frame T_f and the transmission bandwidth B_{TR} represents the available signal resource or the signal domain of the system. The frame structure is repeated every T_f seconds, hence the frame represents a time interval $\{\tau. \; k \leq t \leq (k + 1)T_f\}$, and the time interval $\{\tau. \; kT_f + (m - 1) \, T_f/M \leq t \leq kT_f + mT_f/M\}$ is the slot (m,k), that is, slot m in frame k. The area described by band n and the slot (m,k) is called the signal element.

In a CDMA system each signal is distributed over the complete resource, that is, all of $B_{TR}T_f$ is used. Such signals are characterized by the largest time-bandwidth product $B_{TR}\tau_{sy}$ available in the system, and in Eq. (8.14) it was shown that $B_{TR}\tau_{sy}$ is proportional to the processing gain γ_{PR}. Thus, signals spread over the entire available resource will provide the most processing gain. Such signals are commonly referred to as spread-spectrum signals, and they are used in AJ communications systems.

[18]A simple example of two orthogonal signals are $\sin \omega t$ in the time interval between 0 and 2π. Similarly, the two digital sequences 0110 and 1001 are orthogonal if each symbol is of equal duration.

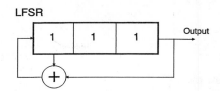

Figure 8.12   Linear Feedback Shift Register

Below we present a summary of two spread-spectrum approaches used for naval SATCOM.

Direct sequence (DS).   Direct sequence systems use signals related to pseudo-random or pseudo-noise (PN) codes. Each PN code word is used to represent one information bit. The ones and zeroes of the code are the transmission symbols and are also referred to as chips. With N chips per bit, the bandwidth of the chip sequence is spread by a factor N, and a progressing gain of $10\log N$ dB is realized.

PN-sequences [GO-1] can be generated by linear feedback shift registers (LFSRs). An LFSR is a shift register where the contents of some of its stages are summed modulo-2 in an exclusive-OR circuit and fed back to the input, as shown in Figure 8.12. The register contains an initial load pattern, which, in conjunction with the feedback arrangement, generates a periodic PN-sequence. The maximum length sequence has $2^n - 1$ bits; in the simple example shown in Figure 8.12, $n = 3$, and the maximum sequence length is seven. Any initial register load pattern except 000 will generate the seven-bit sequence. (The pattern 000 is not allowed as an initial load for the example shown; the pattern 000 can be allowed if a modified summing circuit is used; then an eight-bit sequence would be generated.) For the example shown here, any segment of length seven of the continuous string generated by the LFSR is a PN sequence. Thus, in our example there are seven distinct PN sequences; they are 1110100, 1101001, 1010011, 0100111, 1001110, 0011101 and 0111010.

PN-sequences have the following specific properties: (a) the number of zeroes and the number of ones differ by 1; (b) in every period one-half of the runs have the length one, one-fourth the length two, one-eighth the length three, and so forth as long as the number of runs exceeds one (there will be one run of length n); and (c) the autocorrelation function $C_{j\Delta}$ is two valued. In addition, the modulo-2 sum of the sequence PN and $PN_{j\Delta}$ is $PN_{k\Delta}$, that is, if one of the seven sequences of our example is added modulo-2 to the same sequence shifted by j bits, one obtains the sequence shifted by k bits. Similarly, the modulo-2 sum of PN and $PN_{i\Delta}$ is $PN_{m\Delta}$, that is, when a sequence and its *inverse* is added modulo-2, the result is a $\overline{\text{shifted}}$ *inverse*. Figure 8.13 illustrates the shift-and-add property, and shows the two-valuedness of the autocorrelation function.[19] The sequences are transorthogonal, that is, the value of the autocorrelation outside of the peak is not zero but $-1/(2^n-1)$.

Using the PN sequence correlation function, one can define a discriminator function

---

[19]The symbol $Av(PN)$ used in Figure 8.13 denotes an averaging process for a sequence of ones and zeroes. It is defined by:

$Av(PN) = $ (number of zeroes $-$ number of ones)/(number of zeroes $+$ number of ones).

$$PN \oplus PN_{j\Delta} = PN_{k\Delta} \qquad\qquad PN \oplus \overline{PN}_{i\Delta} = \overline{PN}_{m\Delta}$$

1110100**11101001**110100	1110100**11101001**110100
11101001**11101001**110100	00010110**00010110**001011
101001**1101001**11010	01011**0001011**000101

● Autocorrelation: $C_{j\Delta} = Av(PN \oplus PN_{j\Delta})$

$$Av(PN) = \frac{\text{Number of Zeroes} - \text{Number of Ones}}{\text{Number of Zeroes} + \text{Number of Ones}}$$

$$C_{j\Delta} = \begin{cases} -1/(2^n - 1) & \text{if } j \neq 0 \\ 1 & \text{if } j = 0 \end{cases}$$

Figure 8.13   Properties of PN Sequences

$D_{j\Delta} = C_{\Delta} - C_{-\Delta}$ that can be used for detecting the PN sequence; the correlation and discriminator function is shown in Figure 8.14.

A detector for the received PN-sequence is the delay-locked loop, shown in Figure 8.15. Detection consists of finding a sequence and timing match. The delay-locked loop utilizes the same LFSR that is used for generating the PN-sequence for transmission to generate a replica of the transmitted PN-sequence; the specific PN-sequence (out of the group of $2^n - 1$ PN-sequences the LFSR can generate) is given by the initial loading of the LFSR. The delay-locked loop is an implementation of the discriminator function. If the received PN-sequence is fed into the input of the delay-locked loop, the loop will align the detector's PN-sequence with the received PN-sequence; only for the point $\Delta = 0$ will the loop be stable, and the clock will match the clocking of the received PN-sequence. Since the LFSR is keyed through its initial load pattern to a specific sequence, obtaining stability means the sequence has been detected. If the data stream does not contain the sequence, stability (lock) is not achieved.

The detector can be designed to detect the PN-sequence as well as its inverse (as illustrated in Figure 8.15). Thus, a spread-spectrum system can be constructed where each 1 of the information flow is expressed by a specific PN-sequence, and each 0 of the information flow by the inverse the PN-sequence. By choosing a maximum length PN-sequence of $2^n - 1$ bits, a bandspread factor of $2^n - 1$ is accomplished using an LFSR

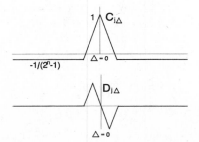

Figure 8.14   Correlator and Detector Function

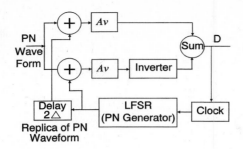

Figure 8.15   PN-Sequence Detector

having n stages. For example, using a 15-stage register, one obtains 32,767 chips per bit, and one can accomplish a bandspread factor of 32,767 corresponding to a 45dB processing gain. Each of the direct sequences bandspread signals uses a different PN sequence and occupies the complete $B_{TR}T_f$ rectangle of the system signal resource plane.

Frequency hopping (FH).   Another method of generating bandspread signals is frequency hopping. Frequency-hopped signals consists of sinusoids of short duration, where each sinusoid occupies one signal element of the time-bandwidth plane; this is illustrated in Figure 8.16. The signals can be chosen such that each signal occupies different signal elements and the sum of all signals occupy the complete time-bandwidth resource. Signals chosen in that manner do not overlap like the CDMA signals described previously; they will not interfere with each other but the regularity of the signal design may afford less protection. Alternately, multiple signals can be chosen such that they are widely spread and each signal pattern is random. Then some overlap is likely to occur, but generally a very wide spread is possible, particularly in the EHF region, and large processing gains can be accomplished.

For military applications the signal hopping rate can vary from low to high. A hopping rate of 256 hops per channel is a high hopping rate. The waveforms used for FH AJ systems are standardized in MIL-STD-188-110 [DD-1] and MIL-STD-1582 [AF-1; BE-2].

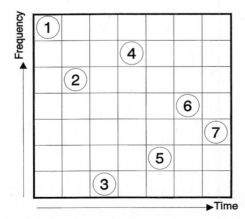

Figure 8.16   Example of Frequency Hopping

Time hopping is analogous to frequency hopping; here the successive signal elements (chips) are distributed randomly over a time frame. A delay of at least one frame plus processing time will be introduced by this method; the delay results from reconstruction of the transmitted signal from the received chips.

8.4 AJ IN NAVAL COMMUNICATIONS SYSTEMS

AJ systems for military use have evolved over the last 30 years using both FH and DS approaches. The early systems, such as the TACSATCOM, relied on FH. During that time, the frequency synthesizers were analog devices and the rate of frequency change was low. FH could use only slow hop rates.

TABLE 8.3   LIST OF AJ SYSTEMS IN NAVAL COMMUNICATIONS SYSTEMS

AJ System	Frequency Band	AJ Technique	ECCM Equipment	RF Radios	Characteristics
HFAJ	HF (2 MHz to 30 MHz)	Frequency hopping	AJ modem	AN/URC-109; ICS-3	Need wideband HF antennas (2–8 MHz, 4–6 MHz, and 12–30 MHz)
SINCGARS	30 MHz to 88 MHz	Frequency hopping	RT-1523(C)/U Receive-Transmitter with ECCM applique	AN/VRC-90 AN/ARC-210	ECCM mode; also single channel mode and airborne relay
HAVE QUICK II	225 MHz to 400 MHz	Frequency hopping	RT-1556 Transceiver with ECCM applique	AN/ARC-164, AN/ARC-182	Slow hop ECCM
JTIDS	965 MHz to 1215 MHz	Frequency hopping	ECCM modem	JTIDS Class 2 terminal	Time slot scrambling
FLTSATCOM	7.9 GHz uplink	Direct sequence	ECCM modem	AN/FSC-64 shore terminal	FLTBROADCAST uplink only from shore station to FLTSATCOM
DSCS II and DSCS III	7.9 GHz to 8.4 GHz	Direct sequence	OM-55 (one channel shipboard AJ modem); interoperable with AN/USC-28 GMF AJ modem	AN/WSC-6 shipboard terminal	Multi-beam antenna with nulling
MILSTAR	44 GHz uplink; 20 GHz downlink	Frequency hopping	ECCM modem	AN/USC-38 terminal	Multi-beam antenna with nulling; high hop rate at 256 hops per sec per channel

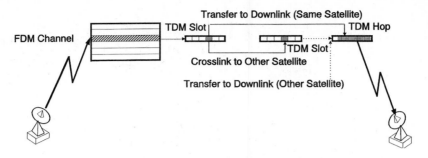

Figure 8.17   Frequency Hopping in the EHF Milstar System

In the 1970s, as result of the development of the PN-code technology used in space vehicles for ranging and communications [GO-2], the technology for DS AJ systems became available. This was also supported by the development of fast miniaturized digital circuits, and high chip-rate DS systems were designed.

Several naval communications systems discussed in the chapters on HF, VHF/UHF, SHF SATCOM, and EHF SATCOM employ AJ techniques. Table 8.3 is a summary of these AJ systems. The FLTSATCOM, the DSCS II, and the DSCS III use the direct sequence technique. All other systems use the frequency-hopping technique with some variations adapted for each application. In SINCGARS and HAVE QUICK II radios, appliques are retrofitted to existing radios, because RF equipment is already deployed in various shipboard and airborne platforms.

The principle of CDMA application in DSCS is straightforward. The information bits are modulated onto PN-type codes. A transmitter and receiver pair have the exact information of the codes used for the CDMA channel in which they operate. Thus, the receiver will decode the transmission encoding and reproduce the transmitted information sequence.

Frequency hopping is used in the EHF MILSTAR satellite. Here only a brief overview is given; details are presented in Chapter 15. The information is coded at the transmitter then recoded at the satellite. The principle of this operation is illustrated in Figure 8.17. The uplink operates as an FDM in the EHF band. For the downlink in the SHF band the signal is translated into a TDM and each FDM uplink channel is allocated a time slot. These slots are converted into frequency-hopped signals and transmitted to the ground receivers. Crosslinking among satellites is possible to provide global coverage without the need for retransmission via a ground station from one satellite footprint area to another one.

As Figure 8.17 shows, the system is complex, and several stages of recoding are used to provide a substantial signal protection.

REFERENCES

[AE–1]   Aein, J. M., "On the Output Power Division in a Captured Hard Limiting Repeater," *IEEE Trans. Comm.*, Vol. COM-14, 1966, pp. 347–349.

[AF–1]   *Satellite Data Link Standards (SDLS); Uplink and Downlink*, Military Standard MIL-STD-1582, Department of the Air Force, Washington, D.C. 20330, 1 Feb. 1989.

[BE–1] Berglund, C. N., "A Note on Power-Law Devices and Their Effect on Signal-to-Noise Ratio," *IEEE Trans. Information Theory*, Vol. IT-10, Jan. 1964, pp. 52–57.

[BE–2] Bennett, R. R., "Theater/User-Dedicated Communications Satellite Concept," *Conference Record of the 1990 Military Satellite Conference*, Monterey, Calif., Sept. 30 to Oct. 3, 1990, pp. 12.1.1–12.1.5.

[BO–1] Bond, F. E. and H. F. Meyer, "Intermodulation Effects in Limiter Amplifier Repeaters," *IEEE Trans Comm.*, Vol. COM-18, Apr. 1970, pp. 127–135.

[CA–1] Cahn, C. R., "A Note on Signal-to-Noise Ratio in Band-Pass Limiters," *IEEE Trans. Info. Theory*, Vol. IT-7, 1961, pp. 39–43.

[DD–1] *Interoperability and Performance Standards for Data Modems*, Military Standard MIL-STD-188-110A, Department of Defense, Washington, D.C. 20350, 30 Sept. 1991.

[DI–1] Dixon, R. C., *Spread Spectrum Systems*, John Wiley & Sons, New York, N.Y., 1967.

[GO–1] Golomb, S. W., *Shift Register Sequences*, Holden-Day Inc., San Francisco, Calif., 1967.

[GO–2] Golomb, S. W., ed., *Digital Communications with Space Applications*, Prentice Hall, Englewood Cliffs, N.J., 1964.

[HA–1] Hancock, J. C., "On Comparing the Modulation Systems," *Proc. Natl. Electron. Conf.*, Vol. 18, Oct. 1962, pp. 45–50.

[JO–1] Jones, J. J., "Hard-Limiting of Two Signals in Random Noise," *IEEE Trans Info. Theory*, Vol. IT-9, Jan. 1963, pp. 34–42.

[SC–1] Schwartz, M., *Information Transmission, Modulation and Noise*, McGraw-Hill, New York, N.Y., 1970.

[SC–2] Schwartz, J. W., J. M. Aein, and J. Kaiser, "Modulation Techniques for Multiple Access to a Hard-Limiting Satellite Repeater," *Proc. IEEE*, Vol. 54, No. 4, May 1966, pp. 763–777.

[SE–1] Sevy, J. L., The Effect of Multiple CW and FM Signals Passed Through a Hard Limiter or TWT," *IEEE Trans. Comm.*, Vol. COM-14, Oct. 1966, pp. 568–578.

[SH–1] Shannon, C. E., "Communication in the Presence of Noise," *Proc. IRE*, Vol. 37, Jan. 1949, pp. 10–21.

[SI–1] Simon, M. K. et al., *Spread Spectrum Communications*, Volumes I–III, Computer Science Press, Rockville, Md., 1985.

9

ELF, VLF, and LF Communications

In this chapter, we discuss Extremely Low Frequency (ELF), Very Low Frequency (VLF), and Low Frequency (LF) communications as they are applied to shipboard communications, particularly to submarine communications.

Radio frequencies between 30 Hz and 300 Hz are called ELF radio, between 3 kHz and 30 kHz, VLF, and between 30 kHz and 300 kHz, LF. The region from 300 Hz to 3 kHz is called Voice Frequency (VF); it is not used for radio communications.[1] As shown in Table 9.1, the wavelengths of the ELF, VLF, and LF bands are on the order of 5000 km (3125 mi), 50 km (31.3 mi), and 5 km (3.1 mi), respectively [HS-1; CA-1]. VLF and LF radios are similar in nature; VLF and LF are also referred to as long-wave; we will use this designation in this chapter [GA-1]. Due to the characteristics of long-wave propagation, the Navy uses the VLF radios to communicate over very long distances, typically halfway around the world. Since long-wave signals can penetrate salt water, long-wave radios are used in submarine communications. ELF communications are in a class of their own; ELF signals penetrate up to 80 m ($\approx$260 ft) into the sea. ELF signals are extremely narrowband and can be used for transmission of very short pre-arranged messages to submerged submarines.

Figure 9.1 shows the frequency allocations for the 30 Hz to 300 kHz region. Be-

---

[1]While VF is not needed for radio communications, it is exploited for underwater sonar communications; see section B.5 of Appendix B.

TABLE 9.1 FREQUENCY AND WAVELENGTH OF LONG-WAVE
FREQUENCY BANDS

Frequency band	Frequency (kHz)	Wavelength (km)
Extremely Low Frequency (ELF)	0.03–0.3	10,000–1,000
Voice Frequency (VF)	0.3–3.0	1,000–100
Very Low Frequency (VLF)	3.0–30.0	100–10
Low Frequency (LF)	30.0–300.0	10–1

sides the Navy frequencies of ELF Sanguine, Shipboard VLF Verdin, TACAMO VLF, and Fleet LF Multichannel Broadcast, two important radio navigation frequencies, the Long Range Navigation (LORAN-C) in the 100 kHz band and the Omega in the 10–14 kHz band, are included in this figure.

9.1 FUNDAMENTALS OF ELF, VLF, AND LF COMMUNICATIONS

Submarine communications use long-wave radios because:

- For long-wave radio propagation, the surface of the earth and the ionosphere act as a waveguide for signals which propagate with relatively small attenuation for ranges of 10,000 km (6,200 miles) and beyond.

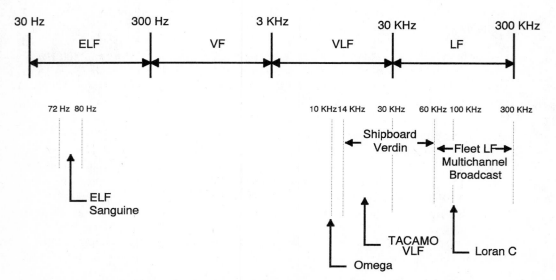

Figure 9.1   Frequency Allocations in the ELF, VF, VLF, and LF Bands

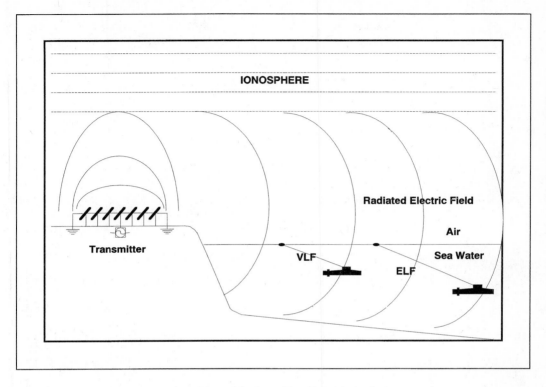

Figure 9.2   Long-Wave Transmission System

- Long-wave radio signals are able to penetrate seawater to as much as 30 m ($\approx$100 ft) which is the periscope depth of submarines (the seawater attenuation is on the order of 0.5 dB per m at a frequency of 20 kHz).
- Long-wave transmissions provide a highly reliable path for communications in northern latitudes.[2]

Figure 9.2 illustrates wave propagation for long-wave radio transmissions. This figure shows that the radiated electric field emanated from a transmitter propagates in a duct formed by the ionosphere and the Earth.

Long-wave radios on submarine platforms act only as receivers for shore transmissions. Signals generated by shore transmitters are strong enough to provide good reception onboard submarines, and receiver antenna matching is generally not required. Long wire, whip, and loop antennas each provide sufficient signal levels to allow receivers to operate satisfactorily. Requirements for very long transmitter antennas make it impossible for submarine platforms to use long-wave signals for return transmissions to shore sites.

---

[2]HF transmissions are not reliable due to the auroral effect in northern latitudes.

TABLE 9.2 VLF TRANSMITTERS

Transmitter	Transmitter Location	Rated Transmitter Output Power	Frequency
AN/FRT-31	Cutler, Maine	2 MW	17.8 kHz
AN/FRT-87	Annapolis, Md.	1 MW	21.4 kHz
AN/FRT-3	Jim Creek, Washington	1 MW	18.6 kHz
AN/FRT-64	Lualualei, Hawaii	1 MW	23.4 kHz
AN/FRT-67	Northwest Cape, Australia	2 MW	22.3 kHz

For long-wave radios, the efficiency of the shore transmitter antenna is proportional to its length. A general design guideline is that an antenna becomes more efficient when its physical size approaches the wavelength. Since the antenna length of long-wave transmitters can never be as large as the signal's wavelength (on the order of 5000 km ($\approx$3100 mi) for ELF and 100 km ($\approx$62 mi) for VLF and LF), long-wave antennas are very inefficient.

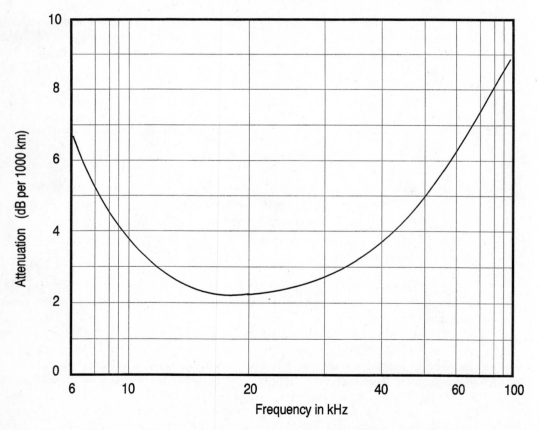

Figure 9.3   Average Long-Wave Attenuation Rate α (in dB/1000 km) for Daytime Conditions and a Mixed Path (20% Land and 80% Sea) (Adapted from [NA-1], Figure 2-26, page 2-32)

Antenna coupling is a very important factor in the long-wave transmitter design. The antenna coupling provides impedance matching for the antenna so that a maximum of radiated power can be produced. One of the problems of building long-wave transmitting antennas is that transmitter power is dissipated in antenna tuning components, in the structure of the antenna, and in the ground.

The receiver antenna length, dictated by the physical constraints of the submarine, is much smaller than the operating wavelength. For a long-wave LF receiver antenna, the efficiency is less critical than for a transmitter antenna. The receiving equipment must have a high enough sensitivity to detect the transmitted signal. If its sensitivity is high enough to overcome atmospheric noise, it is always acceptable for LF communications because the total system noise is essentially the atmospheric noise alone.

When penetrating seawater to reach the submerged antenna, the signal is attenuated at a much larger rate than in free space. In order to obtain a meaningful signal-to-noise ratio for a submarine antenna, the transmitters must be quite powerful; transmitter power must be 1 MW to 2 MW, as shown in Table 9.2 [NA-1].

Propagation loss in the long-wave radio bands is relatively low, typically on the order of 200 dB over 10,000 km ($\approx$6,200 mi). Figure 9.3 shows the average attenuation rate for daytime conditions and a mixed path (20% land and 80% sea) [NA-1]. The basic path loss expected for surface waves propagated over a smooth spherical Earth is shown in Figure 9.4, assuming a conductivity of 0.005 mho/m, dielectric constant of 15 ex-

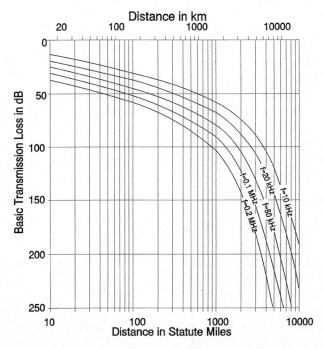

Figure 9.4   Basic Transmission Loss Expected for Surface Waves Propagated over a Smooth Spherical Earth over Land (Adapted from [NO-1], Fig. 8, June 1959)

pressed in esu units, lossless isotropic antennas 10 m (33 ft) above the surface, and vertical polarization [NO-1].

　　Since the transmitter must be capable of producing the required high power, long-
wave radios need amplifiers that can modulate and amplify the signal to be transmitted.
An example of such an amplifier is known as the High Efficiency Solid State Amplifier
(HESSA). HESSAs in long-wave radios must be designed with very special dielectric materials. These dielectric materials must withstand electric breakdown from very high voltages during normal operations. The magnitude of voltage is on the order of 1 MW to 2
MW as shown in Table 9.2; the voltages are several thousands of volts. To drive an antenna from a HESSA, a coupler and a tuner is required, usually consisting of large induc-
tors and capacitors; the capacitors used in these couplers and tuners are on the order of a
few farads. These capacitors use dielectric material that can withstand the very high voltages.

9.2 PROPAGATION

Our primary interest in long-wave signals for submarine communications is the behavior
of long-wave propagation over large distances of 1500 km ($\approx$940 mi) or more.

　　The field strength of long-wave signal (E_N) consists of three components:

$$E_N = E_0 + E_d + E_S,$$

where E_0 is the field strength of the ground wave component that propagates along the
Earth's surface, E_d is the field strength of the direct wave component that propagates directly over line-of-sight distances, and E_S is the field strength of the sky wave component.
The latter radiates from the transmitting antenna at relatively high angles and reaches the
receiver by reflection from ionized layers above the Earth. A VLF and LF ground wave
propagates with relatively low attenuation. A VLF and LF sky wave propagates between
the Earth's surface and ionospheric layers with very little penetration of and absorption by
the ionosphere even at high angles of incidence [NA-1; p. 2-1].

　　A combination of the ground wave (E_0) and the direct wave (E_d) is commonly
called the surface wave. The ground wave is the dominant signal at short distances up to
300 km ($\approx$190 mi). Long-wave signals at distances from 300 km ($\approx$190 mi) to 1500 km
($\approx$940 mi) consist primarily of a combination of ground waves and sky waves.

9.2.1 Propagation Over Long Distances

Surface wave components. Long waves propagate over long distances
mostly as surface waves. The way a surface wave is attenuated depends principally on the
characteristics of the surface over which it propagates. At long-wave frequencies, sub-
strata at depths in excess of 100 m ($\approx$330 ft) can affect the attenuation and the phase of
the signal. Abrupt changes in the magnitude of the ground wave component along the
propagation path are possible. Signals propagated as ground waves are highly stable in
time. Ground conductivity is nearly independent of time, with changes at these frequencies appearing only as very long-term effects [WA-1; WA-3].

The field strength of ground wave contribution at the receiver may be expressed as:[3]

$$E_0 = (mE_u)/d$$

where

E_u is the inverse-distance attenuated field intensity on the Earth's surface, at unit distance from the radiation source;

d is the distance between the transmitter and receiver; and

m is the ground wave attenuation factor, which is a function of frequency, polarization, distance, Earth curvature, the effective conductivity, and the dielectric constant of the terrain over which the signal propagates.

The significance of the above equation is that, for long waves, the field intensity of the ground wave is attenuated approximately inversely to distance. However, experimental measurements show that the attenuation at far fields (1500 km or ≈940 mi) can be larger than the analytic prediction by a factor of about 1.75 [BE-1].

Long-distance propagation.   As a VLF wave front propagates beyond 5000 km (≈3100 mi), it is no longer expanding but converging [NA-1; p. 2-8]. When the wave front approaches the antipodal point on the Earth's surface the effects of convergence and absorption losses balance. Around the antipodal point, the signal strength increases substantially and provides for an additional region of transmission coverage. For example, signals transmitted from Yosami, Japan, converged in the South Atlantic, signals transmitted from Cutler, Maine; Annapolis, Maryland; Jim Creek, Washington; and Balboa, Panama Canal Zone, converge in the Indian Ocean, and signals transmitted from the Northwest Cape, Australia, converge in the North Atlantic.

Effect of the earth's conductivity on propagation.   Ground conductivity affects changes in ground wave components along the propagation path. The conductivity varies according to ground conditions, such as mountainous terrain, large ore deposits, ice fields, and seashores. Examples of conductivity values (in mhos per meter) are 1.0×10^{-3} in Continental United States (CONUS) and 7.5×10^{-6} in Greenland. Table 9.3 lists typical values of dielectric constants and ground conductivities that apply to ELF and VLF propagation [HS-1; p. 28-3]. These dielectric constants and ground conductivities are used in calculating long-wave propagation using a waveguide model.[4]

Long-Wave propagation as a waveguide mode.   In the long-wave regions of the frequency spectrum (3 kHz to 300 kHz), the energy of propagating radio waves is confined principally to the shell between the Earth's surface and the ionosphere,

---

[3]This equation expresses the field intensity of a ground wave signal in terms of E_u, distance d, and ground wave attenuation factor m; E_u is the field intensity on the Earth's surface at unit distance from the radiated source; numerically E_u is equal to the voltage induced at unit distance; see [BE-1; p. 662] for detailed explanation of this relationship.

[4]A description of the waveguide model is presented in the next paragraphs. An example of a long-wave propagation prediction model is presented in section 9.2.5.

TABLE 9.3   GROUND CONDUCTIVITY AND DIELECTRIC CONSTANT
FOR MEDIUM- AND LONG-WAVE PROPAGATION

Type of Terrain	Conductivity σ (mhos/meter)	Dielectric Constant ε (esu)
Salt water	5	80
Fresh water	8×10^{-3}	80
Dry, sandy, flat coastal land	2×10^{-3}	10
Marshy, forested flat land	8×10^{-3}	12
Rich agricultural land, low hills	1×10^{-3}	15
Pastoral land, medium hills and forestation	5×10^{-3}	13
Rocky land, steep hills	2×10^{-3}	10
Mountainous (hills up to 3000 feet)	1×10^{-3}	5
Cities, residential areas	2×10^{-3}	5
Cities, industrial areas	1×10^{-3}	3

and this space is frequently considered as the terrestrial waveguide [GA-1].[5] For long waves, the height of this guide is nearly one wavelength, and the characteristics of wave propagation are determined jointly by the properties of the guide boundaries. There are a number of propagating modes with distinct cutoff frequencies, similar to those present in microwave transmission systems. But unlike the highly conducting waveguides used for the microwave transmission, the upper boundary of the terrestrial waveguide is an absorbent or leaky conductor; the bottom plane (i. e., the Earth's surface) is generally a good conductor.

The height of the terrestrial waveguide for VLF varies from 70 km (44 mi) during the day to about 90 km (56 mi) during the night. In general, the effective waveguide height is less than the free-space ELF wavelength of 1000 km ($\approx$620 mi) to 10,000 km ($\approx$6200 mi); therefore, the waveguide allows only one propagating mode. However, in the VLF band, the effective waveguide height may sometimes exceed the free-space VLF wavelength, that is, from 10 km ($\approx$6 mi) to 100 km ($\approx$62 mi), several modes can propagate in the waveguide. In the LF band, the number of realizeable propagating modes may exceed ten.

Figure 9.5 shows a model for long-wave propagation in a waveguide. The z-axis is the direction of the propagation; the top and bottom planes represent the ionosphere and the Earth in this model. The dimension of the waveguide shown in Figure 9.5 is on the same order of VLF wavelengths; the height ranges from 70 km (44 mi) to 90 km (56 mi).

The VLF and LF bands are below the critical frequencies of the lower layers of the ionosphere. In the VLF and LF bands, only the D layer provides reflected sky waves. The effective reflection height of the D layer varies from about 60 km to 90 km with a daytime average about 70 km and a nighttime average of 85 km. The effective reflection height depends on the ionization levels of the D layer, which are maximum during the sunlight hours when ionization rates are maximum and which decreases at night as the ionization

[5]The ionosphere consists of several ionized layers as discussed in Chapter 10. Only the lowest layer, the D layer, reflects the VLF and LF waves. Higher frequency signals such as MF and HF are absorbed or weakened in the D layer.

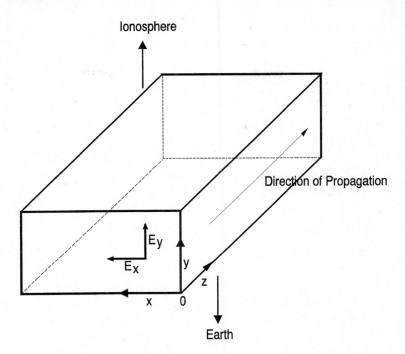

Figure 9.5   Orientation of the Waveguide for VLF/LF Propagation

rate drops and recombination continues [NA-1; p. 2-8] [WA-1; p. 180] [DA-1; pp. 40-41].[6,7]

   The Earth's magnetic field also plays a role in the propagation of the electromagnetic energy contained in a long-wave signal. There are several electric field propagation modes $TE_{m,n}$ and magnetic field propagation modes $TM_{m,n}$. Figure 9.6 illustrates the magnetic and electric fields of the $TM_{1,0}$ mode wave between parallel planes. The $TM_{m,n}$ propagation mode depends on the operating frequency and the waveguide dimension. In the daytime, when the ionosphere is 70 km (44 mi) in altitude, frequencies up to 20 kHz in VLF bands can propagate well in the existing ionosphere-Earth surface waveguide structure. The corresponding $TM_{m,n}$ modes are shown in Table 9.4 [CA-1]. Normally observed modes will be the $TM_{1,0}$ through $TM_{4,0}$ modes; for example, with the frequency of 4.2 kHz, long-wave signals can excite the $TM_{1,0}$ mode, in a waveguide with a height of 70 km (44 mi), the day-time ionospheric height. Similarly, the frequency of 8.6 kHz can excite the $TM_{2,0}$ modes. Doubling the operating frequency (i. e., from 4.2 kHz to 8.6 kHz) results in half the

---

[6]The ionosphere is divided into D, E, and F regions within which layers of electrons may exist. Approximate locations of these layers are: D layer (60 km–90 km), E Layer (100 km–110 km), and F layer (200 km–250 km); see section 10.2 (HF Propagation) for further discussion on the ionospheric layers.

[7]In general descriptions of the ionosphere, it is usually stated that the D layer dissolves at night. However, a layer of weak ionization (remnants of the daytime D and E layers) remains; this weak ionization is sufficient to reflect long-wave signals.

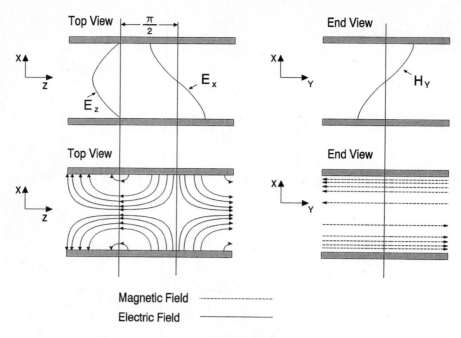

Top View

End View

Top View

End View

Magnetic Field --------------------

Electric Field ————————

Figure 9.6   The $TM_{1,0}$ Mode Wave between Parallel Planes

wavelength (i. e., from 70 km to 35 km). Long waves with frequencies above 21.4 kHz are not used in daytime.[8]

At night, the $TM_{5,0}$ and $TM_{6,0}$ modes will also be excited since the height of the ionosphere increases to 90 km (56 mi); however, there is more attenuation of the higher order modes. The nighttime $TM_{m,n}$ modes corresponding to signals up to 20 kHz are shown in Table 9.4.

9.2.2 Attenuation in Seawater

Another important characteristic of long-wave propagation is its ability to penetrate seawater. Figure 9.7 shows the attenuation of a plane wave in seawater as function of the frequency [TU-1]. At 10 kHz, the attenuation is about 0.3 dB per meter, which becomes 0.5 dB per meter at 20 kHz. Table 9.5 shows the penetration depth of long-wave signals in seawater [CA-1]. The penetration depth may vary slightly with changes in temperature and salinity of the seawater. Long-wave radio signals may be received by an antenna submerged at these penetration depths. Strong VLF signals can penetrate up to 4.6 m (15 ft) below the ocean surface. Typically, submarine antennas can receive VLF signals at a

---

[8]Modes are analogous to standing waves. When viewing Figure 9.6 one sees that the E or H fields show ½ period of a sine wave, determined by the boundary conditions of the waveguide. When a full period, 1½ periods, etc., can be maintained, higher frequencies can propagate. For higher order modes the E field decays more rapidly with distance. For frequencies above 24 kHz the signal decay (or attenuation) is large enough to render the signal of little use. Hence, VLF operation is limited to frequencies below 24 kHz.

TABLE 9.4   ALLOWABLE $TM_{m,n}$ MODES AND CORRESPONDING WAVELENGTH FOR LONG-WAVE SIGNALS UP TO 20 kHz

(a) Daytime; Average Height of D Layer 70 km

Mode \ Freq	4.2 kHz	8.6 kHz	13.0 kHz	17.1 kHz	21.4 kHz	25.9 kHz	30.0 kHz
$TM_{1,0}$	70 km						
$TM_{2,0}$		35 km					
$TM_{3,0}$			23 km				
$TM_{4,0}$				17.5 km			
$TM_{5,0}$					14 km		
$TM_{6,0}$						11.6 km	
$TM_{7,0}$							10.0 km

(b) Nighttime; Average Height of D Layer 90 km

Mode \ Freq	3.3 kHz	6.6 kHz	10.0 kHz	13.3 kHz	16.6 kHz	20.0 kHz	23.5 kHz
$TM_{1,0}$	90 km						
$TM_{2,0}$		45 km					
$TM_{3,0}$			30 km				
$TM_{4,0}$				22.5 km			
$TM_{5,0}$					18 km		
$TM_{6,0}$						15 km	
$TM_{7,0}$							12.9 km

Legend: ▨ Excluded TM Mode

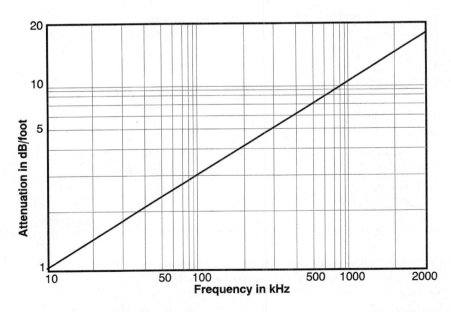

Figure 9.7   Attenuation of a Plane Wave in Seawater (Conductivity = 3 mhos/m) (Reprinted with permission from Robert W. Turner, "Submarine Communications Antenna," Proc. IRE, May 1959, pp. 753–739, Figure 1, Copyright 1959 IEEE.)

TABLE 9.5 PENETRATION DEPTH OF
LONG-WAVE SIGNALS IN SEAWATER [CA-1]

Frequency Band	Penetration Depth (m)
MF	0.14–0.46
LF	0.46–1.4
VLF	1.4–4.6
VF	4.6–14
ELF	14–46[1]

[1]Actual ELF signal penetration depth is not limited to
46 m; it can penetrate up to 80 m.

periscope depth of 10 m (33 ft). However, the ELF signal can penetrate as much as 80 m ($\approx$270 ft) below the ocean surface. This allows deeply submerged submarines to receive ELF signals while operating at depths two orders of magnitude deeper than for submarines receiving VLF signals.

9.2.3 The Effects on Sky Wave Interference

So far, we have discussed signal amplitude reduction as a function of propagation distance and propagation medium. However, the signal amplitude also varies as a function of time. During the day, the signal amplitude is subject to quasi-cyclic variation as the phase of the sky wave component varies relative to the constant phase of the ground wave. The variation of amplitude as a function of time depends on the reflection heights, that is, 70 km (44 mi) to 90 km (56 mi), and reflection coefficients of the ionosphere and on the amplitude and phase of the ground wave. This variation is known as sky wave interference.

9.2.4 Noise and Interference

There are two kinds of variations in the electron density of the ionospheric D layer which affect the signal strength: diurnal changes and effects of the aurora.

Diurnal changes.   Figure 9.8 shows the electron density changes of the D layer of the ionosphere due to the change from day to night [FE-1]. When a signal path crosses the day-night terminator, day-night changes significantly affect the signal strength due to the rapid change of the D layer heights between day and night. The long-wave Path A in Figure 9.8 is an example of a signal path crossing over the day-night terminator. The electron density change over the terminator is fairly constant, but the day-night terminator moves on constantly.

Auroral effects.   Figure 9.8 also shows the changes of the electron density due to the aurora near the North and South Poles. When a signal path crosses over the aurora, electron density changes, and this significantly impacts the signal strength. Figure 9.8 shows the change of the D Layer from 50° N to 70° N and from 40° S to 85° S. The long-wave Paths A and B in Figure 9.8 are two examples of signal paths crossing over auroras. The electron density over the aurora fluctuates randomly.

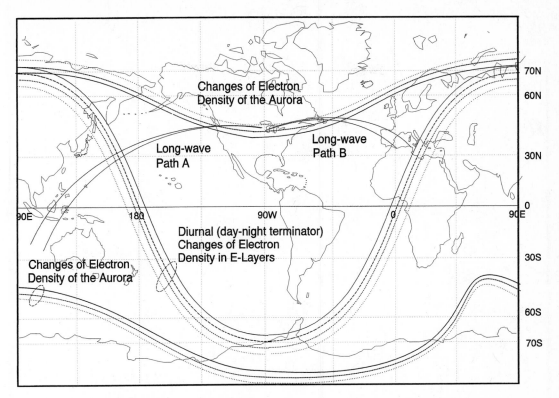

Figure 9.8   Diurnal Changes of D Layers in Inosphere and the Aurora Effects near the North and South Poles [Courtesy of the U.S. Navy, NCCOSC, San Diego]

Noise.   There are three classes of noise in the long-wave bands: atmospheric noise, human-made noise, and cosmic noise.[9] In the long-wave frequency bands, atmospheric noise is predominant. Atmospheric noise is caused by impulses caused by lightning. Lightning will generate transients which travel in the waveguide like radio signals. During local thunderstorms, the noise level is significantly elevated. This noise is called static. Strong impulse interference can be reduced by employing amplitude limiting techniques in the receiver, preferably before bandwidth limiting circuits.

9.2.5 Long-Wave Propagation Prediction Models

Since the 1960s, considerable efforts have been made to develop numerical techniques for the prediction of long-wave propagation and noise. In 1965, the Naval Research Laboratory developed a VLF atmospheric noise prediction model [RH-1; NA-1], which includes algorithms for:

[9]See Figure 7.1 of Chapter 7 for data on atmospheric, man-made, and cosmic noise.

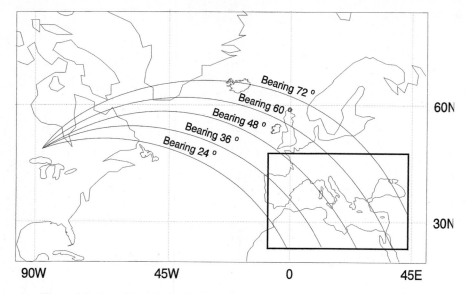

Figure 9.9 Long-Wave Paths Plotted from the Transmitter Site to the Mediterranean Sea, Each Bearing Separated by Ten Degrees at the Transmitter Site (Adapted from [FE-2]; Courtesy of the U.S. Navy, NCCOSC, San Diego)

- Computing the mean and standard deviation values of electromagnetic power radiated from every region of the Earth, using thunderstorm and lightning discharge data.

- Computing the propagation of energy within the Earth-ionosphere waveguide, using the effects of ground conductivity, direction of propagation, latitude, Sun zenith-angle, day or night path conditions, and three most significant modes of propagation.

- Computing noise field intensity, standard deviation, voltage deviations, and value of the angle of signal arrival for any location on the Earth's surface at any frequency from 10 kHz to 30 kHz.

- Generating contour plots, polar plots, diurnal plots, and frequency interpolation plots.

In 1967, the Naval Research Laboratory also developed a computer model for Single Mode Propagation Prediction [NA-2; NA-1]. This model uses the theoretical-waveguide propagation model for single-mode propagation prediction, developed by J. R. Wait [WA-2]. The prediction model uses simple statistical variations of the parameters in Wait's single-mode equation. The predicted signal statistics are derived from an assumed normal distribution of hourly signal levels for day-to-day signal variation occurring during any particular month. This model is a first approximation of the more complex multimode VLF propagation mechanism, which results in a more accurate signal prediction.

In 1989, the Naval Ocean Systems Center[10] developed a long-wave waveguide

---

[10]It is now called Naval Command and Control and Ocean Systems Center (NCCOSC).

propagation model called Long-Wave Propagation Capability-4 (LWPC) [FE-1; FE-2]. This is a computer simulation of long-wave propagation. The simulation covers the frequency range from 30 Hz to 60 kHz and is modeled for a curved Earth waveguide. The path of long-wave propagation from the transmitter site to the receiver is determined by segmentation methods, which treat complicated propagation paths as horizontally homogeneous segments. The parameters of the segments are determined by local values of ionospheric D layer height, geomagnetic field, and ground conductivity. The resultant tracks are on great circle paths. Figure 9.9 shows different bearings plotted from the

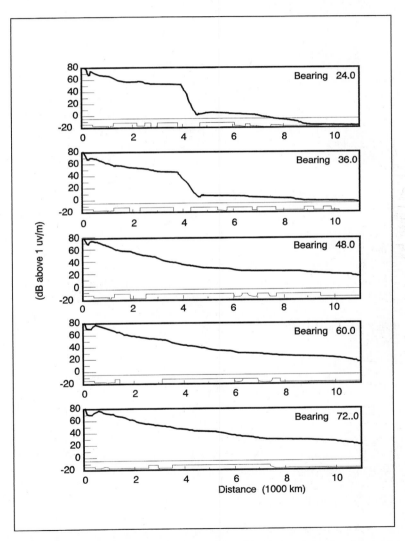

Figure 9.10 Plot of Signal Strength versus Distance (For VLF Paths in Figure 9.8) (Courtesy of the U.S. Navy, NCCOSC, San Diego)

transmitter site, that is, Clam Lake, Wisconsin, to the Mediterranean Sea, and each bearing is separated by 12 degrees at the transmitter site [FE-1]. In this model, both the surface wave and the waveguide propagation as explained in section 9.2.1 are used. Mode computations on a location may not be very important, but as the signal travels on a great circle path, the D layer propagation, ground constants, and geomagnetic field may change the dominant propagation mode at that location.

In this model, the effect of the Earth conductivity on the propagation of the surface wave is included. Also included in this model are: (a) the diurnal variation of propagation due to the change of the D layer height, and (b) the effects of the aurora on the propagation. The aurora near the North and South Poles has a significant impact on the signal strength when a selected path crosses over the aurora in the D layer of the 50° N to 70° N near the North Pole and of the 40° S to 85° S near the South Pole.

The NCCOSC LWPC model was validated over the ocean by measurements made on a ship traveling along a path from the U.S. coast to the United Kingdom. The ship carried a receiver that recorded signal strengths. There were significant changes in the ground conductivity, day-night changes, and the geomagnetic constant changes in the high latitudes. For each bearing a separate path was selected in order to compute the signal strength. Figure 9.10 shows plots of the signal strength versus distance for different bearings [FE-2; pp. 593–597]. Also shown in this figure is the land mass profile along the paths.

9.3 ANTENNA SYSTEMS

Long-wave antennas can be divided into two classes: transmitting and receiving antennas. Long-wave transmitting antennas are either shore-based systems or airborne systems (like the TACAMO trailing as discussed in section 9.4.2).

9.3.1 Transmitting Antennas

Long-wave transmitting antennas are large physical structures with multiple towers several hundred meters high [NA-1]. The wavelength of long-wave signals is on the order of hundreds of meters, and the size of transmitter antenna elements should be on the order of a quarter-wavelength. If the wavelength is not matched, the transmitter efficiency decreases. Furthermore, the transmitter must be capable of handling the required high power capacity, and high RF voltage [JA-1].

The long-wave antenna generally consists of towers across which long wires are mounted. The structure of a typical long-wave transmitting antenna is shown in Figure 9.11. On the top of the towers, an array of five or six long wires is installed. The height of such a tower is typically 300 m (1000 ft) [WA-1].

9.3.2 Receiving Antennas

There are three types of long-wave receiver antennas commonly used on submarine platforms [TU-1]:

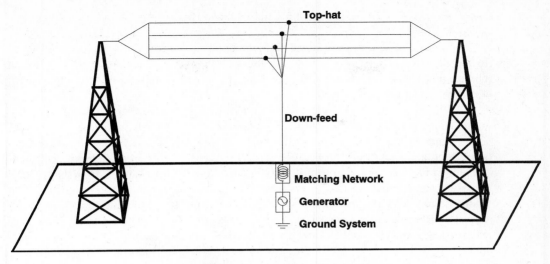

Figure 9.11   VLF/LF Transmitting Antenna

- Towed buoy antenna: A small loop antenna is enclosed in a towed buoy, which is tethered to a submarine.
- Mast-mounted, multifunction antenna: Several antennas are mounted on the mast of a submarine, covering various frequency bands (long-wave, MF/HF, UHF, VHF, and SATCOM).
- Buoyant cable antennas: An antenna element assembly is connected to long RF transmission lines.

Towed buoy antennas.   An antenna frequently used on submarines is the loop antenna; it is housed in a sea-following buoy tethered to a submarine. The buoy rises close to the sea surface when towed by a submerged submarine as shown in Figure 9.12. Typical buoy antennas currently in use are the OE-305, AN/BRR-6, and AN/BRR-8. These antennas provide VLF/LF and MF/HF reception for submarines.

A typical towed buoy antenna system consists of: (a) a towed buoy, (b) a small loop antenna, (c) an antenna amplifier, and (d) a tow cable. Normally a towed buoy is housed

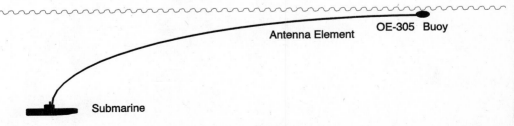

Figure 9.12   Typical Towed Buoy Antenna

in a buoy compartment as shown in Figure 9.13, and is deployed as a sea-following buoy when required.

The tow cable must be built so that it can withstand the tension required to pull the towed buoy at the maximum speed. The tow cable also carries the control signal from a control unit to the loop antenna and the signal received by the antenna to the long-wave receivers via a multi-coupler.

A small loop antenna is shown in Figure 9.14. It is electrically small, that is, the loop diameter is less than one tenth of the wavelength (0.1 λ). Also shown in Figure 9.14 is the figure-eight antenna pattern of the small loop antenna. The radiating pattern of these small loop antennas is directional. The directional antennas may not be very convenient for submarine operation due to difficulty arising from aspect changes. However, with a cross-loop antenna, that is, two loops placed perpendicular to each other, one can obtain an omnidirectional antenna pattern, which is more useful for submarine applications. The operating RF range for loop antenna is from 10 kHz to 180 kHz. The minimum sensitivity of a long-wave receiver with a small loop antenna used in a submarine platform is 7 μV/m at the center of the loops. An electrically small loop antenna induces a sufficiently large signal for the receiver to operate acceptably well [WA-1].

Mast-mounted, multifunction antennas.   Mast-mounted, multifunction antennas are also frequently used on submarines. Unlike on surface ships, submarine antenna systems must combine several functions on a mast. A set of antennas perform receiver antenna functions for communications, navigation, and identification. A specific antenna for a mast system is selected primarily according to physical space availability. A typical mast-mounted, multifunction antenna system consists of: (a) a VLF/LF loop an-

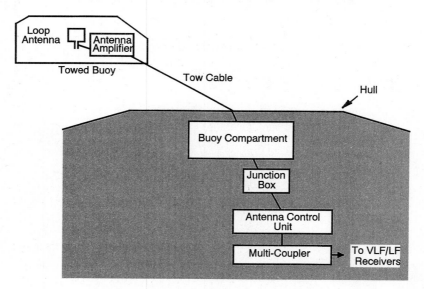

Figure 9.13   Typical Towed Buoy and Its Mechanical Devices

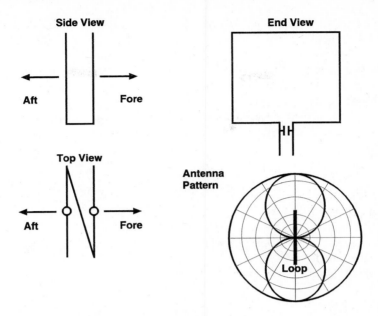

Figure 9.14   A Loop Antenna and Its Radiation Pattern

tenna, (b) a UHF dipole antenna, (c) a retractable HF/MF whip antenna, and (d) a UHF SATCOM helical antenna. These complex antennas are housed in a mast fairing as shown in Figure 9.15. In addition, a separate EHF SATCOM antenna, and a GPS receiver antenna, and LORAN-C receiver antenna are often included in the multifunction antenna system. In this chapter we describe only the long-wave receiving antenna.

Long-wave signals received by the loop antenna are processed and controlled by the antenna amplifier and switch assembly unit, since several multifrequency bands are received by a single antenna. The high sensitivity of a communications receiver does not require high signal field strengths for good reception, and antenna matching is not required. The long-wave receiving antenna is nonresonant. Received signals are connected via an antenna coupler to the receivers.

Note that the mast fairing is colocated with the periscope; there are also other energy emitting devices as illustrated in Figure 9.16. The problem of top-side antenna placement usually becomes a complex EMC problem, and simultaneous operation of several circuits using the multifunction antenna arrangement is very difficult.

Typical buoy antennas currently in use are the AN/BRA-34, AN/BRA-23, OE-207, and OE-176.

Buoyant cable antenna.   Another submarine antenna is the buoyant cable antenna, an antenna element assembly connected to a buoyant transmission line, which is attached to a submarine conning tower through a streaming conduit. The antenna element assembly rises near to the sea surface when towed by a submerged submarine as shown in

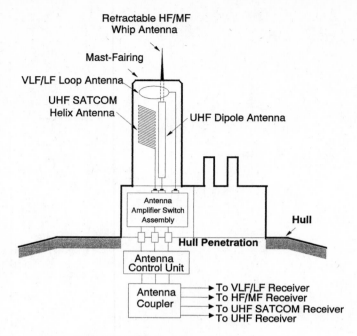

Figure 9.15   A Typical Mast-Mounted Multifunction Antenna

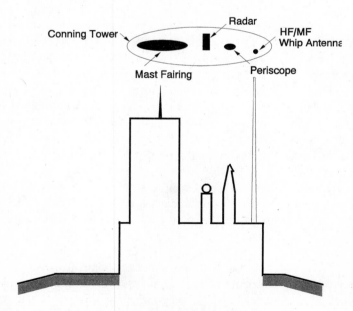

Figure 9.16   Typical Top-Side Arrangement of Conning Tower

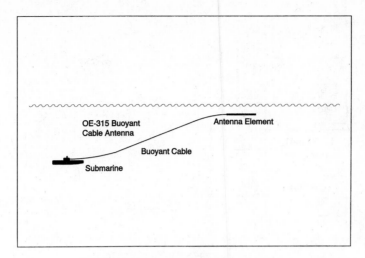

Figure 9.17   Typical Buoyant Cable Antenna

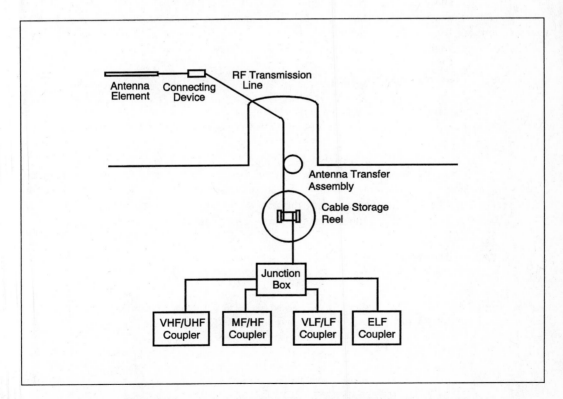

Figure 9.18   Typical Buoyant Cable Antenna and Its Electrical and Mechanical Devices

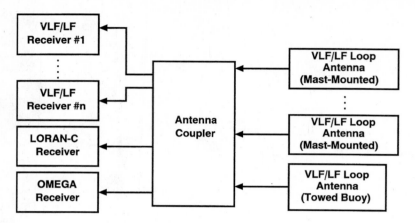

Figure 9.19   Typical VLF/LF Antenna Coupler System

Figure 9.17. Typical buoyant cable antennas currently in use are the OE-315 and AS-2629B. These antennas allow reception of signals from the ELF to VHF bands. They can also be used for transmitting in the MF to HF bands. A typical buoyant cable antenna system consists of: (a) an antenna assembly, (b) an antenna cable transfer assembly, (c) a cable storage reel, and (d) an antenna coupler as shown in Figure 9.18.

9.3.3 Tuners and Couplers

Antenna couplers process and condition long-wave signals received by loop antennas in mast-mounted, multifunction antennas and cross-loop antennas in a towed buoy antenna, and transfer signals to VLF and LF receivers and to navigation receivers such as LORAN-C and OMEGA, as shown in Figure 9.19.

The processing function of the antenna coupler interprets commands from the antenna control unit for selection of proper antennas and pointing in the optimum direction. The conditioning function matches the antenna impedance to that of the receiver side, usually at 50 Ω. Typical tuners and couplers currently in use are listed in Table 9.6.

TABLE 9.6   LIST OF TUNERS AND COUPLERS

Equipment Name	Functional Description
• CU-1441/BRR	• VLF/LF multi-coupler; couples mast, buoyant cable, and whip antenna signals
• CU-1396/BRA-16	• VLF/LF multi-coupler; couples towed buoy, mast, navigation, and buoyant cable antenna signals
• AN/BRA-6	• MF/HF tuner for emergency whip antenna systems (AT-441 and AT-774)
• CU-2364/BRR	• VLF/LF multi-coupler; couples towed buoy, mast buoyant cable, towed buoy auxiliary wire, and E-field antenna signals

9.4 MODULATION SYSTEMS FOR LONG-WAVE RADIO TRANSMISSION

In communications system design, the modulation methods used are determined by factors such as the available bandwidth, the signal strength at the receiver, the noise in the system, and the required output signal quality, and equipment limitations; these factors are discussed in Chapter 7. The need to provide protection of signals from unauthorized access through encryption requires the application of digital modulation.

9.4.1 Long-Wave Modulation System Design Considerations

In the long-wave frequency bands, there are a number of radio systems in operation for communications and navigation. The primary design factors in selecting modulation systems in long-wave radios are bandwidth limitations and size limitations of the transmitter antennas. Effective bandwidth use is essential and adjacent channel interference needs to be reduced by using the minimum bandwidth for transmission of signals. Low bandwidth and high overall system performance are key factors in selecting an appropriate modulation method for a system operating in the long-wave region of the radio spectrum.

The use of narrowband systems at long-wave radio frequencies restricts information transmission to low signal rates and requires that the equipment has a high frequency stability. This explains why either hand-keyed or machined-key Morse at conventional signal speeds are used, or why more modern systems use single-channel telegraphy at 60 words per minute (w/m) or 75 bps.

Practical narrowband digital teletypewriter (TTY) modulation systems suitable for LF radio communications include shaped AM, narrowband FSK, narrowband PSK, and MSK.

9.4.2 Long-Wave Modulation Techniques

CW keying in long-wave radios.   The earliest modulation method for long-wave radio telegraphy, used until the 1960s, was CW keying. CW keying became one of the most effective means of minimum power transmission. In the 1930s and early 1940s, long-haul communications relied on carrier telegraph systems designed for the stable wire medium. The telegraph system sent messages in International Morse Code by wire transmitter keying. This method was adopted for RF telegraphy, where the sending key controlled the output of the radio transmitter. When the key is closed, RF is transmitted, and when the key is opened, the transmitter is cut off. At the receiver, the operator would either listen for the interrupted short and long tone signals and read them as dots or dashes, or the dots and dashes were automatically recorded on a tape. This method was known as Interrupted Continuous Wave (ICW) keying. Characters in the International Morse Code, represented by dots and dashes, are sequences of unequal length of one to five dot-dash combinations. This type of coding does not lend itself readily to the direct

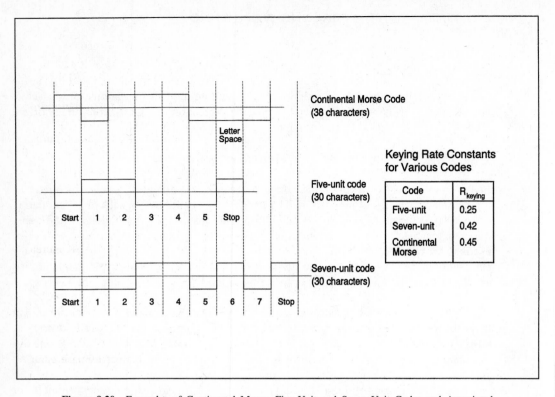

Figure 9.20 Examples of Continental Morse, Five-Unit and Seven-Unit Codes and Associated Keying Rates [WA-1; p. 569]

conversion of received messages from on-off signals to alphabetical characters [WA-1]. Therefore, the five-unit code was adopted, which represents each alphabetical character by five bits. Figure 9.20 shows the five-unit code and two other variants, the seven-unit code and the Continental Morse Code.[11] The five- and seven-unit codes have equal length bits where the ones are transmitted by presence of the carrier and zeroes by absence of carrier; the modulation also became known as on-off keying (OOK) or Amplitude-Shift keying (ASK).

FSK modulation.   A limitation of ICW keying (or ASK), which was designed for the wireline telegraphy, when used over radio circuits, is distortion caused by selective fading and multipath of the constantly varying ionospheric transmission path. Frequency-

---

[11]A good summary of these codes can be found in [JO-1; pp. 122–125] [WA-1; p. 569] [HS-1; pp. 30–36].

Shift Keying (FSK) modulation was introduced to overcome the problems of ICW keying. In FSK two different frequencies are used to designate mark and space, and the shift between them is controlled by the normal on-off telegraph signal. FSK combined with a limiter at the demodulator provides a signal that has less ambiguity between mark and space than ASK.

A number of theoretical and experimental studies have indicated that the optimum receiver bandwidth for TTY FSK transmission in the presence of thermal noise is equal to about 5/3 times the frequency shift. The studies have also indicated that the optimum frequency shift ΔF is equal to about twice the signaling speed. The optimum frequency shift for five-unit code stop-start TTY FSK modulation at a signaling rate of 60 wpm is about 60 Hz.

FSK is widely applied to data transmission at rates of 1200 bps or less. Digital transmission using FSK has the following advantages: (a) the implementation is straightforward, (b) FSK shows a 3- or 4dB performance improvement over ASK against most types of noise (as the frequency shift becomes greater, the advantage over ASK increases in a noisy environment, but the bandwidth also widens), and (c) FSK protects against the effects of multipath fading.

PSK modulation.   Further bandwidth reduction with respect to FSK can be accomplished using phase-shift-keying (PSK) modulation. In PSK the carrier frequency is constant, but the mark and space are expressed by phase changes. Binary PSK, where the mark and space (or binary zero and one) are distinguished by 180-degree phase changes (phase reversals), has a 3-dB performance advantage over FSK; however, PSK is not very resistant to fading and multipath distortions.

MSK modulation.   One of the most effective modulation methods in terms of bandwidth usage is minimum-shift keying (MSK). It is also relatively complex and its implementation requires application of the technological advances that became available in the 1960s and early 1970s.

MSK is akin to FSK where the frequency is shifted in such a manner that the result is a phase increase or decrease of 90° each τ seconds (τ is the duration of one bit). The phase shift is gradual and there is no abrupt phase change. It is accomplished by phase modulating alternate bits of the data stream of ones and zeroes (for example, the stream 00110110) to generate PSK on an in-phase carrier (or I-carrier) and a quadrature carrier (or Q-carrier); the I-carrier is represented by $cos(2\pi ft)$, the Q-carrier is given by $sin(2\pi ft)$. The sum of the modulated I- and Q-carriers are 2τ seconds in duration, but they are offset by the time τ, that is, the duration of one bit. Because of the offset, the bits are staggered in such a manner that there is exactly a 90° phase change for each bit; furthermore, the phase change is gradual during each bit. There are no abrupt phase or amplitude changes, and as a result the sum of the modulated I- and Q-carriers has a minimal spectrum occupancy. A more detailed description follows in section 9.4.5. Figure 9.21 shows the waveforms of ASK, FSK, PSK, and MSK.

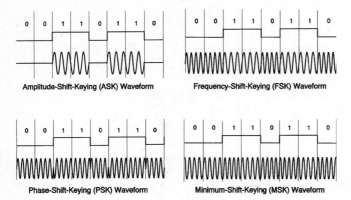

Amplitude-Shift-Keying (ASK) Waveform

Frequency-Shift-Keying (FSK) Waveform

Phase-Shift-Keying (PSK) Waveform

Minimum-Shift-Keying (MSK) Waveform

Figure 9.21 Waveforms Resulting from Different Types of Modulation

9.4.3 Waveform Filtering

From Figure 9.21 it can be seen that there are abrupt changes of amplitude, frequency, or phase at the transitions between ones or zeroes. These abrupt changes widen the bandwidth of the spectrum of the modulated waveforms. In order to improve bandwidth containment (since bandwidth is at a premium), filtering is applied.

The effectiveness of filtering can be measured by a filtering index. The value of the filtering index r ranges from 0 to 1. The filtering index r expresses the degree of filtering accomplished as shown in Figure 9.22. Figure 9.22 shows the frequency response of filters (square, cosine, and raised cosine) and their filter indexes.

Keying waveform filtering. In order to minimize the bandwidth requirements of long-wave radios, several specialized forms of filters are employed [WA-1; NA-1]. Figure 9.23 shows the square wave and its RC and LC filtered waveforms. The pulse width is τ; the bandwidth required to transmit this waveform free of distortion is infinite. However, if the bandwidth is restricted to W_r, a reasonable replica of the square wave can be transmitted.[12] When a filter of bandwidth W is used, the band-filtering factor is τW, where τ is the duration of the keying element, as shown in Figure 9.23. Since $\tau = 1/(2W_r)$ and $W = 2W_{6dB}$ (W_{6dB} is the 6-dB cutoff frequency of the filter used),[13] the band-filtering factor τW becomes W_{6dB}/W_r, or $W_{6dB} = 2W_r \tau W$.

The bandwidth containment will also be a determining factor in the modulation rate,

---

[12]W_r is the first zero crossing of the Fourier transform of a rectangular pulse; the Fourier transform of the rectangular pulse of amplitude 1 and duration τ is $(sin\pi f\tau)/(\pi \tau f)$. The first zero crossing occurs where $\pi\tau f = \pi$; the frequency for this point is designated W_r. W_r is a *one-sided* (low-pass) bandwidth, it corresponds to B_{out} described in Chapters 7 and 8; for transmission in the RF medium a *two-sided* bandwidth of $2W_r$ is required. A rectangular wavetrain has a Fourier line spectrum with the $(sin\pi f\tau)/(\pi \tau f)$ as envelope (see [HS-1; p. 42–44]).

[13]Note that the filter bandwidth is *two-sided* (radio bandwidth) and the bandwidth is defined to be the point at which the filter attenuation has reached 6 dB.

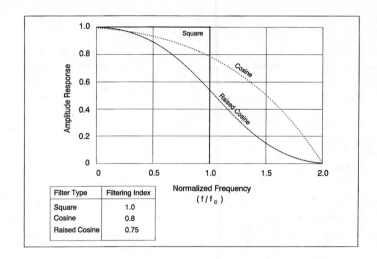

Figure 9.22   Frequency Response of Various Filters (Square, Cosine, and Raised Cosine)

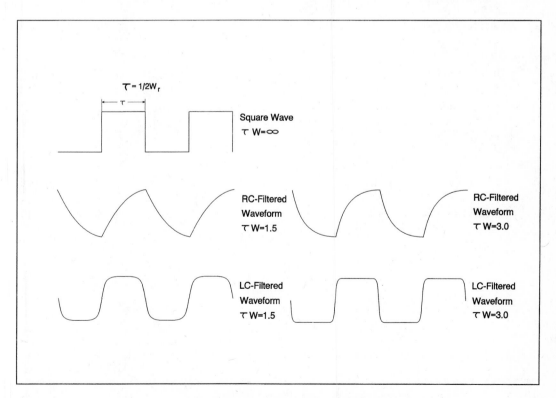

Figure 9.23   Keying Waveforms with Different Filtering

because the filtering will limit the number of pulses per second (pulses/s) that can be transmitted. The modulation rate R_{mod} (pulses/s) is given as keying rate R_{keying} (char/s) times the unit code (pulses/char); the keying rate R_{keying} (pulses/s) is message rate R_W (w/min) times characters per word (char/w) times minutes per second. In TTY transmission, the assumed average number of characters per word is 6. The resultant modulation rate is:

$$R_{mod}[\text{pulses/s}] = R_w[\text{w/min}] \times 6[\text{char/w}] \times 1/60[\text{min/s}] \times \text{UnitCode}[\text{pulses/char}]$$

9.4.4 Examples of Bandwidth Reduction by Filtering for Long-Wave Transmission

Bandwidth requirements for a communications system are determined by data rate, signal coding, and type of modulation employed. For long-wave transmission also the characteristic of the equipment, in particular the receiver antenna bandwidth, needs to be considered. A typical receiver antenna characteristic is shown in Figure 9.24. Numerical examples of the ASK and FSK modulation bandwidths required for long-wave radios are shown below.

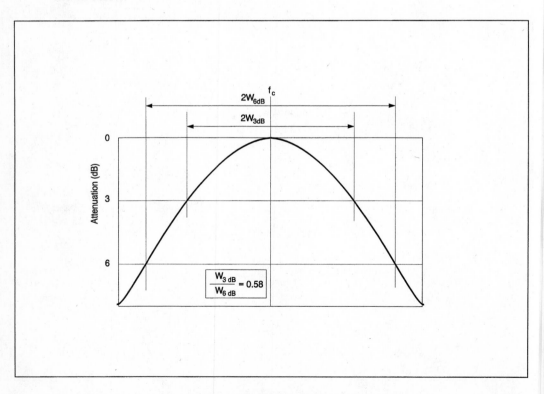

Figure 9.24   Typical VLF Receiver Antenna Characteristics

ASK modulation.   The 6-dB bandwidth for reliable ASK TTY signals is obtained from:

$$W_{6dB} = 2\,W_r \tau W = 2R_{keying}\tau W\,R_W,$$

where R_{keying} = keying rate constant for the unit code used,
τ = keying element length in seconds,
W_{6dB} = 6-dB cutoff bandwidth of the filter used,
R_W = message rate in words per minute, and
τW = filtering characteristic.

Let it be assumed that a VLF signal is transmitted by a single tuned antenna of a 3-dB to 6-dB bandwidth ratio of 0.58 as shown in Figure 9.24. The resulting 3-dB bandwidth can be expressed as [WA-1; p. 569]:

$$W_{3dB} = 0.58W_{6dB} = 0.58 \times 2R_{keying}\tau W\,R_W = 1.16\,R_{keying}\,\tau W\,R_W.$$

As an example, consider VLF TTY ASK transmitted at a rate of 15 wpm, using a waveform filter τW of 3. The keying rate R_{keying} is 0.45 (using Continental Morse Code) as shown in Figure 9.20. Therefore, the 3-dB minimum bandwidth for transmitting VLF ASK TTY is:

$$W_{3dB} = 2 \times 0.58 \times 0.45 \times 3 \times 15 = 24 \text{ Hz.}$$

FSK modulation.   The equation for deriving the minimum bandwidth for FSK TTY via VLF transmission is shown below. The 3-dB bandwidth for the reliable FSK TTY signal can be expressed as [WA-1; p. 609], [NA-1]:

$$W_{3dB} = 1.16\,R_{keying}R_W\,(m + \tau W),$$

where R_{keying} = keying rate constant for the unit code used,
R_W = message rate (wpm),
m = modulation index ($\Delta f/f$),
τ = keying element length in seconds, and
W_{6dB} = 6-dB cutoff bandwidth of the filter used.

As an example, consider VLF FSK TTY transmitted at a rate of 50 (w/min) using a waveform filter τW of 1.42. The keying rate R_{keying} (seven-unit code) is 0.42 as shown in Figure 9.20. The message rate R_W of 50 [w/min] using seven-unit code is:

$$R_W = (60/6) \times 50[\text{w/min}]\,/7 = 71.42\ [\text{w/min}].$$

The modulation index is assumed to be $m = 1$. Thus, the 3-dB minimum bandwidth required to transmit VLF FSK TTY signals is:

$$W_{3dB} = 1.16 \times 0.35 \times 71.42 \times (1 + 1.42) = 70\ [\text{Hz.}]$$

9.4.5 Long-Wave Modulation Bandwidths Comparison

In Figure 9.21 we show the waveforms for four digital modulation methods that are applicable to long-wave radio TTY transmission:

- Amplitude-Shift-Keying (ASK) modulation
- Frequency-Shift-Keying (FSK) modulation
- Phase-Shift-Keying (PSK) modulation
- Minimum-Shift-Keying (MSK) modulation

In the previous sections, we discussed in depth two modulation methods, ASK and FSK, since they have been extensively used. We now examine which of the four modulation methods is optimum for long-wave transmission.

ASK modulation.   This converts two binary values into two different amplitudes of the carrier frequency:

$$y_1(t) = K sin \, (2\pi f_c t + \varphi_0), \qquad binary \; 1,$$
$$y_0(t) = 0, \qquad\qquad\qquad binary \; 0$$

where f_c is the carrier frequency, and K is the amplitude of the carrier signal. (The ASK waveform is included in Figure 9.21.)

FSK modulation.   This converts two binary values into two different frequencies near the carrier frequency:

$$y_1(t) = K sin \, (2 \, \pi \, (f_c - \Delta f) \, t + \varphi_0), \qquad binary \; 1,$$
$$y_0(t) = K sin \, (2 \, \pi \, (f_c + \Delta f) \, t + \varphi_0), \qquad binary \; 0.$$

The FSK waveform is included in Figure 9.21.

Binary PSK modulation.   This converts two binary values into two different phases of the carrier frequency:

$$y_1(t) = K sin(2\pi f_c t + \pi), \qquad binary \; 1,$$
$$y_0(t) = K sin \, (2\pi f_c t), \qquad binary \; 0.$$

The binary PSK waveform is included in Figure 9.21.

MSK modulation.   This can have two implementations, serial MSK and parallel MSK [AM-1; GR-1]. In both cases the spectrum is shaped by filtering to minimize the bandwidth occupancy of the signal. Here parallel MSK is described. The MSK waveform is included in Figure 9.21.

For MSK modulation the data stream of ones and zeroes is separated into two component streams, each consisting of alternate bits; for example, the stream 00110110 is separated into component streams 0101 for the odd (first, third, etc.) bits and 0110 for the even bits. The first component stream modulates the in-phase carrier (or I-carrier), the second component stream the quadrature carrier (or Q-carrier); if the I-carrier is represented by $cos2\pi ft$, the Q-carrier is given by $sin2\pi ft$. The signals on the I- and Q-channel are offset by one bit interval τ. The signals are also modulated by $\pm cos(\pi t/2\tau)$ (for the I-carrier) or $\pm sin(\pi t/2\tau)$ (for the Q-carrier); thus, the PSK signals have a cosine or sine envelope that has a duration of one half period. The sum of the modulated I- and Q-carriers exhibits a constant amplitude and a gradual phase change that will be exactly $+90°$ or $-90°$ at the end of a bit. The overall phase shift accumulates as a function of the bit stream that is transmitted. The sum of the modulated I- and Q-carriers has a minimal spectrum occupancy; with modern detection techniques (a block diagram of a receiver is shown in Figure 9.34) the demodulated signal has a very good signal-to-noise ratio. The bits of the MSK data stream can be represented by sinusoids of the form [ZI-1], [SI-2], [DO-1]:

$$\pm A \; cos \; 2\pi \; [f_0 \pm (1/4\tau)] \; t$$

and

$$\pm A \; sin \; 2\pi \; [f_0 \pm (1/4\tau)] \; t.$$

These sinusoids can be considered as having the frequencies $[f_0 \pm (1/4\tau)]$ Hz; the signs of the expression are such that a continuous waveform results. The quantity $(1/4\tau)$ is the offset of the carrier frequency in the MSK. The frequency $[f_0 - (1/4\tau)]$ is called "mark" frequency and it is the frequency generated by a sequence of either all ones or all zeros. The frequency $[f_0 + (1/4\tau)]$ is called "space" frequency and it is the frequency generated by a sequence of alternating ones and zeros. MSK is phase continuous Offset Quadrature-Phase Shift-Keying (OQPSK) [SP-1].

An illustration of the MSK waveforms is presented in Figure 9.25. The figure shows a data bit stream of zeroes and ones. The data bit stream is separated into an I-channel and Q-channel sequence. In this process the zeroes and ones are transformed into $+1$ and -1. The illustration shows the $+1$ and -1 sequences in the I- and Q-channel. In the I- and Q-channel the PSK waveforms are modulated by either $cos(\pi t/2\tau)$ or $sin(\pi t/2\tau)$; this will create a gradual phase change by $-90°$ or $+90°$ degrees for each bit of duration τ.

The superposition of the I- and Q-channel waveforms creates a bit signal stream with a gradual phase change; this is equivalent to having a frequency $[f_0 + (1/4\tau)]$ or $[f_0 - (1/4\tau)]$. The phase change is continuous, that is, there is no phase jump from one bit to the next. Figure 9.25 shows both the phase change and frequency for each bit duration. The net phase change for the duration 2τ of two bits is either 0, $+180$, or -180 degrees.

The resulting waveforms are also shown in Figure 9.25. If one assumes a bit duration to be 1 ms, and f_o to be 4000 Hz, then $1/4 \; \tau$ is 250 Hz, the MSK "mark" frequency is 3750 Hz and the MSK "space" frequency is 4250 Hz. The waveforms in Figure 9.25 show the subtle frequency changes and the phase and progression change for this set of values.

Channel usage and data rate.   The performance of various modulation methods in digital communications depends on the bandwidth of modulated signal. A measure of the effective use of the RF spectrum is bandwidth effectiveness η_B,[14] which is defined

---

[14]See also Chapters 7 and 8.

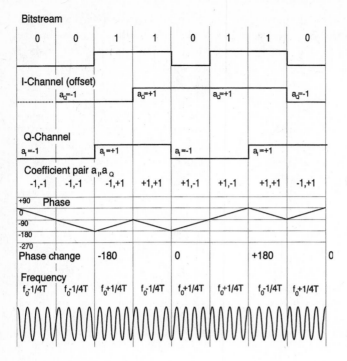

Figure 9.25 An Illustration of MSK Waveforms

as the ratio of data rate to transmission bandwidth.[15] (This ratio is also known as the bandwidth efficiency [ST-1; SK-1], and bandwidth utility [FI-1]).

The bandwidth for ASK is:

$$W_{ASK} = (1 + r) R,$$

and the ratio of the data rate to transmission bandwidth is:

$$R/W_{ASK} = 1/(1 + r),$$

where R is the data rate and the filtering index r is the ratio of the 3-dB to 6-dB bandwidth of the filter used in shaping the spectrum (see Figure 9.22).

The bandwidth for FSK is:

$$W_{FSK} = (f_2 - f_1) + (1 + r)R,$$

where f_1 and f_2 are two FSK frequencies representing ones and zeroes. As limiting case, the bandwidth W_{FSK} for FSK becomes dominated by the term $(f_2 - f_1)$ in the higher carrier frequency and the term $(1 + r)R$ in the lower carrier frequency. The ratio of the data rate to transmission bandwidth for FSK is:

---

[15]The effectiveness of a modulation system for long-wave radio transmission can be expressed as: R/W_{3dB}, where W_{3dB} is the two-sided (RF) filter 3-dB cutoff and R is the data rate.

TABLE 9.7   RATIO OF DATA RATE TO TRANSMISSION BANDWIDTH
FOR VARIOUS MODULATION METHODS [ST-1]

Filtering Index	$r = 0$	$r = 0.5$	$r = 1$
ASK	1.0	0.67	0.5

FSK
   Higher carrier frequency
      $(f_2 - f_1) \gg R$
   Lower carrier frequency
      $(f_2 - f_1) \approx f_c$
	~0
	
	1.0
		~0
		
		0.67
			~0
			
			0.5

PSK
   2-level
   4-level
   8-level
	1.0
	2.0
	3.0
		0.67
		1.33
		2.0
			0.5
			1.0
			1.5

MSK (Equivalent to 4-level PSK or OQPSK)	2.0	1.33	1.0

$$R/W_{FSK} = R(f_2 - f_1) + 1/(1 + r).$$

The bandwidth for multilevel PSK is:

$$W_{PSK} = (1/\log_2 L)\,(1 + r)R,$$

where L is the number of levels. The ratio of the data rate to transmission bandwidth for PSK is:

$$R/W_{PSK} = (1/\log_2 L)/(1 + r).$$

The bandwidth for MSK is:

$$W_{MSK} = 0.5(1 + r)R,$$

since it corresponds to a special case of offset QPSK where the level of signal $L = 4$. The ratio of the data rate to transmission bandwidth for MSK is:

$$R/W_{MSK} = 2/(1 + r).$$

Table 9.7 shows that multilevel PSK modulation provides the best bandwidth efficiency among the four modulation methods (ASK, FSK, PSK,[16] and MSK).

Table 9.8 shows the bandwidth efficiencies for the above four modulation methods (BPSK, QPSK, MSK, and sinusoidal FSK).

Fractional power occupancy bandwidth.   Another definition of the signal bandwidth efficiency is one where 99% of the power is inside the occupied band.[17] V. K. Prabhu studied the difference in performance between PSK and MSK modulation methods by computing the power contained outside the band $[-W, W]$ [PR-1]. Figure 9.26 shows the

---

[16]In Table 9.7, the filtering indices for PSK with 2-, 4- and 8-level are included. Multilevel PSK where $m > 2$ is more effective than MSK in its bandwidth usage. However, multilevel PSK requires more power for the same error rate [SK-1]. Also it requires more complex equipment.

[17]This definition is the standard by which the FCC certifies frequency bandwidth of equipment [AM-1].

TABLE 9.8   THE LOWEST 99% POWER
CONTAINMENT BANDWIDTH

Modulation Type	99% Power Containment Bandwidth
BPSK	20.56
QPSK	10.28
MSK	1.18
Sinusoidal FSK	2.20

spectral occupancy of binary PSK with several filtering methods (rectangular, cosine, and raised cosine) applied.[18] The normalized power contained outside the band $[-W, W]$ is a function of the time-bandwidth product in the radio band $(2WT)$.[19] For example, 99% of the binary PSK signal, passed through the cosine filter, lies within a time-bandwidth product in the radio band equal to 4. Figure 9.27 shows the spectral occupancy of MSK with the same filtering. The modulation uses a smaller bandwidth than binary PSK modulation according to Prabhu.

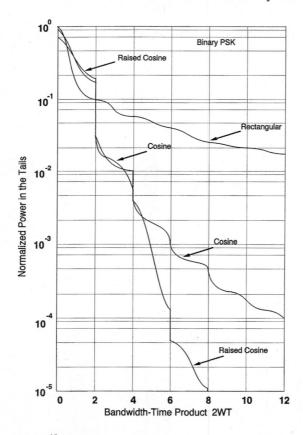

Figure 9.26   Normalized Power Occupancy versus Bandwidth for Binary PSK (Reprinted with permission from V. K. Prabhu, Copyright 1976 AT&T)

[18]The lower bound shown in Figure 9.28 corresponds to the Shannon's ideal channel.

[19]See discussions of the time-bandwidth product 2WT in section 7.1.5 and section 8.3.

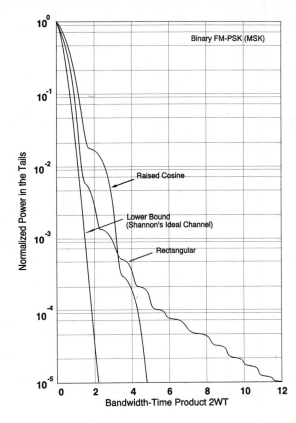

Binary FM-PSK (MSK)

Raised Cosine

Lower Bound
(Shannon's Ideal Channel)

Rectangular

Normalized Power in the Tails

Bandwidth-Time Product 2WT

Figure 9.27 Normalized Power Occupancy versus Bandwidth for MSK (Reprinted with permission from V. K. Prabhu, Copyright 1976 AT&T)

F. Amoroso further studied the bandwidth efficiency of several other modulation methods [AM-1; DE-1; PA-1]. Figure 9.28 shows a comparison of the four modulation methods (BPSK, QPSK, Sinusoidal FSK, and MSK). From the curves shown in Figure 9.28, the 99% power occupancy bandwidths in the radio band for the four modulation methods can be tabulated as and are given in Table 9.8.

In summary, MSK modulation is applied to long-wave communications systems because it requires a minimum bandwidth for a given data rate. A typical circuit which implements MSK modulation is explained in the next section.

9.4.6 Typical Long-Wave Receiving Equipment

VLF/LF shipboard receivers must be devices capable of: (a) fine tuning over a range from 14 kHz to 30 kHz in step of 1 Hz, 10 Hz, or 100 Hz, (b) maintaining frequency stability of 1 part in 10^8 per day, and (c) having signal sensitivity of 0.3 μV.[20] In addition, VLF/LF

---

[20]From the Harris R-2368/URR-79 VLF/LF Receiver [HC-1] and the Rohde and Schwarz Model EK-070 VLF/LF Receiver [RO-1; pp. 45–54].

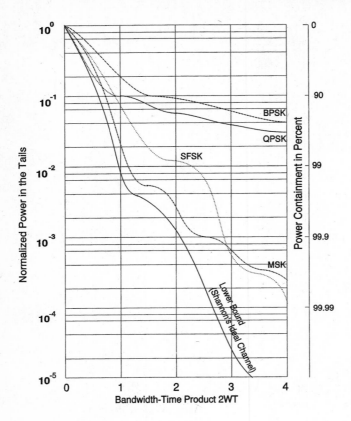

Figure 9.28 Fractional Out-of-Band Power for Various Modulation Methods (Reprinted with permission of F. Amoroso, Copyright 1980 IEEE) [AM-1]

receivers are designed to be able to eliminate high impulse, non-gaussian noise, which is introduced in long-haul VLF/LF propagation channels.

Included in VLF/LF receiving equipment is a demodulator-processor, which transforms received ICW, FSK, or MSK signals into the format needed for subsequent processing. Figure 9.29 shows a block diagram of typical VLF/LF Digital Communication Equipment [RW-1]. Table 9.9 lists shipboard VLF/LF receiver equipment and their characteristics.

9.4.7 Compact (CVLF) Systems

The Compact VLF (CVLF) system was developed to provide compact, lightweight, nuclear-hardened, highly reliable long-wave communications equipment for use on submarines, aircraft, and shore stations [SI-1]. A functional description of the CVLF receiving equipment is shown in Figures 9.30, 9.31, and 9.32 [SC-1]. The first part of the CVLF receiver (Figure 9.30) shows the receiver IF subsystem. The second part of the CVLF receiver (Figure 9.31) shows the noncoherent detection subsystem for FSK modulation. The third part of the CVLF receiver (Figure 9.32) depicts the coherent detection subsystem for MSK modulation [SI-1].

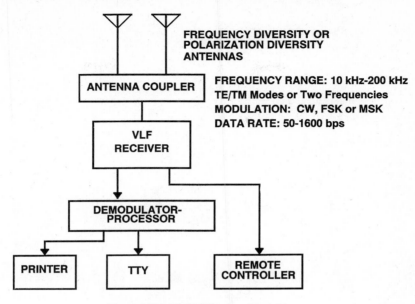

Figure 9.29   VLF/LF Receiver Block Diagram

TABLE 9.9   SHIPBOARD VLF/LF RECEIVING EQUIPMENT

Receiving Equipment	Frequency Band	Modulation	Manufacturer
AN/URR-76 VLF/LF Receiver	14 kHz–175 kHz	CW, FSK, and MSK	Telephonics
R-2368 VLF/LF/MF/HF Receiver (AN/URR-79)	14 kHz–30 kHz (the MF and HF portions are not listed)	CW, FSK, and MSK	Harris RF Communications
AN/BRR-3 VLF Receiver	14 kHz–30 kHz	CW and FSK	
AN/FRR-21 VLF/LF Receiver	14 kHz–600 kHz	CW and FSK	
R800 VLF/LF Receiver	10 kHz–200 kHz	CW, FSK, and MSK	Redifon, Ltd.
TRC-251 VLF/LF Telegraph Receiver	10 kHz–200 kHz	CW, FSK, and MSK (STANAG 3050 Requirements)	Thomson-CSF
EK-070 VLF/LF/MF/HF Receiver	10 kHz–200 kHz (the MF and HF portions are not listed)	CW, FSK, and MSK	Rohde and Schwarz

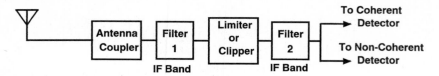

Figure 9.30   Typical VLF Receiver Circuit (Receiver RF and IF Subsystem) (Reprinted with permission from Schwartz, Copyright 1989 IEEE)

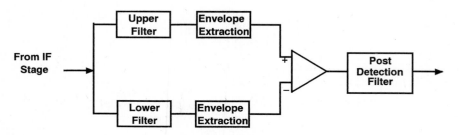

Figure 9.31   Typical VLF Receiver Circuit (Noncoherent Detection Subsystem) (Reprinted with permission from Schwartz, Copyright 1989 IEEE)

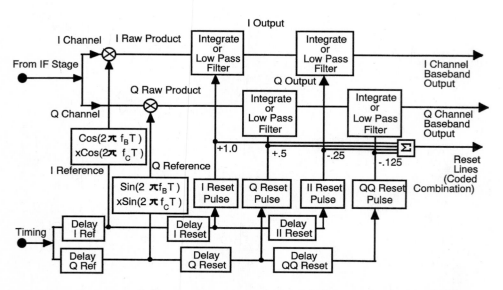

Figure 9.32   Typical VLF Receiver Circuit (Coherent Detection Subsystem) (Reprinted with permission from Schwartz, Copyright 1989 IEEE)

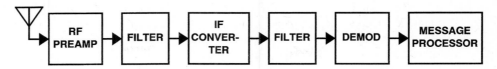

Figure 9.33   Functional Block Diagram of the CVLF Receiver

Frequency range.   The frequency range of CVLF receiver is 14 kHz to 175 kHz. The CVLF system provides each platform with the flexibility to receive communications from a large number of geographically diverse transmitters.

Equipment size.   Figure 9.33 shows the RF and IF circuits of the CVLF receiver. Figure 9.34 shows the AN/URR-76 CVLF receiver set and its subsystems. The receiver set occupies approximately 0.37 m^3 (1.3 cu. ft.); this represents a significant decrease in volume from its predecessors, the shipboard VERDIN AN/WRR-7 and TACAMO Airborne VERDIN AN/ART-52.

Receiver performance.   The CVLF system uses Minimum-Shift-Keying (MSK), which has the advantage of needing less bandwidth and operating at a lower SNR than conventional receivers of binary FSK signals [SI-1].

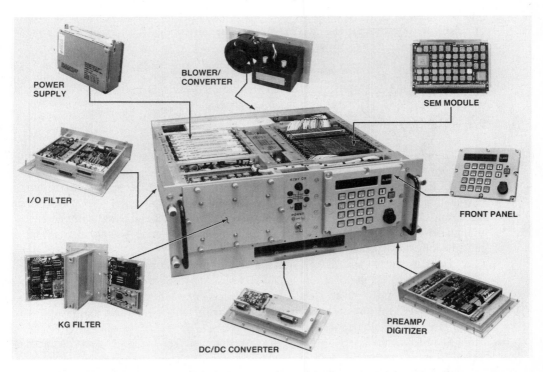

Figure 9.34 Compact VLF Receiver Set AN/URR-76 (Reprinted with permission from Telephonics Corporation)

There are two filters, before and after an IF converter. The second filter is critical to long-wave receiver performance. Opening the bandwidth of the receiver results in a reduction in selectivity and an increased acceptance of the background gaussian noise. The benefit is a faster recovery from impulsive noise, giving low bit error rates during local storms [RE-1; SC-1]. In the CVLF system, the digital filters are tunable over the RF frequency range of 14 kHz to 175 kHz, adjustable in bandwidth from 100 Hz to 10 kHz, and able to be shaped to optimum receiver selectivity.

The heart of the CVLF processor is the Distributed Digital Signal Processor (DDSP), which operates at a rate of 6 mega-operations per second. The DDSP consists of: (a) a 16-bit microprocessor-controller for making decisions and (b) six 16-bit digital processing modules that perform repetitive polynomial calculations for I- and Q-channel signal detection.

The CVLF receiver is not limited by amplifier saturation because the AGC holds amplification in the linear range. Strong signals limit the gain before the noise figure of the analog-to-digital converter ultimately limits dynamic range. The dynamic range increases correspondingly with weak signals, which can be received in the spread-spectrum mode.

The CVLF receiver set correlation capabilities permit the receiver to test a large number of time and phase hypotheses of I- and Q-channels to acquire and track a desired communications signal in the presence of receiver thermal noise and gaussian or highly impulsive atmospheric noise. The CVLF receiver has been statistically tested in a variety of noise environments for true synchronization, the absence of false synchronization, and resultant low character error rate.

9.5 OTHER LONG-WAVE SYSTEMS

9.5.1 ELF Sanguine Systems

An ELF radio system operating at 72 Hz to 80 Hz is used for deeply submerged submarine communications [CO-1; KE-1; JA-1; pp. 146–147] [BU-1]. As discussed in section 9.2.1, ELF propagation characteristics allow the signal to penetrate much deeper into the seawater than VLF. However, the effective usable ELF bandwidth is extremely small (2–3 Hz). The transmitter antennas are inefficient, radiating 2 and 8 watts of power from about 1.36 MW input, but the system is capable of transmitting signals that can be received a long distance away.

The data rate in ELF is very low. An early experiment involving the USS Tinosa (SSN-606) suggested the data rate was 0.003 bps, hence ELF can only carry short, prearranged, coded messages to alert submarines and call them up for a long message on VLF and SATCOM circuits. ELF communications of only two letters can provide over 600 different instruction codes.

The ELF receiver uses the AN/UYK-44 computer. The computer decodes the transmitted ELF bit stream and monitors signal strength. The slow data rate (on the order of a tenth of a letter per minute) makes this system primarily an alert system for Trident submarines to be positioned in a shallower depth in order to receive communications from VLF transmitters, or UHF transmissions via satellite [NA-1; EW-1].

ELF signals are inherently resistant to the effects of EMP radiation created by a nuclear burst. But the transmitters are not designed to survive the blast effects of a direct nuclear attack [EW-1; pp. 55–57].

9.5.2 TACAMO/VLF System

TACAMO (Take-Charge-and-Move-Out) is an airborne survivable, VLF message injection system [KE-1; pp. 76–79], [RW-2]. The TACAMO communications system relays critical operational messages to fleet ballistic missile submarines. TACAMO aircraft operate far at sea where they can receive messages from the shore command authority.

An aircraft with a trailing wire antenna is used as an alternate transmitter platform for submarines. The TACAMO aircraft is the EC-130 and E-6A. A long antenna of 10,000 m, and a short antenna of 1500 m are available for use, but the average transmission calls for an antenna of about 5000 m length. Both long and short antennas are tipped with a terminal drogue that helps the antennas achieve the required verticality. TACAMO is survivable in a nuclear strike. Shore-based ELF/VLF transmitter stations are vulnerable to nuclear attack.

TACAMO broadcasts in the VLF band (14 kHz to 30 kHz range) because the long wavelength of VLF enables sufficient penetration of seawater to allow a submerged submarine to copy message traffic via a wire antenna that is reeled out astern.

The TACAMO communications system generates about 200 kW of power, which translates into 100 kW of effective radiated power. TACAMO receives VLF uplink messages from fixed VLF shore stations and other HF and UHF SATCOM messages from various other stations. The TACAMO downlink retransmits received messages via a VERDIN transmitter terminal.

REFERENCES

[AM–1] Amoroso, Frank, "The Bandwidth of Digital Data Signals," *IEEE Communications Magazine*, November 1980, pp. 13–24.

[BE–1] Belrose, J. S., W. L. Hatton, C. A. McKerrow, and R. S. Thain, "The Engineering of Communications Systems for Low Radio Frequencies," *Proceedings of IRE (now IEEE)* , May 1959, pp. 661–680.

[BU–1] Burrows, M. L., "ELF Communications in the Ocean," Conference Record, *1977 International Conference on Communications*, IEEE Publication 77CH1209-56CSCB, June 1977.

[CA–1] Callahan, Michael B., "Submarine Communications," *IEEE Communications Magazine*, November 1981, pp. 16–25.

[CO–1] Richard, Compton-Hall, *Sub vs. Sub*, Orion Books, New York, N.Y., 1988.

[DA–1] Davis, Kenneth, *Ionospheric Radio Waves*, 2nd Ed, Blaisdell Publishing Company, Waltham, Mass., 1989.

[DE–1] De Buda, Rudi, "Coherent Demodulation of Frequency-Shift-Keying With Low Deviation Ratio," *IEEE Transaction on Communications*, June 1972, pp. 429–435.

[DO–1] Doelz, M. L. and E. H. Heald, *Minimum-Shift Data Communications System*, U. S. Patent 2 977 417, March 28, 1961 (Assigned to Collins Radio Company).

[EW–1] *The C3I Handbook*, EW Communications, Inc., Palo Alto, Calif., 1988.

[FE–1] Ferguson, J. A. and F. P. Snyder, *Long-Wave Propagation Capability Program Description and User's Guide*, NOSC Technical Document 1449, AD-B130-808, NOSC (now NCCOSC RDT&E Div.), San Diego, Calif. 92152-5185, January 1989.

[FE–2] Ferguson, J. A., "Long Wave Propagation Model," *1989 IEEE Military Communications Conference*, October 15–18, 1989, Boston, Mass., pp. 593–597.

[FI–1] R. F. Filipowsky, "The Utility Chart, a Convenient Design Tool for the Evaluation of Modulation Methods," *Proceedings of Fourth Space Congress*, Cocoa Beach, Fla., Apr. 3–6, 1967.

[GA–1] Galejs, Janis, *Terrestrial Propagation of Long Electromagnetic Waves*, Pergammon Press, New York, N.Y., 1972.

[GR–1] Gronemeyer, S. A. and Alan L. McBride, "MSK and Offset QPSK Modulation," *IEEE Transactions on Communications*, Vol. Com-24, No. 8, August 1976.

[HC–1] *Harris Product Brochure on R-2368/URR Series High-Performance General Purpose VLF/LF/MF/HF Receivers*, Harris Corporation RF Communications Group, Rochester, N.Y., April 1990.

[HS–1] *Reference Data for Radio Engineers*, sixth ed., Howard W. Sams & Co., Indianapolis, Ind., 1977.

[JA–1] Jasik, H. ed., *Antenna Engineering Handbook*, McGraw-Hill, New York, N.Y., 1961, p. 19-7.

[JO–1] Jolley, E. H., *Introduction to Telephony and Telegraphy*, Hart Publishing Company, Ind., New York, N.Y., 1968, pp. 122–125.

[KE–1] Keiser, B. E., "Early Development of the Project Sanguine Radiating System," *IEEE Transactions on Communications*, April 1974, p. 366.

[NA–1] *VLF, LF & MF Communication Systems*, NAVELEX Shore Electronics Criteria, 0101,113, Naval Electronic Systems Command (now Space and Naval Warfare Systems Command), Washington, D. C., August 1972.

[NA–2] *Theoretical VLF Multimode Propagation Prediction*, NRL Report 6663, Naval Research Laboratory, Washington, D. C., December 1967.

[NO–1] K. A. Norton, *Transmission Loss in Radio Propagation: II*, National Bureau of Standards (now National Institute of Standards and Technology) Technical Note 12, Figure 7, NIST, Gaithersburg, Md., June 1959.

[PA–1] Pasupathy, Subbarayan, "Minimum-Shift-Keying: A Spectrally Efficient Modulation," *IEEE Communications Magazine*, pp. 14–22, July 1979.

[PR–1] Prabhu, V. K., "Spectral Occupancy of Digital Angle-Modulation Signal," *The Bell System Technical Journal*, April 1976, pp. 429–453.

[RE–1] Reid, C., "LF Receiver Designs for Impulsive Noise," *1989 IEEE Military Communications Conference*, October 15–18, 1989, Boston, Mass., pp. 564–569.

[RH–1] Rhoads, F. J. and W. E. Garner, *An Investigation of the Model Interference of VLF Radio Waves*, NRL Report 6359, Naval Research Laboratory, Washington, D. C., October 1965.

[RO–1] Rohde, U. and Bucher, T. T. N., *Communications Receivers—Principles and Design*, McGraw-Hill, Company, New York, N.Y., 1988.

[RW–1] *Rockwell-International Date Sheet: VLF-600: A Family of Modular Long-Wave/MF Digital Communication Equipment*, Rockwell-International, Richardson, Tex., undated.

[RW–2] *Rockwell-International Brochure on TACAMO Advanced Communications System*, Rockwell-International, Richardson, Tex., undated.

[SC–1] Schwartz, J., "Application of Simulation Based Analysis to Receiver Design for Operation in Impulsive Atmospheric Noise," *1989 IEEE Military Communications Conference*, October 15–18, 1989, pp. 570–576.

[SH–1] Shannon, C. E., "A Mathematical Theory of Communications," *Bell System Technical Journal*, Vol. 27, pp. 379–423, July 1948; pp. 623–656, October 1948.

[SI–1] Sielman, P. F., "CVLF," *1987 IEEE Military Communications Conference*, October 19–22, 1987, Boston, Mass.

[SI–2] Simon, Marvin K., "A Generalization of Minimum-Shift-Keying (MSK)-Type Signaling Based upon Input Data Symbol Pulse Shaping," *IEEE Transactions on Communications*, August 1976, pp. 845–856.

[SK–1] Sklar, B., "Defining, Designing, and Evaluating Digital Communication Systems," *IEEE Communications Magazine*, Vol. 31 No. 11, Nov. 1993, pp. 92–101.

[SP–1] Spilker, J. J. Jr., *Digital Communications by Satellite*, Prentice Hall, Englewood Cliffs, N.J., 1977.

[ST–1] Stallings, William, *Data and Computer Communications*, Macmillan, New York, N.Y., 1988.

[TU–1] Turner, Robert W., "Submarine Communication Antenna Systems," *Proceedings of the IRE (now IEEE)*, May 1959, pp. 735–739.

[WA–1] Watt, A. D., *VLF Radio Engineering*, Pergammon Press, New York, N.Y., 1967.

[WA–2] Watts, James R., "The Mode Theory of VLF Ionosphere Propagation for Finite Ground Conductivity," *Proceedings of IRE (now IEEE)*, Vol. 45, June 1957, pp. 762–767.

[WA–3] Wait, James R., *Electromagnetic Waves in Stratified Media*, Macmillan, New York, N.Y., 1962.

[ZI–1] Ziemer, Rodger E. and Carl R. Ryan, "Minimum-Shift Keyed Modem Implementations for High Data Rates," *IEEE Communications Magazine*, October 1983, pp. 28–37.

10

HF Communications

In this section we discuss HF shipboard long-haul communications. Although the HF band is technically defined as the frequency region between 3 MHz and 30 MHz, we refer to HF as the region from 2 MHz to 30 MHz, since this frequency band is commonly used for long-haul naval HF communications. The 2 MHz to 30 MHz band supports a substantial part of communications between afloat units and shore stations, particularly for platforms which are not equipped with satellite communications. HF long-haul circuits are narrowband; they typically carry 3-kHz voice, 75-bps TTY, or data up to 2400 bps.

Historically, HF radios were one of the primary means for long-haul shipboard communications [BE-1]. HF paths, although subject to ionospheric fluctuations varying with the time of day, the season, the prevalence of sunspots, and multipath and fading, are relatively free from atmospheric noise. In addition, efficient antennas for HF radios became available in the early years of radio. Consequently, the pioneers of shipboard radios concentrated their efforts on developing long-range HF communications. Their efforts resulted in continuing technical refinements that improved the level of attainable HF performance through:

- Improved antenna performance
- Use of single-side-band (SSB) modulation that preserves bandwidth and uses transmitter power more efficiently
- Propagation prediction methods

With the advent of satellite communications, an alternative to long-haul HF communications became available. Shipboard communicators rapidly adopted satellite communication systems to provide global connectivity. Satellite communications became the primary communications means, and HF communications became a back-up system.

As electronic vulnerabilities, high costs, and limitations of satellite communications were better understood, HF radios re-emerged as an important means for long-haul, shipboard communications. It became apparent that no single, long-range communications medium would meet all shipboard communications requirements. A resurgence in HF circuits began in recent years, supported by major HF communications systems programs such as the HF Improvement Program (HFIP), the Link 11 Improvement (LEI) Program, the HF Automatic Link Establishment System, the Communication Support System (CSS), and multimedia communications systems with HF radios.

Improved HF communications systems set new performance benchmarks:

- Higher data rate (i. e., 4800 bps)
- Improved point-to-point link establishment
- Adaptive frequency selection
- Better shipboard radio resource applications
- Anti-Jam (AJ) capabilities
- Low Probability of Intercept (LPI).

10.1 CHARACTERISTICS OF HF COMMUNICATIONS

Table 10.1 summarizes characteristics of HF communications in terms of (a) operational characteristics, (b) practical design considerations, (c) connectivity effectiveness, and (d) operational limitations.

HF signals propagate via two paths: the ground wave and the sky wave. The ground wave supports short-haul communications, the sky wave supports long-haul communications. The sky wave, enabling the long-haul path, results from ionospheric reflection, and gives rise to multipath effects. The latter cause signal interference and fading and require frequent changes of the operating frequency. By using either a prediction method or HF ionospheric sounding, one can find the best operating frequency for an HF circuit [AA-2].

In general, the bandwidth that can be used for HF communications is narrow because of the multipath and dispersion effects. A normal effective bandwidth for the tactical shipboard radio is 3-kHz; tactical shipboard radios can transmit four nominal 3-kHz voice channels simultaneously, for a maximum of a 12-kHz bandwidth.

Figure 10.1 shows that the HF region has unique characteristics: the skywave promotes long-haul communications, but, in contrast to VLF and LF, high antenna gains can be realized, and the atmospheric noise is relatively low. The interaction of these factors (e. g., skywave range, atmospheric noise, and antenna efficiency) provides an indication of how to exploit the HF band in the best manner: (a) HF is the highest frequency band that will be reflected from the atmosphere, (b) smaller antennas can be used and thus operate with greater efficiency, and (c) receiver noise level is less in lower frequency bands.

TABLE 10.1   CHARACTERISTICS OF HF COMMUNICATIONS

Criteria	Characteristics
Operational considerations	• Propagation: Ground wave and sky wave • Ground wave: Short range (noisy) • Sky wave: Long-haul, but subject to multipath • Maximum usable frequency (MUF)
Practical design considerations	• Deep fades require diversity reception and limit data rates • Large antennas needed • Simultaneous operation of HF and VHF difficult • Requires high transmitter power • ECCM (HFAJ) requires antenna coupler for frequency agility • Can be pinpointed by DF • Vulnerable to HEMP • Vulnerable to jamming
Connectivity effectiveness	• Intra-BG • Link 11 TADIL-A (wideband) • Long-haul ship-to-shore (narrowband) • Short-haul via ground wave usually good • HFAJ: Approximately 500 km
Operational limitation	• Data rates: Up to 2400 bps for short-haul • Up to 300 bps for long-haul (90% probability of good reception)

The overall system performance is given by the range equation[1]

$$\frac{S}{N} = \frac{P_T G_T G_R c^2 L_L M}{(4\pi R f)^2 k T_{op} B}$$

where

(S/N) is the signal to noise power ratio at the receiver,
P_T the transmitted power (in W),
G_T the transmitter antenna gain (in the direction of the receiver),
G_R the receiver antenna gain (in the direction of the transmitter),
c the velocity of light (3×10^8 m/s),
L_L additional losses (absorption, system line losses),
M the design margin,
R the range (distance between transmitter and receiver, in m),
f the frequency (in Hz),
k Boltzmann's constant ($1.38\ 10^{-23}$ W K^{-1}Hz^{-1}),
T_{op} the receiver operating temperature (in K), and
B the receiver bandwidth (in Hz).

The factors over which the designer has the most control are the antenna gains, the operating frequency, the transmitter power, the receiver sensitivity, and the design mar-

[1]The range equation is discussed in detail in Chapter 12.

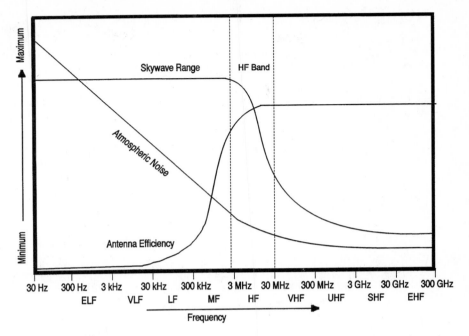

Figure 10.1 Sky wave Properties by Frequency Band (Reprinted with permission from AFCEA *Signal,* November 1984 and Robert H. Sternowski)

gin. These factors, as they apply to HF links, will be discussed in more detail in the paragraphs to follow.

Range.   Low-frequency RF signals (ELF, VLF, LF, and MF) travel long distances (of 8000 km or more) as surface waves. With increasing frequency, absorption becomes more pronounced, and hence, with increasing frequency the range declines; for the low end of the HF band (2 MHz–6 MHz), surface wave propagation of only a few hundred km is typical.

For frequencies below 30 MHz, an HF straight-line ray is generally reflected by the ionosphere. In this manner HF signals propagate long distances with relatively small losses. Above 30 MHz (the lower end of the VHF band) signals penetrate the ionosphere and propagate into space without being reflected.

Atmospheric noise.   Atmospheric noise, both natural and man-made, is a function of frequency. A graph of background noise as function of the receiver operating temperature is shown in Figure 7.1. For the HF band, noise increases with decreasing frequency.

Antenna efficiency.   In order to have an efficient transmitter or receiver, the size of an antenna must be comparable with the wavelength. Figure 10.1 shows that practical antenna efficiency becomes favorable with MF and higher frequencies. For example, at 30 MHz (10-m band) an efficient quarter-wave antenna is 2.5 m ($\approx$8 ft.) long; at 3 MHz a quarter-wave antenna is 25 m ($\approx$83 ft.) long. Typical HF whips range from 5 m to 10 m ($\approx$17 to 33 ft.). Note that an antenna's behavior is closely dependent on the operating frequency, and, therefore, antenna tuners or matching networks are needed for HF antennas.

Practical design considerations.   Multipath causing deep fades in HF operations can be mitigated by using the frequency and space diversity techniques. Additional antennas, transmitters, and receivers are needed to accomplish this.

Shipboard HF antennas, superstructures, and other antennas, such as those used for radars, often cause interference with HF transmission and reception. Simultaneous operation of HF, VHF, and other equipment in shipboard environments requires careful use of topside space to minimize electromagnetic interference.

The transmitter power of shipboard HF radios ranges typically from a few watts to 1 kW. High output power is required for long-haul communications (i. e., ship-to-shore communications). But high power can be a serious radiation hazard to personnel, and adequate precautions must be taken.

HF radios, like other frequency band radios, are susceptible to jamming. Some shipboard HF radios are equipped to mitigate jamming. HF Anti-Jamming (HFAJ) radios need broadband antennas with frequency agility since HFAJ equipment uses frequency hopping AJ techniques.

HF communications have a higher probability of being intercepted than VHF, UHF, and satellite communications systems, and they increase a ship's vulnerability to detection and interception. Hence, shipboard HF radios must be designed to operate in a low probability of interception (LPI) mode.

HF radios are also susceptible to high-altitude electromagnetic pulse (HEMP). The radios and their connections to antennas require careful design that eliminates or minimizes the possibility of damage from HEMP.

Connectivity effectiveness.   Within a Battle Group, Link 11[2] provides very effective, 2400-bps shipboard tactical communications. All long-haul shipboard HF radio communications are narrowband devices with data rates from 75 bps up to 2400 bps; 300 bps is more typical. Very good reliable connectivity can be established for short-haul ship-to-ship communications at a range of up to 160 km (100 mi) by ground wave. HFAJ radios are effective at ranges of up to 500 km ($\approx$310 mi).

Operational limitations.   One operational limitation of HF radios is limited circuit availability.[3] For long-haul shipboard tactical communications, one would expect a circuit availability of 20% at a data rate of 2400 bps. However, at a much lower data rate, (e.g., 300 bps), one would expect a circuit availability of 90%.

---

[2]Three terminologies—Naval Tactical Data System (NTDS), Link 11, and Tactical Digital Information Link-A (TADIL-A)—are often used interchangeably in tactical HF communications. To be accurate, the NTDS is a combat direction system; the term Link 11 is the tactical radio link designation originated from NATO; TADIL-A is the U.S. DoD Tactical Digital Information Link (A) protocol.

[3]Availability is used here to mean availability to conduct communications.

HF radios for long-haul shipboard communications.   HF radios provide a significant part of long-haul shipboard communications, particularly for smaller vessels. Ships, separated by 50 km or more, often rely on HF radios. Satellite communications are more costly and therefore not always available on small ships. HF radios, on the other hand, are relatively low cost items.

10.2 HF PROPAGATION

When an HF signal travels through the Earth's atmosphere from a transmitter at location A to a receiver at location B as shown in Figure 10.2, it may take three different paths:

- Path 1, in which the signal progresses along the line of sight from A to B, is known as the direct wave path.
- Path 2, in which the signal travels from A to B via ionospheric reflection, is called the sky wave path.
- Path 3, which goes from A to B via the ground, is called the ground reflected wave path.

When transmitter and receiver are both at the Earth's surface level, there are two paths: the ground wave path and the sky wave path.

Ranging from 300 km to 600 km, HF propagation is very poor because the take-off angle of signals is too acute for reliable reflection back to Earth, and the range is also beyond the line-of-sight [DA-1]. This region is known as the skip zone. It will be discussed later in section 10.5.5.

There are several ionospheric layers at different heights and of different electron density. They are known as the D, E, F_1, and F_2 layers and are illustrated in Figure 10.3. These layers determine at what height the HF sky wave will be reflected. During the nighttime the D layer and E layer dissolve and the F_1 and F_2 layers combine at a lower altitude (Night F_2). Consequently, during the daytime sky waves mainly reflected at

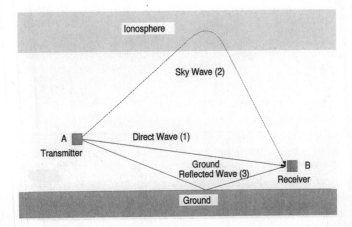

Figure 10.2   HF Wave Propagation

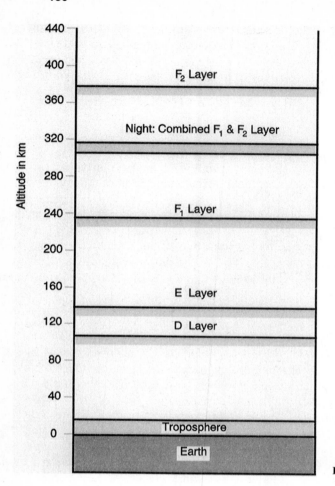

Figure 10.3 Layers of Earth's Atmosphere

the E layer cover a shorter distance than sky waves at night, that are reflected at the F layer.

10.2.1 Fading

While the HF signal reflected via the ionosphere has very useful characteristics for long-haul communications, it also gives rise to undesirable effects. The ionospheric layer, which reflects the signal, is not a perfect reflector, but a layer with variable electron density that depends on the time of day and location. A radio signal, depending on its polarization, can be reflected at different angles in this layer and may take different paths.

Consider a signal that leaves the transmitter at an angle of θ. When the signal reaches the ionosphere, it may take three different courses, depending on its frequency and take-off angle θ: (a) it passes through the ionosphere, (b) it dissipates as heat, and (c) it is reflected back to Earth. When signal arrives at the receiver, it has several time-variant components. The vector sum of the various components reinforce or cancel each other, creating an interference effect known as fading. Short-term fading lasts from several ten-thousandths of seconds to a few thousandths of seconds.

The ionosphere, depending on its electron density, also acts as an absorber, resulting in frequency selective fading. Fading, whether short-term or frequency selective, often causes loss of the signal.

10.2.2 HF Operating Frequency

One of the most important elements for successful operation of an HF system using sky wave propagation is an operating frequency that will assure a reliable path. Due to multi-path fading, selective fading, and delay of signal components, finding such a frequency is difficult. There are good frequencies for operation during daytime, reflecting off the E layer, and other frequencies for nighttime operation, reflecting off the F layer. East-West paths generally have different characteristics than North-South paths. Table 10.2 shows normal ionospheric variations for the E, F_1, and F_2 layers in terms of height, diurnal, sea-

TABLE 10.2   NORMAL IONOSPHERIC VARIATION

E Layer (100 km to 140 km)	F_1 Layer (140 km to 250 km)	F_2 Layer (250 km to 400 km during day; 300 km to 350 km during night)
Diurnal		
Critical frequency follows the zenith angle of the Sun; practically disappears during night hours.	Similar to the E layer; the height is fairly constant; maximum critical frequency occurs around noon, as with the E layer.	Height and density increase during the daylight hours; the critical frequency generally continues to increase after the Sun's zenith angle has reached maximum; there is a complex variation with latitude, longitude, and with respect to the geomagnetic equator.
Seasonal		
Both layers vary slowly according to the Sun's zenith angle; during the summer, with a high zenith angle, the critical frequencies are considerably higher and occur for longer periods than during shorter winter days, when the Sun's zenith angle is considerably lower.		Daytime summer critical frequency values are much lower than those in winter, but summer night critical frequencies are higher than winter night critical frequencies; the F_2 layer height is subject to considerable variation; the geographic and geomagnetic variations are the same as diurnal variations.
Cyclic		
Electron density, and consequently critical frequencies of all layers, both daytime and nighttime, summer and winter, increase with more sunspot activity, and vice versa; good correlation between critical frequencies and sunspot numbers enables predictions of critical frequencies three to six months in advance.		

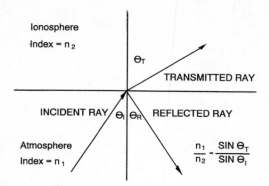

Figure 10.4 Determination of Maximum Usable Frequency (MUF)

sonal, and cyclical changes [DA-1]. Finding a favorable frequency and maintaining a path for a long period is very difficult, and various measurement and prediction tools are now available to assist in maintaining an HF path.

The Central Radio Propagation Laboratory (CRPL)[4] prediction utilizes the maximum usable frequency (MUF), lower usable frequency (LUF), and optimum working frequency (OWF) [or in French, fréquence optimum de travail (FOT)]. The MUF is the highest frequency for a path between a transmitter and receiver site and LUF the lowest usable frequency for a path, it is also the optimum frequency. However, the operating frequency for an HF path changes with time. Thus, when selecting a frequency that is too close to the MUF, the signal may drop below the fade margin, and then one experiences radio link dropouts (or losses of the circuit). In practice, the OWF is chosen to be about 85% of the MUF in the F_2 layer.

Another theoretical method of determining MUF is based on signal ray optics. Figure 10.4 is a simplified representation that shows an HF signal directed at the ionosphere at an incident angle θ_I; this signal is split into two components: one is reflected back to the Earth at an angle θ_R, and the other is transmitted through the ionosphere at an angle θ_T. Snell's Law links the incident, reflected, and transmitted angles θ_I, θ_R, and θ_T, with the indices of refraction n_1 (atmosphere) and n_2 (ionosphere):

$$n_1\sin\theta_I = n_1\sin\theta_R = n_2\sin\theta_T.$$

For all practical purposes the index of refraction n_1 in the atmosphere equals 1, and $n_2 < 1$, since the ionosphere is the optically thinner medium. Thus, the ray is bent away from the normal on the plane of incidence, and $\theta_T > \theta_R$, $\theta_I = \theta_R$. For the incident and transmitted rays, the angles are related as follows:

$$\frac{1}{n_2}\sin\theta_i = \sin\theta_T.$$

[4]CRPL stands for Central Radio Propagation Laboratory. The CRPL became Institute of Telecommunications Sciences, the National Telecommunications and Information Administration, Department of Commerce, Boulder, Colorado.

When the value of $(1/n_2)\, sin\theta_I > 1$, $sin\theta_T$ cannot be real, and the transmitted ray disappears (this phenomenon is also known as total reflection). The limiting condition is $sin\theta_T = 1$.

If the ionosphere is considered as a plasma, an approximation for n_2 is given by [FI-1]:

$$n_2{}^2 \approx 1 - \frac{f_c{}^2}{f^2}$$

where f is the operating frequency and f_c is the so-called plasma frequency, which in turn can be approximated as $9\sqrt{N}$, with N being the electron density in the ionosphere. For the limiting condition the incident frequency for which the ray is totally reflected then becomes the MUF, hence:

$$sin^2\theta_I \approx 1 - \frac{f^2{}_c}{f^2{}_{MUF}},$$

and from this:

$$f_{MUF} \approx \frac{f_c}{\sqrt{(1 - sin^2\theta_I)}} \approx \frac{9\sqrt{N}}{cos\theta_I}.$$

10.2.3 HF Sounding

When a shipboard HF radio operator selects an Optimum Working Frequency (OWF) for a long-haul, point-to-point HF link, CRPL HF propagation forecast provides the operator a guidance for the OWF selection. For more precise OWF selection the operator can use HF sounding equipment.

Ionospheric sounding provides testing ability of selected channels or the entire HF band by emitting a brief beaconlike broadcast test signal. The test signal is then used by other stations to evaluate channels for possible use in HF communications. This broadcast consists of a transmission of HF pulses on discrete frequencies (pulse sounding or channel sounding), or a rapid scan of a sweeping CW signal across the HF band (chirpsounding).

AN/TRQ-35 and AN/TRQ-42 tactical frequency management system. The Tactical Frequency Management System (TFMS) is designed to improve the quality and reliability of HF circuits by providing the operator with instrumentation that continuously measures and displays the best communications frequencies as ionospheric conditions change [WI-1; FE-1]. The Chirpsounder® is a part of the TFMS. The sounder used in the AN/TRQ-35 TFMS is the original militarized Chirpsounder® in use since the late 1970s. A newer system, the AN/TRQ-42, which entered service in the late 1980s, provides automatic ranking of assigned frequencies based upon propagation quality and channel occupancy. The automatic ranking simplifies the operator's task of selecting specific OWFs from a range of propagating frequencies.

The Chirpsounder® transmitter, T-1373/TRQ-35(V) or T-1562(V)/TRQ-42(V), consists of a sweep generator, a power amplifier, and a filter and diplexer. The sweep generator provides: (a) clock and timing controls to initiate frequency sweeps at selec-

five-minute intervals, (b) synthesis of the transmitted FM-CW chirp waveform, and (c) control functions for the filter and diplexer. The chirp transmit waveform is amplified in the power amplifier. The amplifier output signal is passed through a set of sequential, half-octave, low-pass filters. The filters attenuate harmonics to at least 60 dB below the fundamental chirp waveform. The filter and diplexer also combine the transmitted chirp waveform with the output of a communications transmitter.

The Chirpsounder® transmitter transmits a CW test signal which linearly sweeps frequencies from 2 MHz to 30 MHz in 280 seconds. The swept signal is commonly known as an FM CW or chirp signal. The chirp signal can be transmitted separately, or diplexed with a fixed frequency communications signal on a single antenna, using the diplexer available with the Chirpsounder® transmitter as shown in Figure 10.5. The Chirpsounder® transmitter creates negligible interference for other HF spectrum users by using very low output power (typically 10 W). This interference occurs because the chirp signal rapidly sweeps through all communications channels. The chirp signal is "in-band" for only 30 ms, and the chirp signal RF output is muted, when sweeping through frequencies of nearby receivers or other protected frequencies.

The Chirpsounder® receiver, R-2081/TRQ-35(V) or R-2488(V)/TRQ-42(V), consists of three units: a control and display, a receiver, and a power divider. The control and display unit provides: (a) a high-stability frequency standard, (b) clock and timing controls to initiate frequency sweeps at selectable five-minute intervals, (c) controls to generate the local oscillator signal, (d) spectrum analysis and display of the received radio

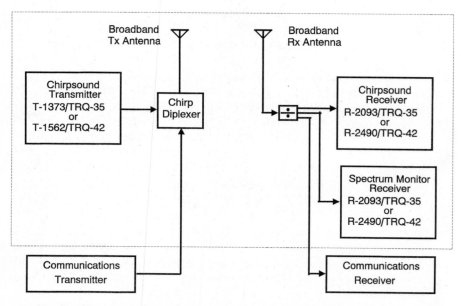

Figure 10.5 Application of the AN/TRQ-35 and AN/TRQ-42 Chirpsounders® (Courtesy of BR Communications Inc.)

propagation information, and (e) numeric display of time instantaneous sweep frequency and cursor frequencies.

The Chirpsounder® receiver automatically and synchronously sweeps with the Chirpsounder® transmitter and provides a real-time measurement of propagation. The receiver generates a Chirpsounder® record which is a graphical display of propagating mode structure (multipath) and signal strength as a function of frequency. From the display, the operator can easily determine the OWF. In the AN/TRQ-42 systems, the instrument also tabulates all assigned communications frequencies ranked from best to worst.

The HF spectrum monitor consists of a control mechanism and a display and a receiver. The monitor scans the entire HF spectrum in 3kHz steps every 11 seconds, then compiles and continuously updates the propagation conditions for the 2 MHz to 30 MHz band in 5-minute and 30-minute time blocks.

Figure 10.6 is an AN/TRQ-42 Chirpsounder® record of a 1000 km path from Salt Lake City, Utah, to Sunnyvale, California. The horizontal axis shows radio frequency from 2 to 30 MHz, the vertical axis is mode structure time delay from 0 to 5 ms, and the bargraph at the top of the display shows signal strength (received signal-to-noise ratio). The dashed cursor marks the Maximum Observed Frequency (MOF) at 22.4 MHz of the 1-hop F layer mode. Note the curvature and hook at the MOF that is typical of medium- to long-path F layer modes. Also visible below the F layer mode is a straight horizontal mode extending from approximately 2 MHz to 12 MHz. This straight line represents sporadic E propagation. Note the high signal strengths between approximately 9 MHz and 13 MHz, and then again between 16 MHz and 21 MHz, separated by a valley of low signal strength around 14 MHz and 15 MHz. The 2-hop F layer mode between 5 MHz and 15 MHz is somewhat spread and would produce signal dispersion on a communications

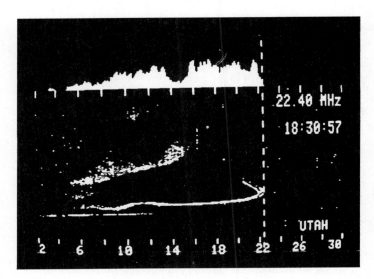

Figure 10.6   Chirpsounder® Record of the Combination of 1-hop and 2-hop F-mode Propagation and Sporadic E-mode Propagation. (Courtesy of BR Communications Inc.)

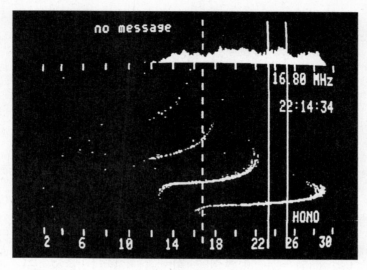

Figure 10.7 Chirpsounder® Record of the Combination of 1-hop, 2-hop, and 3-hop F layer Mode Propagation. (Courtesy of BR Communications Inc.)

channel operating below 15 MHz. Thus, the best frequencies for this path are between 16 MHz and 20 MHz because they provide strong single-mode propagation and are not too close to the MUF so as to assure a good fade margin, as explained in section 10.2.2.

Figure 10.7 is another AN/TRQ-42 Chirpsounder® record of a 4000-km path from Hawaii to San Francisco. Figure 10.7 shows four propagating modes which cause multipath distortion and fading often found on over-water paths. The best band of working frequencies is identified by the two solid cursors between 23 MHz and 25 MHz where only a single mode is propagating (no multipath) and strong signal amplitude (realizing a high signal-to-noise ratio) is obtainable. The dashed cursor at 16.8 MHz marks a frequency in which the signal strength is lower, and three propagating modes of 1-hop, 2-hop, and 3-hop F layers are present with a multipath spread of approximately 2.5 ms. This multipath would create significant fading and thus would render this frequency a poor choice for data communications.

10.3 HF COMMUNICATIONS SYSTEMS

The HF spectrum is a limited resource, and therefore effective use of the spectrum is extremely important. One of the early developments of HF radios was the improved transmitter spectrum and power usage made possible by the introduction of the single-sideband (SSB) modulation.

When an audio input signal is modulated onto a carrier f_c, the resultant spectrum produces two sidebands, an upper sideband (USB) and a lower sideband (LSB); this is shown in Figure 10.8(a). In this illustration, the spectrum of audio input signal is represented by a triangle (A1 and B1), and the signal normally has a bandwidth of 3 kHz. In double-sideband (DSB) modulation both the upper and lower sidebands are transmitted. The total transmission bandwidth (designated B_{TR} in Chapters 7 and 8) is twice the low-pass bandwidth (designated B_{out} in Chapters 7 and 8). In DSB, the carrier is usually retained.

When the audio signal is modulated by an SSB modulator, the resulting spectrum at the HF modulator's output produces a single sideband as shown in Figure 10.8(b). In this illustration, the audio input signal is represented by a triangle (B1), which normally has a bandwidth of 3 kHz. The SSB transmitter generally uses one half of the power and occupies one half of the bandwidth of the DSB transmitter. SSB modulation may be produced by using a bandpass filter, which completely eliminates all signals on one side of the carrier frequency as shown in Figure 10.8(b). In order to be most effective, the carrier f_c is also eliminated. This is known as suppressed carrier SSB.

Similarly, four sidebands can be grouped into an HF channel as shown in Figure 10.8(c). This is called Independent Sideband (ISB) modulation. In the illustration, the four audio signals represented by triangles (A1, A2, B1, and B2) form a single channel of 12 kHz.

One of the most important considerations of using suppressed carrier SSB modula-

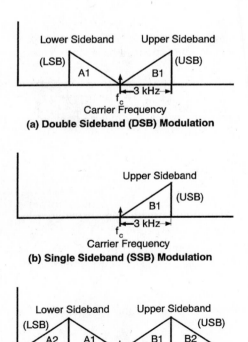

Figure 10.8 Various HF Emission Types: (a) DSB Modulation, (b) SSB Modulation and (c) ISB Modulation

HF Communications   Chap. 10

tion is accurate and stable generation and reinsertion of the carrier frequency at the receiver. The carrier is generated locally and must be as close as possible to the original suppressed carrier frequency. The accuracy of carrier generation and reinsertion is an extremely important design criterion of an SSB receiver. If the reinserted carrier frequency is off even by a few Hz, the performance will begin to degrade, and the audio signal will not be accurately reproduced.

Navy tactical data links.   Tactical communications carry information to and from afloat units; the information typically contains action coordination, commands, surveillance data, combat system control, and weapons system control. The required range for tactical communications on HF links is from a few kilometers to several hundred kilometers.

The HF band is used for Navy tactical data links because it can provide for a longer range than the VHF and UHF bands. The HF links transmit data on both the ground wave and the sky wave. The Navy currently uses several tactical data links as described in section 5.1.11.[5] Link 11 operated in the HF band for long-haul communciations. Link 11 may also operate in the UHF band in an LOS mode. Link 11 provides HF and UHF communications for tactical data exchange for all Navy shipboard, airborne, and shore needs. Link 11 supports the Naval Tactical Data System (NTDS); the protocol used is TADIL-A. Link 11 operates as netted circuit.[6]

Link 11 and the NTDS system were developed by the U. S. Navy in the 1960s. These systems brought about several innovations in naval communications:

a. The first shipboard tactical data systems to use digital computers called the AN/USQ-17, the AN/UYK-7, and later the AN/UYK-43.

b. The first shipboard systems to use automatic computer-to-computer data exchange from ship-to-ship, and from ship-to-aircraft.

c. The basis for subsequent U. S. and NATO message structure and protocol standards for data links [SW-1].

Link 11 is used for anti-submarine and anti-surface warfare control by carriers, cruisers, destroyers, frigates, attack submarines, amphibious assault ships, and by S-3B Viking carrier-based anti-submarine warfare aircraft, P-3 Orion long-range, shore-based maritime patrol aircraft, and E-2C Hawkeye carrier-based surveillance aircraft [DD-2].

The number of platform equipment with Link 11 is limited to those with NTDS systems. NTDS handles and displays combat status and target information. Link 11 uses a polling technique to access various units and exchange target information [SC-1].

NTDS characteristics are summarized in Table 10.3 [LO-1]. The NTDS digital data

[5]In Table 5.2, seven data links (Link 11, Link 16, Link 4A, Common High Band Data Link, LAMPS Data Link, Link 1, and Link 14) are described.

[6]Link 16 is a UHF L_x band tactical data exchange system for shipboard, aircraft, and shore use. Its protocol is TADIL-J. When fully deployed, Link 16 will replace Link 11.

TABLE 10.3   NTDS CHARACTERISTICS

Technical Parameters	Characteristics
Data Rates	• 1364 bps (slow) • 2250 bps (fast)
Tone Library	• 605 Hz Doppler tone and 15 data tones: 935 Hz, 1045 Hz, 1155 Hz, 1265 Hz, 1375 Hz, 1485 Hz, 1595 Hz, 1705 Hz, 1815 Hz, 1925 Hz, 2025 Hz, 2145 Hz, 2255 Hz, 2365 Hz, and 2915 Hz
Modulation	Duo-Binary Phase-Shift-Keying (PSK) Modulation: *Date Value Phase Shift* (1,1) $-45°$ (0,1) $-135°$ (0,0) $-225°$ (1,0) $-315°$
NTDS Interface between Computer I/O Channel and DTS	Type A Interface (NTDS Slow) per MIL-STD-1397 [DD-4]
Operational Modes	• Transmit and receive • Receive only
Net Modes	• Roll call • Broadcast • Short broadcast • Net synchronization • Net test
Station Modes	• Net control station • Picket station

are transferred at two data rates: a slow rate (1364 bps) and a fast rate (2250 bps). The baseband digital data is encoded by a tone library of 15 phase-modulation audio signals. The data terminal set (DTS) modulates each audio tone into a duo-binary PSK signal. The DTS and the NTDS computer (CP642A/B, AN/UYK-7, and AN/UYK-43) are interfaced via an NTDS-Slow channel (1364 bps, Type A interface), which is specified in MIL-STD-1397 [DD-4].

The Link 11 net has two types of participants, the net control station and the picket station.[7] The most frequently used mode of net operation is the roll-call mode. Each station on the Link 11 net is identified by a unique address code. Details of the net modes are explained in section 19.2, where the TADIL-A protocol is also described. Table 10.4 is a list of Link 11 interoperable HF and UHF radios [LO-1].

There are several types of data terminal sets (DTSs) which are deployed on various surface ships, aircraft, submarines, and shore nodes as shown in Table 10.5 [LO-1]. The handling capacity for picket stations varies from 4 to 80. Except for the NTDS DTS

[7]See Chapter 4, Figure 4.2 through 4.6.

TABLE 10.4 INTEROPERABLE LINK 11 RADIOS

Radios	Interoperable Link 11 Radios
HF Radios	• AN/SRC-16 • AN/SRC-23 • AN/URC-75 • AN/URC-109 • R-1903/T-1322 • AN/URT-23/R-1051G • AN/URT-23/R-1051H
UHF Radios	• AN/URC-93 • AN/WSC-3 (V) 6 • AN/URC-85 • AN/URC-83

AN/USQ-36, all DTSs can be used in the net synchronization, net test, roll-call, broadcast, and short broadcast operational modes.

10.4 TYPICAL HF EQUIPMENT

HF radios are widely used in shipboard long-haul communications. There are over thirty shipboard HF transmitter, receiver, and transceiver types in the current U.S. Navy inventory. Among these, six HF radios are described in this section, that is, the R-1051(G) HF

TABLE 10.5 LINK 11 DATA TERMINAL SETS

Data Terminal Set	Years Introduced	Modes of Operations	Maximum No. of Picket Stations	Platform Application
AN/USQ-36	1964–66	NS,[2] NT, & RC	20	Surface ship
AN/USQ-59, -63, & -79[1]	1971–79	NS, NT, RC, BC, & SBC	15	AN/USQ-79 for aircraft
AN/USQ-74	1981-87	NS, NT, RC, BC, & SBC	61	Surface ship
AN/USQ-76	1983	NS, NT, RC, BC, & SBC	4	Submarine
AN/USQ-83	1986–87	NS, NT, RC, BC, & SBC	61	Surface ship
AN/USQ-92	1984	NS, NT, RC, BC, & SBC	4	Surface ship
MX-512P	1990	NS, NT, RC, BC, & SBC	80	Surface ship

[1]AN/USQ-59, AN/USQ-63, and AN/USQ-79 are grouped together since they are all based on the same architecture.

[2]NS = Net Synchronization; NT = Net Test; RC = Roll Call; BC = Broadcast; SBC = Short Broadcast.

receiver, the R-2368 HF receiver, the AN/URT-23(D) and AN/URT-23(E) HF transmitter, the HF-80 transceiver, the AN/USQ-83 HF tactical data terminal, and the AN/URC-109 Integrated Communications System (ICS-3).

10.4.1 R-1051(G) /URR HF Receiver

One of the most widely used shipboard HF receivers is the R-1051(G)/URR [WI-2]. It is a triple-conversion, superheterodyne SSB, DSB, and ISB communications receiver designed not only for shipboard and submarine use, but also for shore station applications. The entire 2 MHz to 30 MHz frequency range is tunable by a synthesizer in 100 Hz increments, permitting rapid selection of any one of 280,000 communications channels. Operating modes provide reception of voice, CW, TTY, and simultaneous voice/TTY signals. Figure 10.9 shows a block diagram of the R-1051(G) HF receiver.

10.4.2 The R-2368/URR HF Receiver

The R-2368/URR multiband receiver, shown in Figure 10.10, is a part of the AN/URR-79 radio receiver set, and it covers VLF, LF, MF and HF frequency bands [WI-3]. The AN/URR-79 also includes the C-11891/G remote control unit. The R-2368/URR is a militarized version of the RF-590A general-purpose receiver. The URR-79(V) receiver will eventually serve as a replacement for the AN/WRR-3 VLF receiver, the AN/SRR-19 LF receiver, and the R-390/R-1051/R-2174 HF receivers for ship, shore, and transportable shelter applications. The R-2368 uses solid-state technology; it measures 13.3 cm (5.2 inches) high, 48.3 cm (19 inches) wide, and 19.5 cm (7.7 inches) deep, and weighs 18 kg (40 lbs). A block diagram of the R-2368/URR is shown in Figure 10.11.

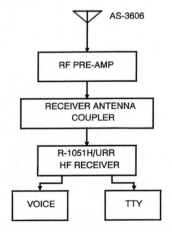

SHIP, SUBMARINE, AND SHORE APPLICATIONS
IN SERVICE WITH THE U.S. NAVY
FREQUENCY REGION 2 - 30 MHz

Figure 10.9   Block Diagram of the R-1051(G) HF Receiver

Figure 10.10   R-2368/URR VLF/MF/HF Receiver (Courtesy of Harris RF Communications Group)

10.4.3 AN/URT-23 HF Transmitter

The AN/URT-23 HF transmitter is a general-purpose shipboard HF transmitter. The AN/URT-23(D) is an earlier version; the AN/URT-23(E) is a newer version of the AN/URT-23(D) [WI-4].

The AN/URT-23(D) contains a linear amplifier, with an RF output of 1 kW peak envelop power (pep). The AN/URT-23(D) is an ISB HF radio transmitting set designed

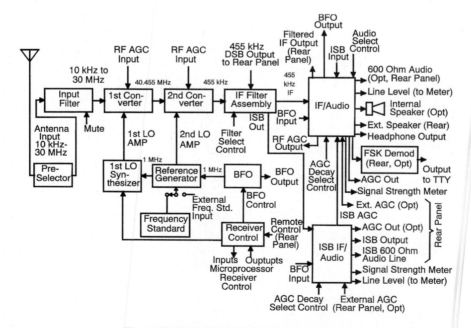

Figure 10.11   R-2368/URR Receiver Block Diagram

for reduced and suppressed carrier applications, with frequency coverage from 2 MHz to 30 MHz. Its operating modes include AM voice, CW, and radioteletype (RATT). The AN/URT-23(D) is designed for use with the antenna coupler group AN/URA-38 for automatic antenna tuning with a remote control system. The AN/URT-23(D) can be used for ship, fixed shore station, and transportable station applications. Figure 10.12 shows the AN/URT-23(D) HF transmitter. Figure 10.13 shows a block diagram of the AN/URT-23(D).

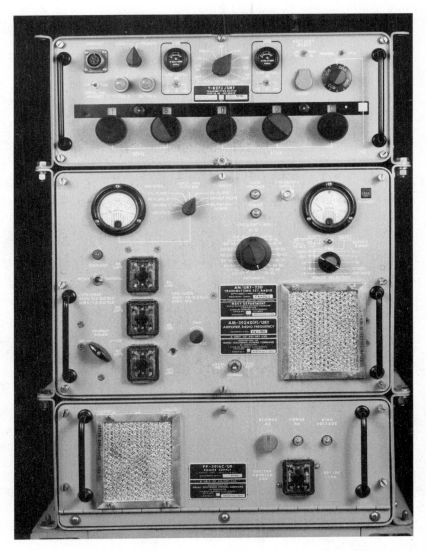

Figure 10.12   AN/URT-23(D) HF Transmitter (Reprinted courtesy of Stewart-Warner Electronics Corporation)

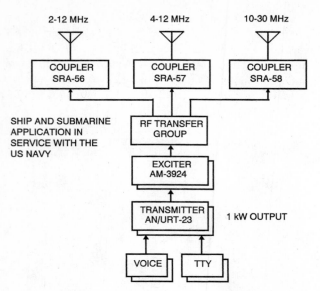

Figure 10.13   Block Diagram of the AN/URT-23(D) HF Transmitter

The AN/URT-23(E) contains several new features such as Automated Link Establishment (ALE), improved coding and waveforms, and a MIL-STD-1553B bus [DD-3; DD-5]. ALE is a method of establishing HF point-to-point communications; it works by performing selective calling, handshaking, scanning, sounding, polling, and coordinating and managing HF networks. ALE is discussed further in section 10.6. MIL-STD-188-141A [DD-1] defines the equipment guidelines and standard algorithms governing ALE.

10.4.4 HF-80 HF Transceivers

The HF-80 HF transceiver family is designed and built with modular flexibility so that each can be configured as a single station, manually controlled by one operator, or several can be configured as a multisite, processor-controlled, remote system[8] [WI-5]. Table 10.6 lists various system configurations of HF-80 radio equipment. Its RF output power ranges from 1 kW to 10 kW. The system can be configured to be a transmitter, receiver, or transceiver. Operating modes include CW, AM, FM, TTY, Voice, and Data.

Figure 10.14 shows a block diagram of an HF-80 transceiver configuration. The HF-8020 power amplifier has three power output options of 1 kW, 3 kW, and 10 kW, with a VSWR of up to 3-to-1. The HF-8014 exciter permits control of up to four input channels. The AC-8075 antenna switch couples the transmitter and receiver into a single antenna with a frequency range from 1.5 MHz to 30 MHz. SELSCAN is a special-purpose device which monitors frequency activities in the HF band. The SELSCAN alerts the operator when a signal of interest is detected. The receiver/exciter control unit permits

---

[8]In a remote system, the transmitter is located away from the main HF installation in order to avoid RF interference.

TABLE 10.6 HF-80 SYSTEM CONFIGURATIONS

HF-80 System \ HF-80 Equipment	HF-8010 Exciter (L)	HF-8010A Exciter (L/R)	HF-8070 Rec-Exc (L)	HF-8070A Rec-Exc (L/R)	HF-8020 1 kW PAmp	HF-8030 1 kW PSup	HF-8021 3 kW PAmp	HF-8022 10 kW PAmp	HF-8050	HF-8050A	HF-8091	HF-8090 Tx Control	HF-8092 Tx-Rx Ctrl
kW Transmitter (Local Control)	■				■	■							
1 kW Transmitter (Local/R Control)		■			■	■						■	
1 kW Transceiver (Local Control)			■		■	■							■
1 kW Transceiver (Local/R Control)				■	■	■							■
3 kW Transmitter (Local Control)	■						■						
3 kW Transmitter (Local/R Control)		■					■					■	
3 kW Transceiver (Local Control)			■				■						■
3 kW Transceiver (Local/R Control)				■			■						■
10 kW Transmitter (Local Control)	■							■					
10 kW Transmitter (Local/R Control)		■						■				■	
10 kW Transceiver (Local Control)			■					■					■
10 kW Transceiver (Local/R Control)				■				■					■
Receiver HF-8050 (Local)									■				
Receiver HF-8050A (Local/Remote)										■			
Receiver HF-8091 (Control)											■		■

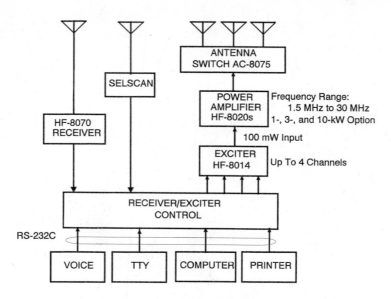

Figure 10.14   Block Diagram of the HF-80 HF Transceiver

complete control of frequency, mode, RF gain, AGC, power level, keying, and ISB channels. The HF-8070 receiver is a stand-alone receiver, normally operated in a manual control mode, either at a receive-only site or a full communications station. The digital input and output devices are connected to the reciever exciter control via RS-232 digital interfaces.

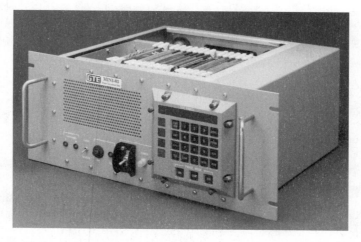

Figure 10.15   AN/USQ-83 NTDS Data Terminal Set (Courtesy of GTE Government Systems Corporation)

TABLE 10.7   TECHNICAL CHARACTERISTICS OF AN/USQ-83 NTDS DATA TERMINAL SET

Parameter	Characteristics
Frequency Band	HF and UHF
Functional Modes	• Link 11/TADIL-A • Voice Frequency Carrier Telegraph (VFCT) (AN/UCC-1) • ANDVT (Data A) • MIL-STD-188C
Data Rate and Interrogation Interval (HF Link 11 Applications)	• 2250 bps (9.09 ms) • 1364 bps (18.18 ms)
Doppler Frequency Correction	±75 Hz
NTDS Interface	8-bit Parallel Type A NTDS Slow
Input Power	110 VAC; 47 Hz to 400 Hz
Size	35.5 cm (14 inches) H × 48.6 cm (19 inches) W × 48.6 cm (19 inches) D
Weight	121.5 kg (270 lbs)

10.4.5 AN/USQ-83 Tactical Data Link System

The AN/USQ-83 is a data terminal set for the Naval Tactical Data System (NTDS) [WI-6]. It provides a two-way interface between digital data systems and a ship's available transceiver. The AN/USQ-83 is shown in Figure 10.15. Table 10.7 lists its technical characteristics. The equipment is designed to handle multimedia (HF and UHF) and multimodes (TADIL-A, clear voice, secure voice, and MIL-STD-188C input/output) [DD-6].

10.4.6 ICS-3 Integrated Communication System (AN/URC-109)

The ICS-3 Integrated Communication System (AN/URC-109) [AA-1; WI-7] is an HF Anti-Jamming (HFAJ) system. Figure 10.16 shows a block diagram of the AN/URC-109 HFAJ System. The HFAJ system employs frequency hopping technology.[9]

In order to operate the HFAJ system, a wideband antenna is required with the capability to transmit a hopping carrier. This kind of antenna is subjected to interference by collocated emitters, and this is a dominant concern for the system design and installation.

A new feature in the ICS-3 is the application of broadband techniques that enable all HF transmission to be amplified and radiated simultaneously using a single bank of amplifiers and three antennas, without any RF electromechanical units. The broadband antenna used in the ICS-3 is an inclined fan-type antenna. A twin fan-type antenna covers

---

[9]Both direct sequence and frequency hopping anti-jamming techniques are discussed in section 8.4.

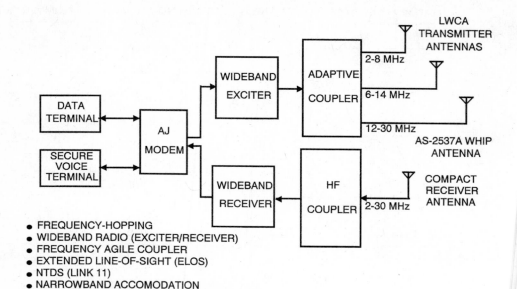

- FREQUENCY-HOPPING
- WIDEBAND RADIO (EXCITER/RECEIVER)
- FREQUENCY AGILE COUPLER
- EXTENDED LINE-OF-SIGHT (ELOS)
- NTDS (LINK 11)
- NARROWBAND ACCOMODATION
- ANTENNA (WIDEBAND LONG WIRE - PASSIVE)
- ICS3 (WIDEBAND WHIP - ACTIVE)

Figure 10.16   Block Diagram of the AN/URC-109 HFAJ System

the 2 MHz to 14 MHz frequency band; a twin whip antenna covers the 12 MHz to 30 MHz frequency band. A general view of broadband twin fan-type antennas used for the AN/URC-109 is shown in Figure 10.17. Figure 10.18 shows a close-up view of the same broadband twin fan-type antenna.

The design of the broadband antennas is dependent on constraints imposed by the

Figure 10.17   General view of Broadband Fan Type Antennas (Courtesy of GEC-Marconi, Ltd.)

Figure 10.18   Details of a Broadband Fan-Type Antenna Used in Conjunction with the AN/URC-109 (Courtesy of GEC-Marconi, Ltd.)

dimensions of the available structure and the proximity of other topside systems in a ship. However, in most instances, the HF band can be covered by two antennas. The key characteristics are the frequency agility of the exciter and the broadband antenna.

10.5 HF ANTENNAS

Typical shipboard HF antennas are nearly omnidirectional such that HF reception can be independent of the ship's heading.[10] Since shipboard HF antennas must be placed in space-limited areas, they often are: (a) single pole whips, (b) twin pole whips, (c) long-wires, or (d) fans [NA-1]. Other HF antennas, such as discones, helix monopoles, and directional log-periodic arrays, are described in Chapter 16.

In choosing an HF antenna, one must consider the following parameters: (a) bandwidth, (b) space requirements, (c) gain, and (d) the Voltage Standing Wave Ratio (VSWR);[11] Table 10.8 summarizes a comparison of these parameters for single pole whip, twin pole whip, long wire, and fan-type antennas.

---

[10]The general discussion and an overview of shipboard communications antennas is given in Chapter 16. In this chapter details of shipboard HF antennas resulting from their complexity in certain applications are discussed.

[11]The VSWR is an index of the impedance match between radio equipment and antennas. The VSWR is defined as the ratio of V_{max} to V_{min} of a standing wave on connecting equipment of a transmission line. A large VSWR indicates power would be reflected back toward the source of the wave. If a significant amount of power is reflected back to the source (the transmitter), equipment damage can result.

TABLE 10.8   COMPARISON OF CHARACTERISTICS OF SHIPBOARD ANTENNAS

Parameters	Monopole Whip	Twin Pole Whip	Long-Wire	Fan-Type
Bandwidth	1 Octave (2:1)	1–2 Octaves (between 4:1 and 2:1)	1–2 Octaves (between 4:1 and 2:1)	2–3 Octaves (between 8:1 and 4:1)
Physical Dimension	6 m to 10 m High	6 m to 10 m High	15 m to 23 m Long	15 m to 23 m Long
Polarization	Vertical	Vertical	Horizontal	Vertical
VSWR	2:1	2:1	3:1	3:1

10.5.1 HF Whip Antennas (Monopole Whips)

For shipboard HF communications, vertical whip (monopole) antennas are widely used. Whip antennas are self-supporting, thin, rigid, aluminum or fiberglass poles, supported by an insulator at the base as shown in Figure 10.19. Their length is usually limited such that they represent a quarter wave antenna at 7.5 MHz (40-meter band);[12] longer whips present mechanical difficulties. The length is approximately 6 m to 10 m ($\approx$20 to 33 ft.) [WI-8].

An HF whip antenna is vertically polarized, which results in an antenna pattern as shown in Figure 10.20. The solid line pattern is for an ideal ground plane, the dashed line pattern is for an average shipboard ground plane. The angle ψ_B is the Brewster angle, the angle at which the reflected wave is 90° out of phase with respect to the direct wave. The Brewster angle for a particular location is a function of ground plane, the soil conductivity (salt water at sea), the dielectric constant, and the operating frequency. For seawater, it is approximately 14° [AR-1; pp. 3.2–3.5].

Monopole antennas are narrowband, and below 5 MHz they are electrically short and rather inefficient. With atmospheric noise as the limiting factor at these frequencies, HF receiving antennas can be lossy without excessively degrading the effective receiving sensitivity and short antennas are not a detriment. Aboard a ship this will tend to reduce the interference of local transmission; in other words, a lossy receiver antenna affords a degree of receiver electromagnetic interference (EMI) protection.

However, for a transmitter, the situation is quite different. To cover the entire 2 MHz to 30 MHz range, the narrowband monopole must be provided with a base tuner to allow optimum power transfer (by optimum impedance matching) between transmitter and antenna. A shortcoming of tuned monopoles is that only a limited number of them can be installed onboard a ship with sufficient spatial separation to prevent excessive mutual coupling interference between adjacent base tuners [OR-1]. For example, a minimum of 13 m is required between transmit whips using the standard Navy 2 MHz to 30 MHz AN/URA-38 base coupler.

---

[12]This frequency is also known as the natural frequency of an antenna; the natural frequency is the lowest resonance frequency of an antenna without added inductance or capacitance.

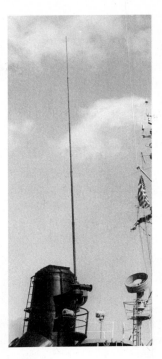

Figure 10.19 AS-2537 10-m Fiberglass Whip Antenna (Courtesy of the U.S. Navy; Naval Sea Systems Command)

10.5.2 Twin Whip Antennas

An alternative method of broadbanding monopole antennas is to use a pair of 10-m whips as a single unit connected to a single feed. Twin whip antennas, so connected and operating as a single antenna for better performance, are widely used aboard ships.

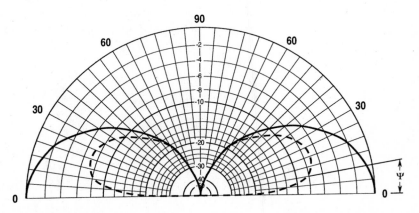

Figure 10.20 Vertical-Plane Radiation Patterns for a Ground-Mounted Quarter-Wave Vertical Whip (Courtesy of *The ARRL Antenna Book,* Figure 3-2, p. 3-2)

Figure 10.21   AS-2537A Shipboard HF Twin Whip Antenna (Courtesy of the U.S. Navy; Naval Sea Systems Command)

Twin whip antennas provide a good impedance match over a wide frequency range and thus have an advantage over narrowband antennas. The advantage of a broadband antenna is that it needs fewer tune-up adjustments, and if used for an ECCM system, a broadband antenna system can support a high rate frequency hopping.

In a twin configuration, the individual antennas are usually mounted on a common platform that may be either stationary or able to be tilted. If the distance between the antennas is relatively close, the antenna may be connected with a crossbar which is then fed at the center point from a matching network using a flexible strap. Alternatively, a wire cable from each antenna to the matching network may be used, provided the wires neither place undue strain on the matching network insulator nor come into contact with the ship's structure during antenna motion in seasway. Figure 10.21 is a picture of a twin whip, 10-m high in a vertical installation including the base tuner, used for 4 MHz to 12 MHz communications.

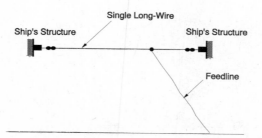

Figure 10.22   Single Long-Wire Antenna

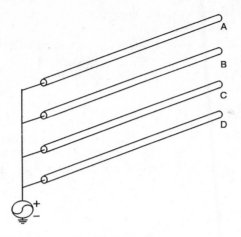

Figure 10.23　Four-element Dipole
Antenna with Negligible Coupling

10.5.3 Long-Wire Antennas

The term long-wire antenna means any configuration where the antenna length exceeds the wavelength, and not simply a straight-wire antenna. Figure 10.22 shows a typical long-wire antenna configuration in which a single wire is stretched between two parts of the ship's structure. In general, a modest gain is obtained by a longer wire, but the bandwidth of the antenna becomes narrower. Where space for the antenna is limited, a multi-element array antenna such as the fan type can be used. Multi-element array antenna yield a wider gain and wider bandwidth than a single long-wire antenna. For example, consider four dipoles with negligible coupling between elements as shown in Figure 10.23. The gain of this antenna surpasses that of a long-wire antenna. Table 10.9 shows a comparison of gains for multiple dipoles as shown in Figure 10.23.

 Long-wire antennas are often chosen despite their deficiencies, because the construction of long-wire antennas is simple, both electrically and mechanically, and there are no critical dimensions or adjustments required. The long-wire antenna works well and gives satisfactory gain and directivity over a 2-to-1 frequency range; in addition, it accepts power and radiates well on any frequency for which its overall length is not less than approximately half the wavelength.

 Long-wire antennas have directive patterns that are pronounced in both the horizon-

TABLE 10.9　COMPARISON OF GAINS OF VARIOUS DIPOLES WITH NEGLIGIBLE COUPLING

Dipoles	Relative Output Power	Relative Input Power	Power Gain	Gain (in dB)
A Only	1	1	1	0
A and B	4	2	2	3
A, B, and C	9	3	3	4
A, B, C, and D	16	4	4	6

tal and vertical planes, and they tend to concentrate radiation at low vertical angles that are most useful at the higher frequencies.

For explanation of a long-wire antenna pattern, consider an elementary dipole's pattern in the plane containing the wire axis as shown in Figure 10.24. The radiation intensity of the dipole is greatest at right angles to the direction of the antenna wire, and it decreases as the direction becomes more nearly in line with the antenna wire. The solid pattern of the elementary dipole is a torus (doughnut), shown in Figure 10.25.

The long-wire antenna pattern is similar to that of a dipole antenna. The ground plane generates an electrical image of the antenna. The resultant pattern is a superposition of the long wire and its image. This will result in a pattern that looks somewhat like a toroidal pattern, distorted and tilted. The exact shape depends on wavelength (operating frequency) and other distortions from the ship's structure.

10.5.4 Fan-Type Antennas

Dedicating an antenna to a single device (transmitter, receiver, or transceiver) is an inefficient way of using space on the deck and superstructure of a ship. To be more efficient,

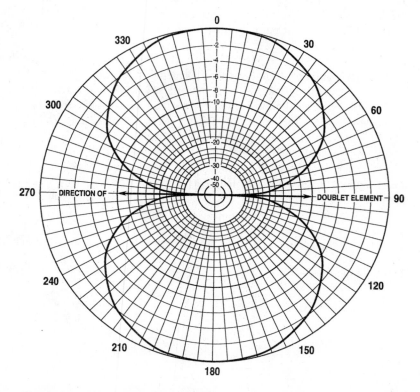

Figure 10.24 Antenna Pattern of an Elementary Dipole in the Plane Containing the Wire Axis (Courtesy of *The ARRL Antenna Book*)

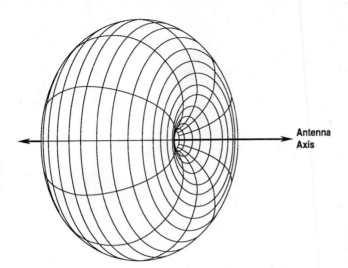

Figure 10.25 Solid Pattern of an Elementary Dipole (Courtesy of *The ARRL Antenna Book*)

multicouplers are used for combining signals from transmitters and receivers for one or more antenna(s). Multicouplers for transmitters generally are connected to a broadband antenna. A wire-rope, fan-type antenna [LI-1] operates over a 4-to-1 frequency range with a power loss of only 20 percent due to mismatch. Wire-rope, fan-type antennas are the most frequently used type of broadband HF antennas; they are used for the Naval Tactical Data System (NTDS) and long-haul HF communications circuits.

Figures 10.26 and 10.27 show drawings of typical wire-rope fan installations. The first is a three-wire, single-fan type and the other is a three-wire, twin-fan type. Wire-rope fan antennas are standardized for the lower portions of the HF range, particularly the low band (2 MHz to 6 MHz), the mid-band (4 MHz to 12 MHz), and the full-band (2 MHz to 30 MHz), with a base tuner. Fan-type antennas are used mostly for transmitting, but they can also be used for receiving. Fan-type antennas are designed to be near omnidirectional, taking into account the distorting effects of structural interference from a ship's superstructure.

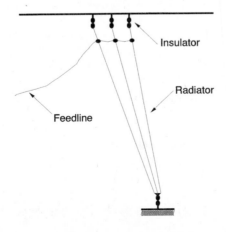

Figure 10.26 Typical Wire-Rope, Fan-type Antenna (Three-Wire, Single-Fan)

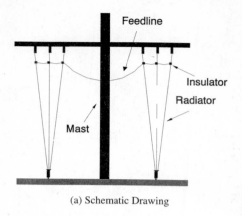

(a) Schematic Drawing

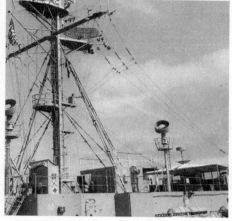

(b) Top-side of a ship (Courtesy of the U.S. Navy; Naval Sea Systems Command)

Figure 10.27   Three-Wire, Twin-Fan Type, Light Weight Communications Antenna (LWCA)

In general, wire-rope antennas are heavy for shipboard environments and vulnerable to shock and vibration. The standard 7.9-mm diameter wire rope is so heavy that the fan antenna center of gravity is typically 33 m (100 ft) or more above the baseline of the ship. Also, all insulators used in the construction of the existing antennas are made of porcelain, and as a consequence, they are susceptible to fracture from shock. To overcome these deficiencies, the U. S. Navy has developed broadband antennas called Light Weight Communications Antennas (LWCA) [BI-1].

A thermoplastic polymer (Delrin) is used instead of ceramic insulators to improve shock resistance. The advantage of the Delrin over ceramic compounds lies in its superior toughness and lower weight. In order to reduce overall weight, the LWCA used 4.8mm diameter phosphor-bronze wire instead of 7.9-mm diameter phosphor-bronze wire.[13] The weight is reduced from 284 kg (511.1 lbs) to 142 kg (315.6 lbs) for the FFG-7 U.S.S. Oliver Perry class installation. Table 10.10 shows the weight comparison of existing antennas and LWCA (FFG-7 Class) [BI-1]. The LWCA is installed on DDG-51 U.S.S. Arleigh Burke class, FFG-7 U.S.S. Oliver Perry class, LHD-1 U.S.S. Wasp class and other ships.

10.5.5 Near-Vertical Incident Skywave (NVIS) Antennas

As discussed in section 10.2, there is a skip zone of 300 km to 650 km in poor HF signal propagation when typical antennas (such as long-wire and whip antennas) are used. An antenna that is placed in a horizontal position can effectively overcome poor HF signal propagation in the skip zone. This type of antenna is known as a Near-Vertical Inci-

---

[13]Phosphor-bronze wire is used because of its corrosion-resistant and nonmagnetic properties.

TABLE 10.10 WEIGHT COMPARISON—EXISTING ANTENNAS
AND LIGHT WEIGHT COMMUNICATIONS ANTENNA (LWCA)
FOR FFG-7 CLASS

Part	Existing Antenna	Light Weight Antenna
Antenna Wire	83 kg (184.4 lbs)	36 kg (80.0 lbs)
Upper Support Assembly	99 kg (220.0 lbs)	35 kg (77.8 lbs)
Feedwire Assembly	76 kg (168.9 lbs)	30 kg (66.7 lbs)
Apex Assembly	23 kg (51.1 lbs)	41 kg (91.1 lbs)
Total	284 kg (511.1 lbs)	142 kg (315.6 lbs)

dent Skywave (NVIS) antenna. An NVIS antenna is a horizontally polarized antenna, which provides for efficient operation of HF signals at high take-off angles [NS-1; HA-2; TO-1].

In order to cover the skip zone, an NVIS antenna has a pattern with high take-off angle as illustrated in Figure 10.28. For comparison, the pattern of a whip antenna is also shown in Figure 9.28. Figure 10.29 shows the geometry and the ray pattern of an NVIS antenna. NVIS antennas, typically horizontal, direct the signal upward rather than toward the horizon, as whip antennas do [NS-1; ST-1]. NVIS antennas operate on the lower side of the HF band, that is, 2 MHz to 6 MHz.

One can place a whip antenna in a horizontal position in order to achieve NVIS-like antenna effects for shipboard applications. Another NVIS antenna is a "towel bar" antenna, shown in Figure 10.30.

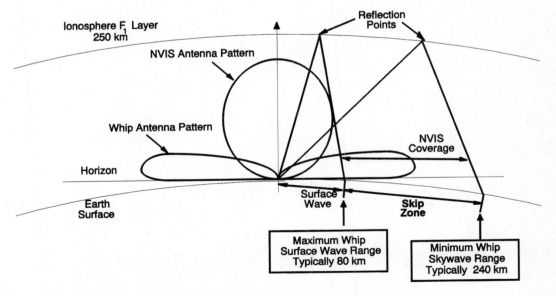

Figure 10.28   NVIS HF Antenna Pattern

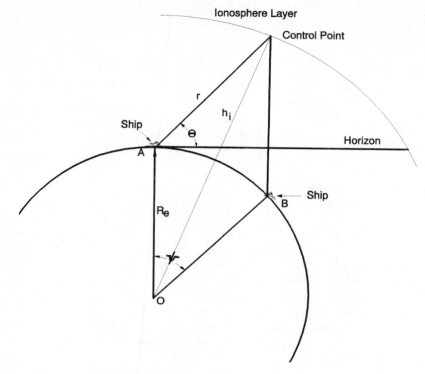

Figure 10.29   NIVS HF Antenna Geometry

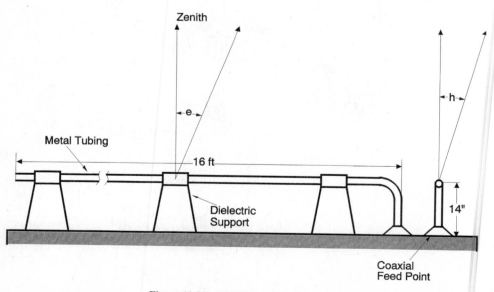

Figure 10.30   NVIS "Towel Bar" Antenna

10.6 AUTOMATIC LINK ESTABLISHMENT TECHNIQUES

Over the past few years HF radios staged a comeback as an alternative means of shipboard long-haul communications, since HF communications feature portability, versatility of location, and long-distance connectivity. In addition, users became aware of problems associated with satellite communications due to high costs and congested satellite channels. Newly automated and microprocessor-controlled radios now offer capabilities previously not available for automatic HF link establishment (ALE).

10.6.1 Description of ALE Method

Figure 10.31 shows a typical ALE radio system, which includes an antenna, an RF coupler, a transmitter, a receiver, an ALE controller, and user terminals and computers. The ALE is a method of automatically initiating contact between calling and called MF and HF radio stations [HA-1]. In order to facilitate ALE, a methodology of operational rules is needed; these rules are listed below:

- Availability of a separate receiver for listening
- Constant listening for the ALE signal
- Constant response to the ALE protocol
- Constant scanning
- Noninterruption of an active ALE channel
- Constant exchange of link quality analysis data with participating nodes when requested, and constant monitoring of the signal quality of others
- Response in preset, derived, or directed time slot for nets, groups, or special calls
- Constant maintenance of connectivity
- Minimum channel occupancy time
- Minimum power usage

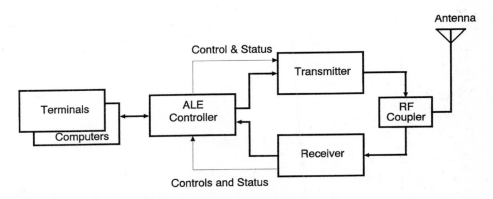

Figure 10.31   Typical ALE HF Radio System

10.6.2 HF Radio ALE Standards

A key factor in facilitating HF ALE technique is the standardization of HF equipment. Without a standard one cannot achieve the operational rules delineated above. All HF ALE radios must adhere to standards, such as MIL-STD-188-141A [DD-1] and FED-STD-1045 [FS-1]. MIL-STD-188-141A establishes technical parameters, in the form of mandatory standards and optional design objectives, which are necessary to ensure interoperability of new long-haul and tactical radio equipment in the MF and HF bands [DD-7].

FED-STD-1045, which is similar to MIL-STD-188-141A, also establishes technical criteria to facilitate interoperability between HF telecommunications facilities and systems of the federal government and compatibility of the radio interface in these facilities with the federal government's data processing equipment. These two standards address automated HF radio features such as frequency scanning, selective calling, link quality analysis, and sounding.

Figure 10.32 shows the control requirements and system applications of levels for HF radio equipment for facilitating the interoperability.

Level 1 of interoperability applies to common basic HF radios; level 1 requires a standardized set of operational procedures, a national HF directory, and a standardized baseline HF radio. Table 10.11 shows a summary of HF interoperability standard baseline radio characteristics.

Level 2 of interoperability requires selective calling and handshaking, that is, digital address selective calling, followed by a handshake consisting of a response signal. At

HF ALE System Implementation Methods

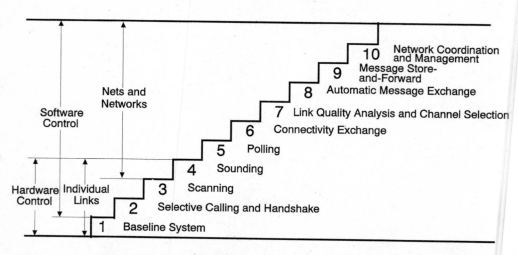

Figure 10.32   Control Requirements and System Applications of Levels for HF Radio Equipment (Courtesy of The MITRE Corporation)

TABLE 10.11   CHARACTERISTICS OF HF ALE INTEROPERABILITY STANDARD
BASELINE RADIO

Parameter	Value	Remarks
Frequency • Low Limit • High Limit • Step Size • Accuracy • Control	• 1.6 MHz • 29.999 MHz • 100 Hz • ±30 Hz • Synthesis	Both Transmit and Receive
Modulation • Voice, Audio, and FSK • Keying	• USB and LSB • CW	• Selectable SSB • Manual Morse
• Mode	• Simplex	• Push-To-Talk
RF Power • Voice • Keying	• 100 W PEP • 100 W Average	• Minimum, SSB Voice • Continuous, CW and FSK
Receiver Sensitivity	0.5 µV	10 dB (S+N)/N, or better
Bandwidth • Voice, FSK • Keying	300–2700 Hz	Minimum Range
Antenna	Basic Antennas	General CONUS orientation; combination of antennas providing high gain, omni-directional coverage, and rapid tuning where needed

level 2 or higher levels, all nodes are always listening for calls, even though they may operate in radio silence, and they automatically respond, unless inhibited by the operator or a special command. Figure 10.33 shows the individual calling protocols in level 2.

Level 3 requires that all available HF nodes continually scan through their receivers

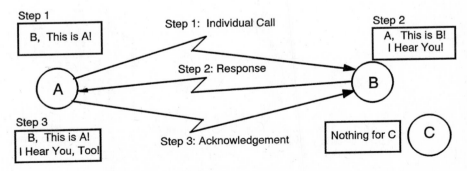

Figure 10.33   Relationship of Three HF ALE Stations Performing Level 2 Functions (Selective Calling & Handshaking) (Courtesy of The MITRE Corporation)

on all channels, seeking calls. Note that level 3 involves a simple, mechanical approach to establishing a working path between two nodes, and it does not require any special information except the frequencies (channels) to be used and the digital address of the called node. Essentially, level 3 assists nodes to find and use any and all working HF paths, regardless of channel conditions such as noise and fading levels.

Level 4 uses sounding, which provides the ability to test each path to every node on each channel, by sending a very brief identifying, beaconlike broadcast signal. Sounding enables one node to assist other nodes with tracking of propagation, connectivity, and station availability. Because sounding signals use one-way transmission, they should not be confused with polling techniques, which consist of two-way exchanges of interrogations and responses.

Level 5 requires polling, which gives the system the ability to track actual available connectivity with other nodes on each channel, to store the information in relationship to time, and to select effective calling channels. At this level of interoperability, the ALE radio system maintains a connectivity matrix, shown in Figure 10.34, that indicates the nature of connectivity between a node and other nodes on the available channels. Each polling node must have memory which contains the matrix, plus a control system to scan, monitor, and identify nodes which are heard.

Level 6 requires the connectivity exchange, which provides a capability to share information about connectivity and to identify possible relay paths. This relay and alternate path information may improve total system connectivity.

Figure 10.32 also shows levels 7 through 10. These additional levels have been identified for future implementation. Level 7 supports link quality analysis and channel selection, which measure and choose the best channel for calling. Level 8 is the automatic message exchange, which enables a station to store an originating message intended for transmission to another station which temporarily does not have connectivity with that storing station. Level 9 is the message store and forward function that enables a station to originate a relay message and specify its routing for indirect relay through other stations

o Connectivity Matrix Between Station A, and Other
  HF Stations (B, C, D, E and F)
    00 = No Connectivity; Failure (Not Heard
        When Polled)
    01 = No Connectivity; Unknown (Not Heard
        and Not Polled)
    10 = Connectivity; One-way (Heard or
        or Sounding)
    11 = Connectivity; Two-way (Handshake)

o When a Station is to be Called, Station A checks its
  Connectivity Matrix to find Which Channel(s) Have
  Worked Recently. The Higher the Numerical Value,
  The Higher the Priority in Channel Selection

	Channels				
Stations	1	2	3	4	5
B	10	11	10	00	11
C	01	11	10	11	01
D	11	01	11	00	10
E	01	10	00	11	11
F	01	00	00	10	01

Figure 10.34 HF ALE Radio Connectivity Matrix Required for Level 5 Functions (Polling Stations) (Courtesy of The MITRE Corporation)

and to automatically accept other stations' messages for direct or indirect relaying. Level 10 contains network coordination and management, which enables an entire network of stations to automatically or, under operator guidance, optimize its functions and allocate its resources to compensate for variations in propagation and operational environment and maximize the reliability and throughput of the network as a system.

Figure 10.35 shows how the five federal standards dealing with HF radios interrelate. Federal Standard 1045 deals with levels 1 through 4 and a portion of level 7 (link quality analysis). Levels 5 through 10 will be contained in Federal Standards 1046, 1047, 1048, and 1049.

10.6.3 Shipboard ALE Using AN/URT-23 Radio Equipment

Some Navy shipboard HF ALE radio equipment has been designed using the AN/URT-23 transmitter [SA-1]; the ALE conforms to MIL-STD-188-141A. Figure 10.36 shows a block diagram of the ALE equipment for shipboard use. The system contains a remote transmitter, a remote receiver, an antenna coupler, an ALE modem remote controller, and user terminals such as the AN/ASQ-69 and computers such as the AN/UYK-43. A MIL-STD-1553B bus is used for connecting the various baseband signals of transmitters, receivers, and ALE modem remote controllers.

In a shipboard environment, HF radios supporting various services such as the NTDS, HF voice, and HF radio TTY, are collocated with the HF ALE. The main HF ALE radio then performs the role of the master link station to which HF radios for the NTDS, voice, and TTY services are connected via a MIL-STD-1553B bus. This arrangement represents integrated shipboard HF radio, which meets MIL-STD-188-141A requirements for HF interoperability and ALE. This integrated system is also fully compatible and interoperable with existing Navy shipboard HF equipment and radio operations.

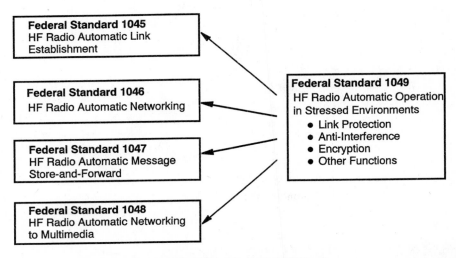

Figure 10.35 Federal Standards for HF Radios

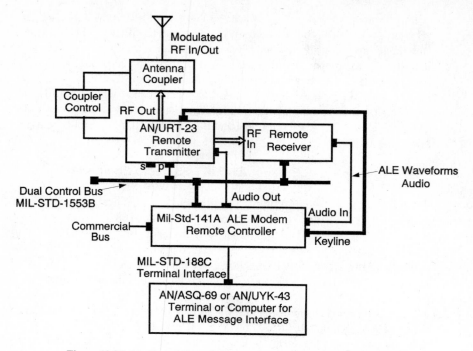

Figure 10.36   Audio and Control Signal Flow Diagram of ALE Master Station

Figure 10.37 shows an OSI communications model for the HF ALE using the AN/URT-23 radio. The right side of Figure 10.37 shows the functions of the AN/URT-23 HF ALE transmitter and remote receiver. The ALE functionality applies to the two lowest levels of the OSI protocols.[14] One finds in this column the physical and link levels of the OSI model. They include:

- RF propagation paths and HF spectrum
- Antennas, RF switches, and couplers
- Radio equipment
- Digital Signal Processing (DSP) modulators and demodulators
- Forward error correcting devices
- Bit interleavers
- Encoders and decoders (e. g., Golay code)
- Link quality analysis devices
- Link data communications controllers
- Radio controllers

[14]The OSI protocols are discussed in Chapter 19.

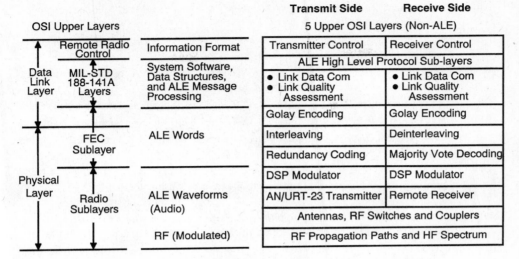

Figure 10.37   OSI Communications Model for HF ALE

Overlayed are the 5 upper layers of the OSI protocols that do not contain ALE functions. The second column shows the signal formats for the ALE in the two lower OSI layers, which include:

- RF waveform and ALE audio waveform
- ALE words
- System software, data structure and ALE message processing per MIL-STD-188-141A [DD-1]
- Other signal/information formats for user peculiar OSI layers

The left column shows how the OSI layers correspond to the ALE HF radio and signal formats.

In summary, the AN/URT-23(E) is an excellent example of how the existing HF radio is adapted to the ALE method by using automation, a microprocessor control, and MIL-STD-1553B bus technology.

REFERENCES

[AA–1] Aarons, J., "General Concepts of Modern HF Communications," *Proceedings of 1986 IEEE Military Communications Conference, October 5–9, 1986, Monterey, Calif.,* pp. 14.1.1–14.1.5.

[AA–2] Aarons, J., *Propagation Impact on Modern HF Communications System Design,* AGARD Lecture Series No. 145, NTIS Document No. ADA-171300, NATO Advisory Group For Aerospace Research and Development (AGARD), Neuilly sur Seine, France, 1986.

[AR–1] The American Radio Relay League, *The ARRL Antenna Book*, 15th Edition, Newington, Conn., 1988, pp. 3.2–3.5.

[BE–1] Bell, C. R. and R. E. Conley, "Navy Communications Overview," *IEEE Transaction on Communications*, September 1980, p. 1573.

[BI–1] Biondi, R., R. Pride, H. Murray, and P. Wheeler, "Lightweight Broadband Communications Antenna," *Naval Engineers Journal*, May 1985, p. 279.

[BR–1] *BR Communications, Inc. Data Sheet on AN/TRQ-35 and Chirpsounder*, Sunnyvale, Calif.

[BR–2] *BR Communications, Inc. Data Sheet on AN/TRQ-42 Chirpsounder*, Sunnyvale, Calif.

[DA–1] Davis, Kenneth, *Ionospheric Radio Waves*, 2nd Edition, Blaisdell Publishing, Waltham, Mass., 1989.

[DD–1] MIL-STD-188-141A, *Interoperability and Performance Standards For Medium and High Frequency Radio Equipment*, Department of Defense, Washington, D. C., September 15, 1988.

[DD–2] MIL-STD-188-203-1A, *Interoperability and Performance Standards for Tactical Digital Information Link (TADIL) A*, Department of Defense, Washington, D. C., March 1987.

[DD–3] MIL-STD-1773, *Fiber Optic Mechanization of An Aircraft Internal Time Division Command/Response Multiplex Data Bus*, Department of Defense, Washington, D. C., October 2, 1989.

[DD–4] MIL-STD-1397A (Navy), *Input/Output Interface, Standard Digital Data, Navy Systems*, Department of Defense, Washington, D. C., January 7, 1983.

[DD–5] MIL-STD-1553B, *Digital Time Division Command/Response Multiplex Data Bus*, Department of Defense, Washington, D. C., September 8, 1986.

[DD–6] MIL-STD-188C, *Military Communication System Technical Standards*, Department of Defense, Washington, D. C., November 1969.

[DD–7] DoD Directive 4640.11, *Mandatory Use of Military Telecommunications Standard in MIL-STD-188 Series*, Department of Defense, Washington, D. C., December 21, 1987.

[FE–1] Fenwick, R. B., "Real-Time Frequency Management for Military HF Communications," *BR Communications Technical Note*, No. 2, BR Communications Inc., Sunnyvale, Calif., June 1981.

[FI–1] Filipowsky, R. F. and E. I. Muehldorf, *Space Communications Systems*, Prentice Hall, Englewood Cliffs, N.J., 1965.

[FS–1] Federal Standard 1045, *Telecommunications: HF Radio Automatic Link Establishment*, General Service Administration, Washington, D. C., January 1990.

[HA–1] Harrison, G. E., *Functional Analysis of Link Establishment in Automated HF Systems*, MITRE Technical Report 86W00015, The Mitre Corporation, McLean, Va., December 1985.

[HA–2] Hagan, George and Jan E. Van Der Laan, "Measured Relative Response Toward the Zenith of Short-Whip Antennas on Vehicles at High Frequency," *IEEE Transactions on Vehicular Technology*, August 1970, pp. 230–236.

[KA–1] Kandoian, K. G., "Three New Antenna Types and Their Applications," *Proceedings of IRE (now IEEE)*, February 1946, pp. 70w–75w.

[LI–1] Linton, R. L., Jr., "The Folded Fan As a Broadband Antenna," *Proceedings of IRE (now IEEE)*, November 1951, pp. 1436–1444.

[LO–1] *Understanding Link 11—A Guidebook for Operators, Technicians, and Net Managers*, Logicon, San Diego, Calif., April 1990.

[NA–1] *HF Radio Antenna Systems*, Naval Shore Electronics Criteria, NAVELEX 0101,104, Naval Electronic Systems Command (now Space and Naval Warfare Systems Command), Washington, D. C., June 1970.

[NS–1] *Communications for Future Over-the-Horizon (OTH) Amphibious Operations*, Appendix H—Near-Vertical Incident Skywave (NVIS) HF System, National Security Industry Association, Washington, D. C., April 1988.

[NT–1] NATO Standardization Agreement (STANAG) 5511, *Tactical Data Exchange—Link 11*.

[OR–1] Orem, J. B., "The Impact of Electromagnetic Engineering on Warship Design," *Naval Engineers Journal*, May 1987, p. 210.

[SA–1] *Stewart-Warner Corporation Brochure on Automatic Link Establishment for Shipboard HF Radio Equipments*, Stewart-Warner Corporation, Chicago, Ill., undated.

[SC–1] Schoppe, W. J., "The Navy's Use of Digital Radio," *IEEE Transaction on Communications*, December 1979, p. 623.

[ST–1] Sternowski, R. H., "Growth in HF C3 Capability," *Signal*, November 1984.

[SW–1] Swenson, E. N., E. B. Mahinske, and J. S. Stoutenburgh, "NTDS—A Page in Naval History," *Naval Engineering Journal*, May 1988, p. 53.

[TO–1] Toher, Martin E., "Tactical HF Enters the Skip Zone," *Proceedings of the Naval Institute*, April 1993, pp. 109–111.

[WA–1] Walsh, E. J., "Navy's Tactical Data Link Advances Fleet Technology," *Signal*, February 1990.

[WI–1] Williamson, J., ed., "AN/TRQ-35(V)-2 Tactical Frequency Management System," *Jane's Military Communications 1991–1992*, Jane's Information Group, Inc., Alexandria, Va., 1991, pp. 112–114.

[WI–2] Williamson, J., ed., "R–1051/URR Tactical Data Information Link 11/Tadil A Digital Data Receiving Set," *Jane's Military Communications 1989*, Jane's Information Group, Inc., Alexandria, Va., 1989, p. 297.

[WI–3] Williamson, J., ed., "R-2368/URR VLF/LF/MF/HF Radio Receiver," *Jane's Military Communications 1991–1992*, Jane's Information Group, Inc., Alexandria, Va., 1991, p. 270.

[WI–4] Williamson, J., ed., "AN/URT-23D HF Transmitter," *Jane's Military Communications 1991–1992*, Jane's Information Group, Inc., Alexandria, Va., 1991, p. 268.

[WI–5] Williamson, J., ed., "HF-80 HF Communication Systems," *Jane's Military Communications 1991–1992*, Jane's Information Group, Inc., Alexandria, Va., 1991, pp. 123–124.

[WI–6] Williamson, J., ed., "AN/USQ-83 Date Terminal Set," *Jane's Military Communications 1991–1992*, Jane's Information Group, Inc., Alexandria, Va., 1991, p. 468.

[WI–7] Williamson, J., ed., "ICS3 Integrated Communication System," *Jane's Military Communications 1991–1992*, Jane's Information Group, Inc., Alexandria, Va. 1991, pp. 738–740.

[WI–8] Williamson, J., ed., "AS-2537A/SR Naval Whip Antenna," *Jane's Military Communications 1991–1992*, Jane's Information Group, Inc., Alexandria, Va., 1991, p. 628.

11

VHF/UHF Line-of-Sight Radio Communications

The radio frequency spectrum between 30 MHz and 300 MHz is referred to as VHF, and between 300 MHz and 3 GHz as UHF. VHF/UHF communications support ship-to-ship, ship-to-shore, and ship-to-air line-of-sight (LOS) communications. VHF and UHF radios are similar, and they are discussed together in this section.

Figure 11.1 shows the military's VHF/UHF band frequency assignments. They are: (a) from 30 MHz to 88 MHz and from 108 MHz to 156 MHz for tactical communications, (b) from 156 MHz to 174 MHz for maritime radiophones, (c) from 225 MHz to 400 MHz for tactical LOS communications,[1] and (d) from 950 MHz to 1150 MHz (the L_x band) for anti-jamming radios such as the Joint Tactical Information Distribution System (JTIDS) and HAVE QUICK II radios.

Historically, early radio experiments were carried out in the VHF/UHF frequency band. The VHF band was used for Hertz's and Marconi's earliest experiments before the beginning of the twentieth century [HO-1]. Shortly after the First World War, Marconi built a 150-MHz AM transmitter using a V24 vacuum tube with a dipole antenna and parabolic reflector. The first VHF antennas were the inverted "L" type, but resonant antennas were later introduced and yielded much better results. In 1928, Yagi and Uda developed their well-known directional antenna using director elements, and the Yagi-Uda antenna was to have a significant influence on radio communications in VHF and UHF bands,

---

[1]In this band, frequencies are also assigned to the Navy's UHF SATCOM; UHF SATCOM is discussed in detail in Chapter 13.

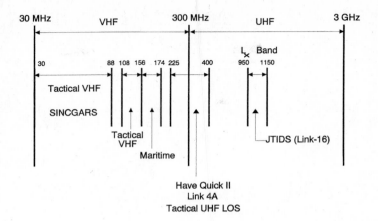

Figure 11.1   Navy Tactical Frequency Allocation (VHF and UHF)

where, in the early years of radio, it was not easy to generate high RF power. They carried out experiments at 6 GHz, obtaining successful radio transmission over 10 km to 30 km.

For Navy shipboard communications the VHF/UHF bands are attractive because (a) smaller antennas can be used than for HF and LF radios, (b) line-of-sight (LOS) propagation results in low detectability, (c) the availability of wider bandwidths allows spread-spectrum application, (d) very low transmission power can be used for limited transmission range, and (e) short and efficient aerodynamically shaped VHF/UHF antennas with low drag can be used for high-performance fighter aircraft.

Before World War II, VHF radios were widely used in shipboard naval communications [HO-1; pp. 513–546]. However, after the war, much of the VHF band was assigned to television, commercial FM radio, civil air traffic control, radio astronomy, and amateur radios. Today, shipboard tactical VHF radios use segments of the VHF band from 30 MHz to 88 MHz and from 108 MHz to 156 MHz for ship-to-shore communications in amphibious operations and for land-mobile shore communications, and they use a segment overlapping the VHF and UHF bands from 225 MHz to 400 MHz for principal tactical ship-to-ship and for ship-to-aircraft Electronic Counter-Counter Measure (ECCM) circuits. The 950 MHz–1150 MHz band is for JTIDS and HAVE QUICK II as well as some satellite radio relays [NT-1; NT-2; NC-3].

11.1 CHARACTERISTICS OF VHF/UHF COMMUNICATIONS

Table 11.1 summarizes the properties of VHF/UHF communications in terms of: (a) operational characteristics, (b) practical design considerations, (c) connectivity effectiveness, and (d) operational limitations. In the VHF/UHF bands, atmospheric noise decreases below the thermal noise.[2] In the VHF and UHF bands, factors affecting receiver perform-

---

[2]Atmospheric and thermal noise are discussed in Chapter 7 (Figure 7.1), and Chapter 12 (Figures 12.3 to 12.5).

TABLE 11.1   CHARACTERISTICS OF VHF/UHF LOS COMMUNICATIONS

Criteria	Characteristics
Theoretical considerations	• Atmospheric noise decreases with increasing frequency (below thermal noise) • Propagation not affected by the ionosphere; line-of-sight propagation • Wider bandwidth available than in HF (up to 1000 times more than HF) • Less susceptible to jamming than HF
Practical design considerations	• Suitable for ship-to-aircraft and ship-to-ship communications • Smaller antennas
Range	• For aircraft at 3000 m altitude, the range can be extended to 240 km
Operational limitations	• Short-range LOS communications (surface communications limited to 40 km) • Co-site interference on shipboard antennas [IS-1; IS-2; RE-1] • In amphibious operations, over-the-horizon communications limitations

ance differ from those in the HF band. It is necessary to pay considerable attention to noise reduction within the equipment itself, since for frequencies above 100 MHz, the noise generated by the receiver front-end amplifiers becomes dominant. Receivers are designed so to perform at as high a signal-to-noise ratio as possible.

Propagation in the VHF/UHF band is not affected by ionospheric reflection and radio signals travel in straight lines. However, there is a multipath effect for transmission between an aircraft and a ship in the VHF/UHF band by signal reflection over water. Consequently, most shipboard VHF radio transmits relatively short-range, LOS signals via elevated antennas.[3]

In the VHF and UHF bands, the wavelength is small (below 1 m for frequencies above 300 MHz), and a ship's structures may cause reflections. Usually, reflected signals interfere with direct LOS signals, and cause multipath effects that may weaken the signal. In shipboard applications, where one or both communications terminals may be at a relatively low altitude, for example, ship-to-aircraft communications, signals may be reflected on water creating multipath signals. The resulting interference will cause the received signal strength to vary from one location on a ship to another.

At higher frequencies (above 400 MHz), transmission closely resembles the transmission of light, with shadowing by obstacles and reflection from terrain features, structures, and vehicles with sufficient reflectivity.

Practical design considerations.   Most VHF/UHF transmissions cover relatively short ranges, and use LOS transmission with elevated antennas, at least at one end of the path. VHF/UHF radios are suitable for ship-to-aircraft, ship-to-ship, and short-

---

[3]Although at the lower end of VHF band (50 MHz and below), signal transmission by ionospheric reflection can occur occasionally, it is not the primary means of transmission. Only around times of very high sunspot activity the ionosphere's electron density increases to the extent that an ionospheric reflection mode for frequencies up to 50 MHz can exist. Ionospheric reflection is rarely used in VHF radio communications [GR-1; p. 2.18]. For all practical purposes, the propagation follows a straightline in the VHF and UHF bands.

range ship-to-shore communications. Table 11.2 lists the output power of ship-to-aircraft and ship-to-ship or ship-to-shore VHF and UHF radios. Airborne radios generally have less output power than shipboard radios.

Much wider signal bandwidths can be used in the VHF and UHF bands than in the HF band, and spread-spectrum techniques can be applied in shipboard communications and navigation. Some LOS radio relay systems make use of this capability. The smaller wavelengths allow the use of antenna diversity even on a relatively small ship. It is also possible to combine inputs from multiple receivers and antennas adaptively to discriminate against interferers or jammers. With wider signal bandwidth and adaptive equalization, much higher data transmission rates can be achieved at UHF, using a wide variety of modulation methods.

Shipboard VHF and UHF receivers are designed for high sensitivity, for a very high range of signal strengths, and for combatting interfering signals. Most VHF/UHF radios used in tactical ship-to-aircraft, ship-to-ship, and short-range ship-to-shore communications employ some spread-spectrum techniques. Also, UHF radios are often used for digital data transmission.

Range.   For many VHF/UHF communications applications the range is dictated by LOS propagation. Figure 11.2 shows the range versus height in VHF/UHF communications. The range of ship-to-aircraft communications is a function of the aircraft height, transmitter power, and antenna gain. VHF/UHF radios provide a large part of shipboard tactical communications connectivities.

To overcome the LOS range limitation, airborne relays are used. For example, an amphibious ship may launch its operations at a distance that LOS communications cannot accommodate. To extend the range, relay platforms such as aircraft, Unmanned Aerial Vehicles (UAVs) and even surface ships using communications payloads are used. Using compatible radios, the connectivity between a ship and an aircraft can be extended to a distance of up to 150 km.

TABLE 11.2   AIRBORNE AND SHIPBOARD VHF/UHF RADIOS

Radios	Frequency Band	Output Power
Ship-to-aircraft		
• AN/ARC-101	VHF	20 W
• AN/ARC-159A	VHF/UHF	30 W
• AN/ARC-182	VHF/UHF	15 W
• AN/ARC-201	VHF	10 W
Ship-to-ship or ship-to shore		
• AN/GRT-21	VHF	30 W
• AN/URC-93	UHF	100 W
• AN/VRC-88	VHF	20 W

ANTENNA HEIGHT	RANGE
SHIP (30 m) TO SHIP (30 m)	25 - 50 km
SHIP (30 m) TO HELO (300 m)	60 - 90 km
SHIP (30 m) TO AIRCRAFT (3000 m)	200 -240 km

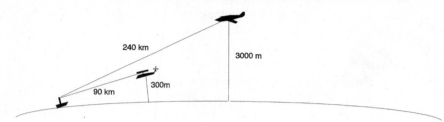

Figure 11.2   Range versus Height in VHF/UHF LOS Communications

11.2 VHF/UHF PROPAGATION

The transmission distance between an airborne VHF/UHF radio and a shipboard VHF/UHF radio is a function of transmitter power, antenna gain, operating frequency, and receiver sensitivity to meet a specified performance requirement, for example, the bit error rate at a given bandwidth or a transmission rate. The relationship of these parameters is given in the range equation. Chapters 7 and 12 discuss the range equation in detail. Figure 11.3 shows a graph of typical minimum transmitter power versus distance for VHF/UHF aircraft-to-ship radio links.

Signals in the VHF/UHF bands can propagate in the following modes: (a) direct line-of-sight (LOS), (b) tropospheric scatter, (c) meteor burst, and (d) via satellite relay. The main propagation mode used in shipboard communications is LOS and satellites. Tropospheric scatter is a viable means of communications for shore sites, but it is not suitable for shipboard applications since it requires large antennas and high transmitter power to offset propagation losses. Signals may also propagate on an intermittent basis by means of scatter from short-lived ionization trails resulting from meteorites entering the Earth's atmosphere and becoming heated to incandescence by friction. Meteor burst applications for shipboard communications are limited by reception zone, burst duration, and reflecting and scattering geometries. However, there is a move toward meteor burst communications for military applications. A draft standard, MIL-STD-188-135, deals with meteor burst communications [DD-2].

Multipath interference in the VHF/UHF band.   In Chapter 10, the multipath effect to HF signals due to ionospheric reflection is discussed. A similar multipath effect occurs to VHF/UHF signals due to reflection over water. When an aircraft transmits

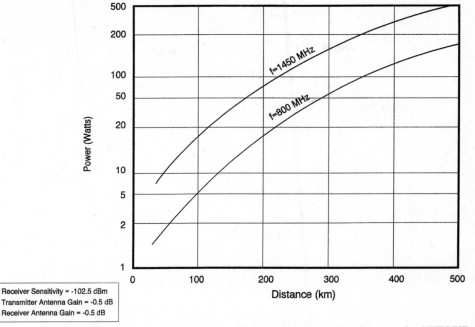

Figure 11.3 Typical Minimum Transmitter Power versus Distance for VHF/UHF Aircraft-to-Ship Radios

signals to a ship, the direct wave and the reflected wave interfere with each other as the two waves arrive with different phases.

Consider an aircraft at the height h_t (= h_{at} + h_{bt}) and ship's antenna at the height h_r (= h_{ar} + h_{br}) as shown in Figure 11.4.

The electric field strength E is related to the direct wave electric field strength E_d by [HS-1]:

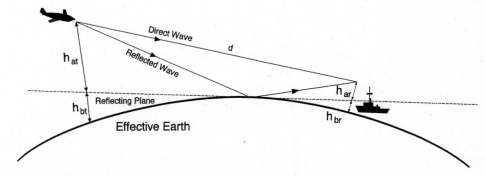

Figure 11.4 Geometry of Direct Wave and Reflected Wave over Water

$$E = 2E_d \sin[2\pi(\frac{\delta}{2\lambda})]$$

where δ is the propagation length difference between direct and reflected waves, and

λ is the wavelength.

The difference δ in the length of the propagation paths can approximately be expressed by [HS-1]:

$$\delta = \frac{2h_{at}h_{bt}}{d}$$

where h_{at} is the height of the transmitter antenna above a reflecting plane tangential to the effective Earth,

h_{bt} is the height of the receiver antenna above a reflecting plane tangential to the effective Earth, and

d is the distance between the transmitter and receiver.

The received power P is related to the direct wave received power P_d by:

$$P = 4P_d \sin^2[2\pi(\frac{\delta}{2\lambda})].$$

Figure 11.5 shows the variation of received power resulting from the multipath effect over water. This curve represents the variation of received power for an airborne transmitter and shipboard radio operating at 1500 MHz, and the aircraft and shipboard receiver antenna heights are 1500 m and 25 m, respectively. Note that the peak power envelope decreases by 20 dB per decade of distance.

As an example, consider a shipboard receiver that has a sensitivity of -134 dBW. For distances of up to 20 km communications are not much affected by the multipath effect, because the received power is well above the receiver sensitivity as shown in Figure 11.5. For distances of more than 20 km, communications are adversely affected by the multipath effect, since disruptive interference reduces the received power, which, for some ranges of the distance, falls below the receiver sensitivity. When that happens, the receive is likely to experience signal drop-out.

Mitigation methods for multipath interference.   There are several methods of mitigating multipath interference on received power in the VHF/UHF band.

The first method is a frequency diversity technique where the signal is modulated onto two carrier frequencies, and the phase separation of the two carriers is equal to a half wavelength ($\lambda/2$) in order to obtain the maximum compensation. An approximate value of the frequency separation is given by:

$$\textit{Frequency separation } (\lambda_{separation}) = \frac{4h_{at}h_{bt}}{d}.$$

As an example, consider an airborne transmitter and a shipboard receiver separated by 100 km, operated at the frequency of 1500 MHz. The aircraft (transmitter) antenna

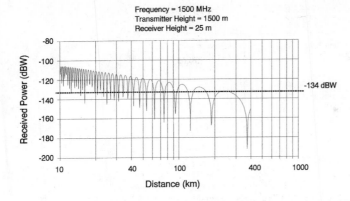

Frequency = 1500 MHz
Transmitter Height = 1500 m
Receiver Height = 25 m

Figure 11.5   Multipath Effect to Received Power as Function of Distance and Frequency

height is assumed to be 1500 m; the shipboard (receiver) antenna height is assumed to be 25 m. The radio frequency separation is then:

$$\lambda_{separation} = \frac{4h_{at}h_{bt}}{d} = 150 \text{ MHz}.$$

The second method is a space diversity technique, in which two receiver antennas are separated in height, and the height of separation is a half wavelength ($\lambda/2$); this provides a means of compensating to a certain extent for changes in electrical path differences between direct and reflected rays. An approximate value of the receiver height separation is given by:

$$\textit{Reciever height separation } (h_{separation}) = \frac{\lambda d}{4h_{at}}.$$

As an example, consider an airborne transmitter and a shipboard receiver separated by 150 km, operated at the frequency of 1500 MHz. The aircraft height is assumed to be 1500 m. The receiver antenna height spacing is then:

$$h_{separation} = \frac{\lambda d}{4h_{at}} = 5.0 \text{ m}.$$

Use of frequencies above 10 GHz for LOS links.

In this chapter we describe LOS links which in most cases use VHF and UHF transmission. For some shipboard LOS links, frequency in the X-band (9.7 GHz to 15.5 GHz) and Ku-band (14.4 GHz to 15.5 GHz) are used, primarily as results of crowding in the VHF and UHF band, and some Navy links requiring wide bandwidth for high-speed LOS data transmission. For completeness, Table 11.3 summarizes the frequency assignments for the United States in the microwave region from 1.71 MHz to 15.25 GHz.

Figure 11.6 shows a frequency designation chart from 0.1 GHz to 100 GHz. Included in this figure are: (a) the official International Telecommunications Union (ITU) frequency bands from VHF to EHF, (b) the frequency designations used during World War II, (c) the commonly used microwave and radar frequency designations, and (d) the official Joint Chiefs of Staff (JCS) frequency designations.

TABLE 11.3 MICROWAVE FREQUENCY ASSIGNMENT
FOR THE UNITED STATES

Assigned User	Frequency (GHz)
• Military	1.710–1.850
• Operational transmission link	1.850–1.990
• Studio transmission link	1.990–2.110
• Common carrier	2.110–2.130
• Operational fixed	2.130–2.150
• Common carrier	2.150–2.180
• Operational fixed	2.180–2.220
• Operational fixed (TV only)	2.500–2.690
• Operational fixed (space)	3.700–4.200
• Military	4.400–5.000
• Common carrier (space)	5.925–6.425
• Operational fixed	6.575–6.875
• Studio transmission link	6.875–7.125
• Military	7.125–7.750
• Military	7.750–8.400
• Common carrier	10.87–11.7
• Operational fixed	12.2–12.7
• CATV-studio transmission link	12.7–12.95
• Studio transmission link	12.95–13.2
• Military	14.4–15.25

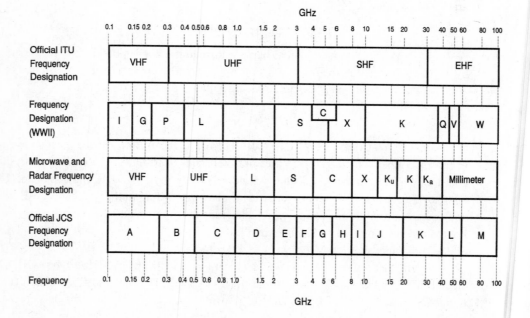

Figure 11.6 Frequency Band Designations

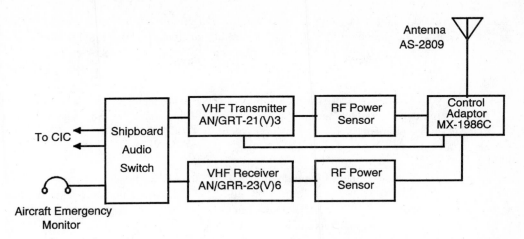

Figure 11.7   Block Diagram of the VHF Aircraft Emergency Monitor System

11.3 VHF/UHF TRANSMITTERS AND RECEIVERS

The AN/GRT-21(V)3 VHF/UHF transmitter [WI-5] and AN/GRR-23(V)6 VHF/UHF receiver [WI-6] are used for transmitting and monitoring aircraft distress communications in the 116.0 MHz to 151.975 MHz frequency band. This transmitter-receiver is used in conjunction with its companion receiver set, the AN/GRR-1986C(V)6, the AS-2809/SRC antenna, and the MX-1986C/SRC control adapter, as shown in the block diagram in Figure 11.7. This transmitter is also used for air traffic control in the 225.0 MHz to 399.97 MHz band.

11.4 VHF/UHF TRANSCEIVERS

The most common equipment used for shipboard VHF and UHF communications is the transceiver, that is, a combination of transmitter and receiver. Transceivers generally are compact, portable, and usually share a common antenna.

11.4.1 SINCGARS

The U.S. Army, Air Force, Navy, and Marine Corps utilize a tactical VHF radio called the Single Channel Ground and Airborne Radio System-VHF (SINCGARS-V), which is a frequency hopping (FH), frequency modulated (FM), spread-spectrum system covering the 30 MHz to 88 MHz frequency band in 25 kHz wide discrete channels [WI-7]. The SINCGARS is designed to provide secure voice and data communications in jamming environments [KI-1; p. 51]. In the single-channel mode, SINCGARS is also compatible with the radio of the existing family of 50 kHz, single-channel radio equipment, such as the AN/VRC-46, operating in the 30 MHz to 76 MHz band [GR-1; pp. 43–45] [WI-7].

The shipboard VHF SINCGARS is a modified version of the ground-based SINCGARS. The shipboard SINCGARS VHF radio must be integrated into the ship's exterior communications system, the available antenna systems, and the Single Audio System (SAS),[4] which is an integrated secure and nonsecure voice interior communications system.

SINCGARS radios consist of transmitter-receiver equipment, antenna/antenna coupler equipment, Communications Security (COMSEC) and Transmission Security (TRANSEC) equipment, and baseband equipment [SN-2]. The SINCGARS radios operate in three modes in tactical situations:

- The single-channel (SC) mode, in which SINCGARS radios are used as ordinary non-ECCM radios; the family of radios, which are compatible with single-channel mode SINCGARS radios, are the AN/PRC-77 (man-pack), the AN/VRC-12 (vehicular short-range), the AN/VRC-46, and the AN/VRC-49.

- The ECCM mode, in which SINCGARS radios are used as ECCM radios; the family of radios, which are compatible with ECCM mode SINCGARS radios, are the AN/PRC-119 (man-pack), the AN/VRC-88 (vehicular long-range dismountable), the AN/VRC-89 (vehicular long-range/short-range), the AN/VRC-90 (vehicular long-range), the AN/VRC-91 (vehicular long-range dismountable and short-range), and the AN/VRC-92 (vehicular dual long-range for retransmit).

- The airborne relay mode, in which SINCGARS radios are used as ordinary ECCM or non-ECCM airborne relay radios; the family of radios which are compatible with airborne relay mode SINCGARS radios are the AN/ARC-201 (airborne transceiver), and the AN/ARC-210.

Shipboard SINCGARS radios have two configurations, type I for ships with single circuit requirements, and type II for ships with up to four circuit requirements. Figures 11.8 and 11.9 show block diagrams of shipboard Type I and Type II SINCGARS VHF radios. Type I shipboard SINCGARS radios consist of a transmitter-receiver with a single antenna, and they are used in ships with a relatively low VHF traffic. Type II shipboard SINCGARS radios consist of up to three transmitters and a single receiver with multiple antennas, and these radios are used in ships that handle heavy traffic. Figure 11.10 shows a picture of the SINCGARS RT-1523 Receiver-Transmitter.

SINCGARS RF and antenna equipment.   Shipboard SINCGARS (type II) radios have up to four antennas (AS-3226A) that can support four channels. Three antennas are for transmitters, and one is for the receiver. When operated in the single-channel (SC) mode, the four antennas are connected to a multicoupler (TD-1289). When operated in Frequency Hopping (FH) mode, the four antennas are connected to four separate RF power sensors in order to support the frequency hopping, spread-spectrum operation.

[4]The Single Audio System (SAS) is also discussed in section 18.3.

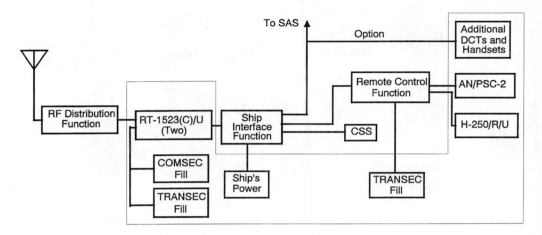

Figure 11.8   Block Diagram of the Shipboard SINCGARS VHF Radio (Type I)

Shipboard SINCGARS radios may encounter problems with antenna co-site inter-ference due to simultaneous operation of the four antennas [IS-1; IS-2; RE-1]. Unless carefully engineered, the receiver antenna may become the victim of the three transmitter antennas.

SINCGARS receiver-transmitter equipment.   The receiver-transmitter (RT) of the shipboard SINCGARS includes four RT-1523 units modified to utilize a ship's power system. The RT-1523 is a part of the AN/VRC-90 SINCGARS radio, and it was originally developed for the Army's vehicular long-range SINCGARS radio.

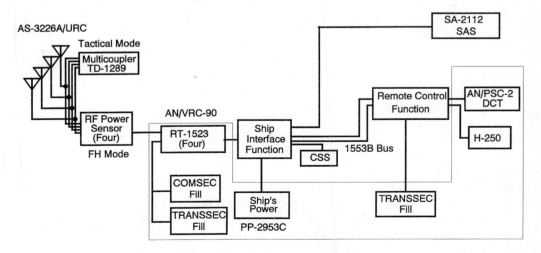

Figure 11.9   Block Diagram of the Shipboard SINCGARS VHF Radio (Type II)

Figure 11.10   SINCGARS Receiver-Transmitter RT-1523 (Courtesy of ITT Aerospace/Communications Division)

SINCGARS COMSEC/TRANSSEC equipment.   The COMSEC/TRANSSEC equipment of shipboard SINCGARS includes a communications encryption device for user traffic and a transmission encryption device for frequency hopping ECCM. For transmission encryption, shipboard SINCGARS provides slow frequency hopping ECCM capability. The ECCM unit of a shipboard SINCGARS radio contains a key generator that controls the frequency hopping pattern of a radio.

In order to initialize and synchronize the shipboard SINCGARS radio, certain information is required. The first item is the Word-of-Day, which provides a unique key for a day/hour/minute variable. The second is the net ID, which provides a unique key for a designated net on which the radio operates. The third is the hopset, which defines the frequencies on which the radios will hop during frequency hopping mode. The number of frequencies in the hopset and their distribution within the band determine the measure of the ECCM performance. In order to synchronize the ECCM unit, a high-precision clock accuracy, to one part in ten million, is required. The clock used in shipboard applications employs a temperature-compensated crystal oscillator that can maintain sufficiently accurate time for the duration of a tactical mission.

SINCGARS baseband equipment.   The baseband equipment of the shipboard SINCGARS includes:

- The AN/PSC-2 Digital Communications Terminal (DCT), a portable communications message processor for Marine Corps Air Defense Early Warning, infantry tactical support, and gunfire support.
- The H-250 handset for voice communications.
- The Single Audio System (SAS), which provides access to the ship's secure and nonsecure voice system via an SA-2112 switch.

Interconnection of the baseband equipment to RT-1523A is established either via a 1553B data bus or an RS-232C interface. The MIL-STD-1553B data bus is capable of trans-

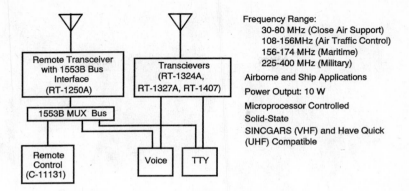

Frequency Range:
    30-80 MHz (Close Air Support)
    108-156MHz (Air Traffic Control)
    156-174 MHz (Maritime)
    225-400 MHz (Military)
Airborne and Ship Applications
Power Output: 10 W
Microprocessor Controlled
Solid-State
SINCGARS (VHF) and Have Quick
(UHF) Compatible

Figure 11.11   Block Diagram of the AN/ARC-182 VHF/UHF Radio

mission of high data rates. It was originally developed for aircraft electronic equipment, using a TDM mode of operation, and a data rate of 1 Mbps. Technical details of the 1553B bus can be found in MIL-STD-1553B (Digital Time Division Command/Response Multiplex) [DD-3].

As shown in Figures 11.8 and 11.9, Shipboard SINCGARS can interface with the Communication Support System (CSS). The CSS is a communications system architecture that enhances battle force communications connectivity, flexibility, and survivability through multimedia access and media sharing [NC-2; SN-1].[5] The CSS permits users to share total network capacity on a priority demand basis in accordance with a communications plan.[6]

11.4.2 The AN/ARC-182 VHF/UHF Radio

Figure 11.11 shows a block diagram of the AN/ARC-182 radio. The AN/ARC-182 VHF/UHF radio is widely used on ships [WI-8]. The frequency covered by this radio ranges from 30 MHz to 400 MHz, as shown in Figure 11.12. This radio is a multiband/multimode tactical radio for close air support, air traffic control, maritime radio-telephone, and military and NATO use.

With 25 kHz channel spacing in VHF/UHF bands, this radio can accommodate 11,960 channels. The baseband can be interfaced via a 1553B data bus or an RS-232 serial data channel. Figure 11.13 shows the AN/ARC-182 VHF/UHF transceiver equipment.

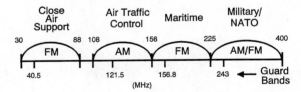

Figure 11.12   Frequency Allocations of the AN/ARC-182 VHF/UHF Radio

[5]See section 20.1.7.

[6]The use of a communications plan implies that some circuits are assigned and/or dedicated to certain services, and some circuits are used as resources. Should a dedicated circuit be temporarily available for other services, it can be added on a priority basis to the resource pool. SINCGARS radio uses the CSS circuit assignment method.

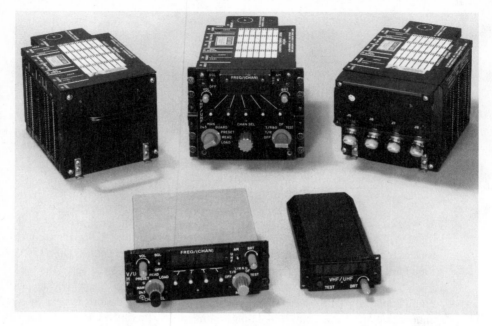

Figure 11.13 The AN/ARC-182(V) VHF/UHF Transceiver (Courtesy of Rockwell-International, Collins Avionics and Communications Division)

11.4.3 The AN/VRC-40 Series VHF Radios

The AN/VRC-40 series VHF transceiver supports short-range, two-way communications in the 30 MHz to 76.0 MHz range. These radios are used in vehicles such as jeeps, armored personnel carriers and tanks, and in Navy ships [WI-3].

A typical AN/VRC-40 series radio consists of the RT-246A receiver-transmitter, the R-442 receiver, the audio amplifier AM-1780, and a frequency control, as shown in the block diagram of Figure 11.14. Voice and intercommunications control can be connected to the AN/VRC-40 radio. Other configurations can be assembled from these basic equipment items to meet a specific user requirement.

11.4.4 The AN/WSC-3(V)6 UHF LOS Radio

The AN/WSC-3(V)6 is a Navy standard shipboard UHF line-of-sight in the 225 MHz to 400 MHz range. A typical AN/WSC-3(V)6 UHF LOS radio consists of an antenna coupler, an RT-1217/WSC-3 receiver-transceiver, and a QPSK modem, as shown in Figure 11.15 [WI-4]. Voice, data, and TTY user terminals can be connected to the AN/WSC-3. The AN/WSC-3(V) radio is extensively used for the UHF SATCOM system which is discussed in section 13.3.1, where an overview of the complete family of the AN/WSC-3(V) is given, as well as a picture of the AN/WSC-3(V) radio.

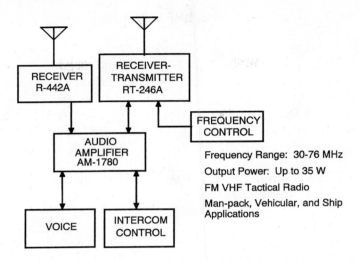

Frequency Range: 30-76 MHz

Output Power: Up to 35 W

FM VHF Tactical Radio

Man-pack, Vehicular, and Ship Applications

Figure 11.14   Block Diagram of the AN/VRC-40 Series VHF Radio

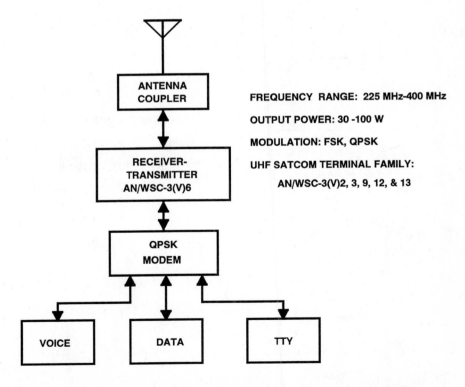

FREQUENCY RANGE: 225 MHz-400 MHz

OUTPUT POWER: 30 -100 W

MODULATION: FSK, QPSK

UHF SATCOM TERMINAL FAMILY:

   AN/WSC-3(V)2, 3, 9, 12, & 13

Figure 11.15   Block Diagram of the AN/WSC-3(V) 6 UHF LOS Radio

11.4.5 The AN/URC-93 VHF/UHF LOS Radio

The AN/URC-93 is a Navy shipboard VHF/UHF LOS radio that operates in the 225 MHz to 400 MHz range. A typical AN/URC-93 radio consists of an antenna coupler, an OR-176(V) receiver-transmitter, and a remote controller, as shown in the block diagram in Figure 11.16. Local or remote voice, TTY, and data terminals can be connected to the AN/URC-93. Several configurations of the AN/URC-93 radios are available for user applications such as voice, ECCM, Low Probability of Intercept (LPI), data, and wideband communications [WI-2].

11.4.6 The Position Location Reporting System (PLRS) AN/TSQ-129(V)

The Position Location Reporting System (PLRS) AN/TSQ-129(V) supports amphibious operations. It provides communications and Military Grid Reference (MGR) navigation capabilities to support battlefield operations. The PLRS communications function also supports free text message exchange [HA-1]. In amphibious operations, the PLRS master

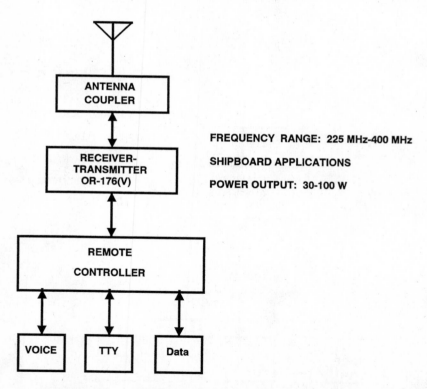

Figure 11.16   Block Diagram of the AN/URC-93 UHF LOS Tranciever

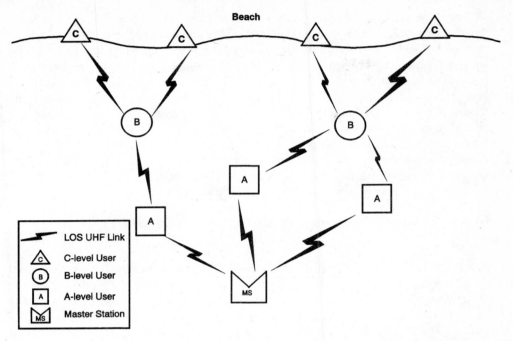

Figure 11.17 PLRS Automatic Multilevel Relay Connectivity (Courtesy of Hughes Aircraft Company)

station is placed in an amphibious command ship as shown in Figure 11.17. PLRS mobile units are carried by embarked force units.

The PLRS is based on synchronized radio transmissions in a network of users controlled by a master station. Major PLRS nodes include airborne, surface vehicular, shipborne, and manpack users as well as the PLRS master station. Characteristics of the PLRS are summarized in Table 11.4.

Figure 11.17 illustrates how the PLRS achieves reliable RF coverage over large de-

TABLE 11.4   PLRS SYSTEM CHARACTERISTICS

Parameters	Characteristics
• Operating frequency	• UHF (420 MHz to 450 MHz)
• System architecture	• TDMA
• ECCM	• Spread-spectrum, frequency hopping, time slot scrambling
• COMSEC/TRANSEC	• Internal cryptographic unit
• Network management	• Automatic, centrally controlled
• Relay capability	• Automatic, four-level
• Typical location accuracy	• 400 user units per master station in a 50 km × 50 km area
• Power output	• 200 W effective radiated power

ployment areas, like in off-shore theaters up to 160 km. This PLRS coverage area is referred to as a PLRS master station community. Several levels of relaying may be utilized by PLRS. Coupled with the PLRS closed-loop adaptive control characteristics, PLRS's integrated, automatic relaying method provides network adaptability, jam-resistance, and battle damage tolerance. The PLRS is a synchronous TDMA system. Time is organized into fixed, cyclic intervals as shown in Figure 11.18. The sixty-four-second epoch is the longest time interval and also the longest user reporting period. The frame has a duration of 0.25 s, which is the shortest reporting period. Each frame is further subdivided into 2-μs time slots. Each user is assigned one or more time slots, and only one PLRS user unit transmits within each master station community at any given time to achieve this multiple user synchronization. A rubidium time standard in the master station is used to update local user clocks and to assure synchronization within a PLRS net.

11.4.7 The Joint Tactical Information Distribution System (JTIDS)

The JTIDS is a high-capacity TDMA system providing integrated communications, navigation, and identification of friend or foe (IFF) capabilities [RI-1; pp. 243–246], [TO-1; SC-1; WI-9]. The JTIDS waveform consists of a series of 6.4-μs pulses in the 960 MHz to 1215 MHz band with anti-jamming capability using frequency hopping techniques [NC-1]. The JTIDS provides the ECCM capability for airborne and surface ship plat-

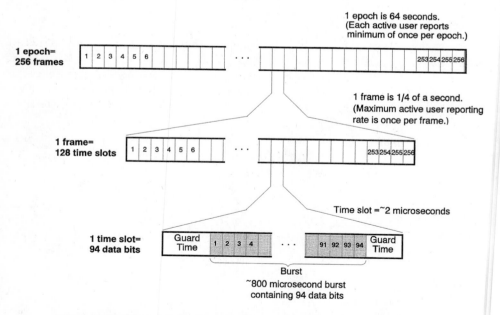

Figure 11.18   PLRS Time Division, Multiple Access (TDMA) Message Structure (Courtesy of Hughes Aircraft Company)

TABLE 11.5   JTIDS CHARACTERISTICS

Parameters	Characteristics
• Operating frequency	• UHF (965 MHz to 1215 MHz)
• Data rate	• 28.8 kbps, 57.6 kbps, or 115.2 kbps
• System architecture	• TDMA
• ECCM	• Spread-spectrum, frequency hopping, time slot scrambling
• COMSEC/TRANSEC	• Embedded cryptographic unit
• Network management	• Automatic, centrally controlled
• Message types	• Formatted or unformatted
• System coverage	• 3640 user units per net in a 480 km × 960 km area
• Power output	• 220 W effective radiated power

forms, extended range of communications, and over-the-horizon communications between surface ships with an airborne relay platform. The JTIDS is designed to accommodate both the digital information associated with Links 14, 4A, and 11 and secure voice. Table 11.5 summarizes the characteristics of the JTIDS.

In order to accommodate multi-user data streams, the JTIDS uses a TDMA scheme as shown in Figure 11.19. The bit stream is subdivided into epochs that contain sixty-four frames each, and an epoch lasting 12.8 minutes. One frame (12 seconds long) has 1563 time slots. Each time slot of 7.8125 ms contains 450 data bits. Each slot is separated by a synchronization burst and guard period to prevent interference that could result from multiple users operating over the same range of operation.

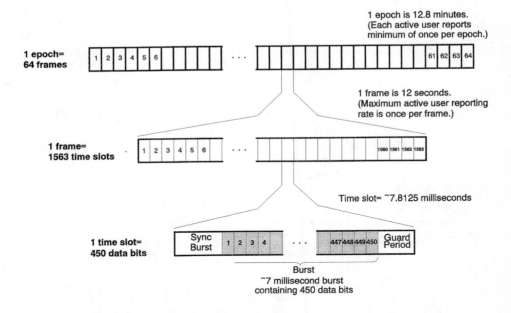

Figure 11.19   JTIDS TDMA Message Structure

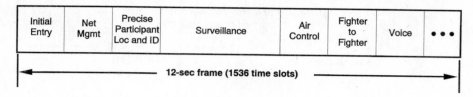

| Initial Entry | Net Mgmt | Precise Participant Loc and ID | Surveillance | Air Control | Fighter to Fighter | Voice | ••• |

◄——————— 12-sec frame (1536 time slots) ———————►

Figure 11.20   Basic JTIDS Message Structure

Each JTIDS message contains various JTIDS message sections called Network Participant Groups (NPGs), as shown in Figure 11.20. In the figure, several NPGs such as initial entry, net management, precise participant location, and identification, surveillance, air control, fighter-to-fighter, and voice are shown. There are nineteen NPGs currently implemented in TADIL-J protocol[7] [NC-1]. NPG definitions are listed in Table 11.6.

Three classes of JTIDS terminals are in use: (a) Class 1 for large aircraft such as AWACS, surface ships, and gateway facilities to ground-based networks, (b) Class 2 for fighter aircraft and small ships, and (c) Class 3 for mobile ground units and small UAVs.

11.4.8 Link 4A

The Link 4A is a half-duplex or full-duplex aircraft control link used by all carrier-based aircraft [WA-1; BE-1; SC-1]. The Link 4A was introduced to support an automatic landing system and later grew into a means to coordinate the E-2C Hawkeye early warning aircraft and F-14A Tomcat fighter aircraft by exchanging status and target data. The Link 4A is also used for aligning the Carrier Aircraft Inertial Navigation System (CAINS).

Technical characteristics of the Link 4A are summarized in Table 11.7. The frequency band allocated for Link 4A ranges from 225 MHz to 399.975 MHz; the RF channels are spaced at 25-kHz increments over the frequency range. The Link 4A uses a time division multiple access (TDMA) technique on a single frequency to connect various units and to exchange target information. The data rate for tactical aircraft control and target information is 5 kbps; the data rate for CAINS is 10 kbps.

A typical Link 4A terminal consists of a UHF radio, modem, cryptographic device, data processor, and user interface device [DD-1], as shown in the block diagram in Figure 11.21. There are two types of terminals in a Link 4A net: the control station terminal and the aircraft terminal. Link 4A terminals control station terminal and aircraft terminal operate in half-duplex mode. However, the control station terminal must be able to operate in full-duplex mode. The other half-duplex channel performs the on-line performance monitoring function.

The Link 4A system exchanges tactical information by using prescribed message

---

[7]Three terminologies JTIDs, Link 16, and TADIL-J, are used interchangeably in the Joint Tactical Information Distribution System. However, to be accurate, the terminologies mean slightly different items: the JTIDS is associated with the system and equipment; the term Link 16 is the tactical radio link designation originated by NATO; TADIL-J is the U.S. DoD Tactical Digital Information Link (J) protocol.

TABLE 11.6 DESCRIPTION OF JTIDS NETWORK PARTICIPANT GROUP (NPG) MESSAGES

Network Participant Group (NPG)	Description
Initial Entry (NPG 1)	• Initial net entry • Coarse synchronization
Round Trip Timing (RTT)-A (NPG 2)	• Active fine synchronization of all units in the network by Addressed Round Trip Timing interrogation
RTT-B (NPG 3)	• Active fine synchronization of all units in the network by Broadcast of Round Trip Timing interrogation
Net Management (NPG 4)	• Command and control • Management of overall network • Dissemination of network management messages to all participants
Precise Participant Location & ID (PPLI) Status A (NPG 5)	• Passive fine synchronization and relative navigation • Exchange of fighter ID, location, and status information • High update rate
PPLI Status B (NPG 6)	• Passive fine synchronization and relative navigation • Exchange of unit ID, location, and status information • Normal update rate
Surveillance (NPG 7)	• Exchange of tactical information (track and track management data) • Support for management of air, surface ship, submarine, land, EW, and intelligence surveillance information
Mission Management (NPG 8)	• Support for control, reporting, and transfer of assets
Air Control (NPG 9)	• Uplink of mission assignments, vectors, and target reports • Downlink of target and status report
Electronic Warfare (NPG 10)	• Support for dissemination of ESM/ECM/ECCM orders and data among EW-capable units
Voice Group A (NPG 12)	• Provision for 16 kbps secure voice, AJ communications in a JTIDS network
Voice Group B (NPG 13)	• Provision for 16 kbps secure voice, AJ communications in a JTIDS network
Indirect PPLI (NPG 14)	• Support of identification and position reporting of nonparticipating units
Weapons Coordination (NPG 18)	• Support of battle group weapons coordination
Fighter-to-Fighter (NPG 19)	• Exchange target information and status between fighters

TABLE 11.7   TECHNICAL CHARACTERISTICS OF LINK 4A

Communications Parameters	Technical Characteristics
• Frequency Band • Data Rate • Connectivity • Information Transfer • Transmission Mode	• from 225 MHz to 399.975 MHz • 5 kbps or 10 kbps • Between aircraft and aircraft carrier • Transfer aircraft control and target information • Carrier Aircraft Inertial Navigation System (CAINS) alignment • Half-duplex for information transfer, TDM on a single frequency • Full-duplex for on-line performance monitoring

formats, defined by TADIL-C. The control station terminal generates a TADIL-C control message during the transmit 14-millisecond frame interval consisting of a synchronization burst, guard interval, start pulse, data, and transmitter un-key signal, as shown in Figure 11.22. The aircraft terminal is designed to recognize the control message and to generate a TADIL-C reply message during the 18-millisecond receive interval, which consists of a synchronization burst, guard interval, start pulse, data, and transmitter unkey signal as shown in Figure 11.22.

11.4.9 HAVE QUICK II

HAVE QUICK II is a modification of several existing tactical UHF radios that provide ECCM capability in the 225 MHz to 400 MHz band. Modifiable tactical UHF radios are the AN/WSC-3, AN/ARC-164, AN/ARC-171, AN/GRC-171, and AN/ARC-182 radios [MG-1]. For shipboard applications, AN/WSC-3 UHF LOS radios are modified by adding an ECCM applique, as shown in Figure 11.23. Original radios such as the AN/WSC-3 and AN/ARC-182 are converted by inserting an ECCM modification kit (applique) between the audio amplifier and the user voice terminal.

The ECCM capability of HAVE QUICK II is accomplished by using slow frequency hopping. HAVE QUICK II radios can operate in ECCM mode as well as in a single-channel (non-ECCM) mode. The ECCM Unit of a HAVE QUICK II system contains a key generator that controls the frequency hopping pattern of the radio. Initialization of the HAVE QUICK II radios requires the same information as used for PLRS, that is, the Word-of-Day, net ID, and hopset.

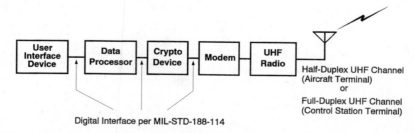

Figure 11.21   Typical Link 4A UHF Half-Duplex or Full-Duplex Terminal

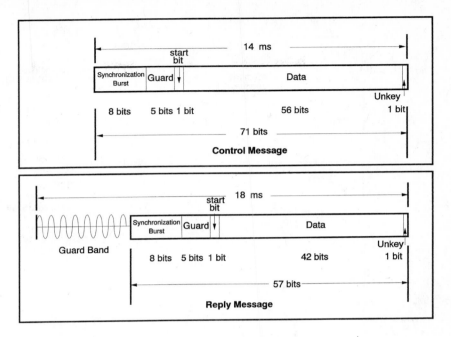

Figure 11.22   TADIL-C Message Formats

Frequency accuracy to 10^{-7} is required; this is accomplished by using a temperature-compensated crystal oscillator. The SG-1192/TRC Reference Signal Generator with rubidium oscillator maintains and disseminates HAVE QUICK time. A TRANSIT satellite receiver provides the time reference.

11.4.10 The Common High Bandwidth Data Link (CHBDL) AN/USQ-123

A Navy shipboard communications system using a microwave LOS link is called the Common High Band Data Link—Ship Terminal (CHBDL-ST) AN/USQ-123. The CHBDL-ST supports imagery data communications from airborne platforms such as reconnaissance aircraft, to ships. The CHBDL-ST system provides two full-duplex microwave digital data links between shipboard and airborne terminals [SN-1]. It uses X- and Ku-band frequencies (9.7–10.5 GHz and 14.4–15.5 GHz). Although the X- and Ku-band used for the CHBDL-ST system are beyond the VHF/UHF band, the CHBDL-ST is included in this section because it is a radio link system that uses LOS microwave radios.

The CHBDL is a tactical microwave digital data link between shipboard terminal user, and airborne platforms [BO-1; SN-1; PA-1; NA-1]. The earlier version of the CHBDL is known as the Modular Interoperable Data Link (MIDL), and its shipboard terminal is known as the Modular Interoperable Surface Terminal (MIST) [PA-2; WI-1]. The CHBDL is an automated communications node used aboard aircraft carriers for acquisition of sensor signals from airborne reconnaissance aircraft and other data link-equipped aircraft. The

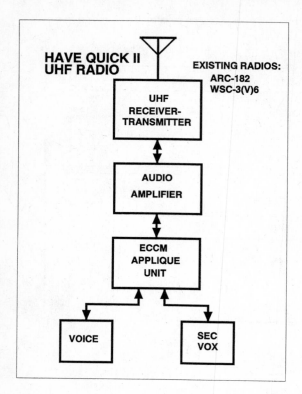

HAVE QUICK II UHF RADIO

EXISTING RADIOS:
ARC-182
WSC-3(V)6

UHF RECEIVER-TRANSMITTER

AUDIO AMPLIFIER

ECCM APPLIQUE UNIT

VOICE

SEC VOX

SINGLE CHANNEL VHF RADIO

AIR FORCE PROGRAM (USED ALSO BY ARMY, NAVY, MARINE CORPS, AND NATO)

CONVERSION BY USING AN APPLIQUE TO EXISTING RADIOS

FREQUENCY RANGE: 225 MHz–400 MHz

PROGRAMMABLE

FREQUENCY HOPPING

MAN-PACK, VEHICULAR, SHIP APPLICATIONS

OUTPUT POWER: 5W–50 W

Figure 11.23   Block Diagram of the Have Quick II UHF Radio

CHBDL is a full-duplex link; the data rates of the downlink from an airborne platform to a shipboard terminal are 10.71 Mbps, 137 Mbps, and 274 Mbps, and the data rate of the uplink from a shipboard terminal to an airborne platform is 200 kbps. Table 11.8 lists technical characteristics of the CHBDL AN/USQ-123(V) transceiver.

Figure 11.24 shows block diagrams of the shipboard CHBDL-ST and the airborne MIDL, which are connected by an X- or Ku-band radio link. The downlink transfers vari-

TABLE 11.8   TECHNICAL CHARACTERISTICS OF THE CHBDL AN/USQ-123 (V) TRANSCEIVER [PA-1]

- Frequency: X-band, Ku-band
- Data Rate
  Uplink: 200 kbps
  Downlink: 10.71 Mbps, 137 Mbps, and 274 Mbps
- Modulation
  Uplink: BPSK
  Downlink: Offset QPSK
- Antenna: Two 1-meter dish antennas

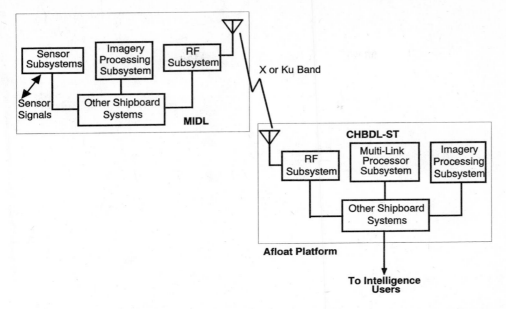

Figure 11.24   Block Diagram of the Common High Bandwidth Data Link (CHBDL) Shipboard Terminal (ST) and Airborne Terminal

ous data acquired in airborne sensor systems, such as electro-optics, infrared (IR), Synthetic Aperture Radar (SAR), and signal intelligence (SIGINT). The downlink data rate is high since the downlink carries bulk data. The uplink data rate is much lower because the uplink transmits control information to airborne terminals. The downlink and uplink of the CHBDL are carried on the Ku- or X-band, which can accommodate a bandwidth from 250 MHz to 350 MHz. The CHBDL-ST shipboard terminal consists of several subsystems: the antenna and RF subsystem, the Multi-Link Processor (MLP) subsystem, the imagery processing subsystem, and other shipboard systems, as shown in Figure 11.24.

11.4.11 The Light Airborne Multi-Purpose System (LAMPS) Data Link AN/SRQ-4

The Light Airborne Multi-Purpose System (LAMPS) Data Link is a tactical digital data link between ships and LAMPS helicopters (SH-60B Seahawk) [SN-3; NA-1; LA-1; BL-1]. The LAMPS downlink radio transfers radar and sonar data acquired in airborne sensors systems to the LAMPS parent ship.

Figure 11.25 shows a block diagram of the LAMPS data link radio terminal set AN/SRQ-4. The AN/SRQ-4 consists of a directional antenna with an AS-3274 radome, an AS-3275 omni-directional antenna, a C-10425 antenna controller/monitor, an OR-209 receiver-transmitter, and a KG-45 cryptographic device. The user baseband terminals which are connected to the AN/SRQ-4 are several secure voice terminals, a radar data processor, and an AN/SQQ-28 sonar signal processor.

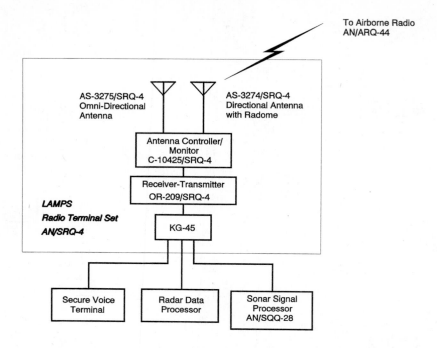

To Airborne Radio
AN/ARQ-44

AS-3275/SRQ-4
Omni-Directional
Antenna

AS-3274/SRQ-4
Directional Antenna
with Radome

Antenna Controller/
Monitor
C-10425/SRQ-4

Receiver-Transmitter
OR-209/SRQ-4

LAMPS

Radio Terminal Set

AN/SRQ-4

KG-45

Secure Voice
Terminal

Radar Data
Processor

Sonar Signal
Processor
AN/SQQ-28

Figure 11.25   Block Diagram of LAMPS Shipboard Radio Terminal Set AN/SRQ-4

Figure 11.26 shows the complete LAMPS Data Link System AN/SRQ-4, including the AS-3274 directional antenna with radome, the OR-209 receiver-transmitter unit, the C-10425 antenna controller/ monitor, and the AS-3275 omni-directional antenna.

The downlink and uplink of the LAMPS data link is carried on the 4 GHz to 6 GHz

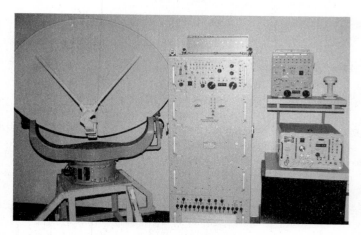

Figure 11.26   LAMPS Data Link System AN/SRQ-4 (Courtesy of Lucas Aerospace Communications and Electronics, Inc.)

Synchronization Word	Computer Data	Range Data	Voice Data	Voice Data

(a) Uplink Data Format

Synchronization Word	Radar Data	Computer Data	Radar Data	Range Data	Sonar Data	Voice Data	Radar Data	Voice Data	Radar Data

(b) Downlink Data Format

Figure 11.27   Ship-to-Aircraft Link and Aircraft-to-Ship Link Data Formats for the LAMPS Data Link

G-band. The LAMPS data link is a full-duplex link, with FSK modulation at the rate of 25 Mbps. Figure 11.27 shows the uplink and downlink data formats for the LAMPS data link.

11.5 VHF/UHF ANTENNAS AND COUPLERS

Shipboard antennas are discussed in Chapter 16. In this section, supplemental information is provided that is specific to VHF and UHF band antennas but is not included in Chapter 16. VHF/UHF LOS radio antennas are typically dipole, whip, discage, log-periodic, or stubs-on-ground-planes. Some examples are the AS-1729 VHF antennas used with AN/VRC-40 radios. UHF antennas are also dipoles or coaxial stubs. A typical shipboard UHF antenna frequently used is the AS-390 (at the end of the yardarm in Fig. 11.28). Chapter 16 includes a list of shipboard communications antennas.

VHF and UHF antennas are small; the dimensions of radio antennas are 1.5 m to 3 m for the VHF band and 60 cm to 1.5 m for the UHF band. These antennas can be placed on a ship's topside, on mast platforms and yardarms, as illustrated in Figure 11.28. In the figure, four dipole antennas (AS-1735/SRC) are mounted around the mast. These antennas radiate RF signals at 225 MHz to 400 MHz; the power rating is 1 kW. Also shown are several AS-390 antennas.

11.5.1 VHF/UHF Multicouplers

Often the number of usable shipboard communications circuits and links may be limited by the number of available antennas, because of topside space limitation in shipboard environments. Hence, one chooses antennas designed to serve several transmitters and receivers simultaneously. In this case, impedance of transmitters and receivers must be matched to impedance of the single antenna. Devices that match the impedances for a single, multifunction antenna are antenna multicouplers.[8]

VHF/UHF multicouplers can be used for transmitters and receivers. The close chan-

---

[8]A general discussion of antenna multicouplers can be found in section 18.1.2.

Figure 11.28 Typical Placement of Shipboard VHF/UHF Antenna Systems AS-1735/SRC (Courtesy of the U.S. Navy, Naval Sea Systems Command)

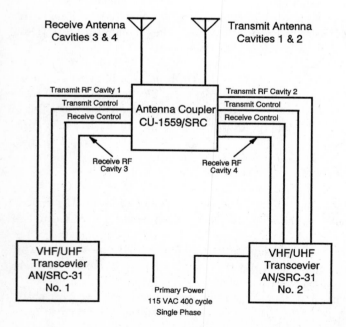

Receive Antenna
Cavities 3 & 4

Transmit Antenna
Cavities 1 & 2

Transmit RF Cavity 1
Transmit Control
Receive Control

Antenna Coupler
CU-1559/SRC

Transmit RF Cavity 2
Transmit Control
Receive Control

Receive RF
Cavity 3

Receive RF
Cavity 4

VHF/UHF
Transcevier
AN/SRC-31
No. 1

VHF/UHF
Transcevier
AN/SRC-31
No. 2

Primary Power
115 VAC 400 cycle
Single Phase

Figure 11.29 Interconnecting Diagram of the VHF/UHF Antenna Coupler CU-1559/SRC

nel spacing desired during operation requires multiple-resonator filters to achieve the necessary isolation between channels. VHF/UHF multicouplers are designed for use with antennas that have a VSWR[9] of 2:1 or less throughout the operating frequency range of a multicoupler. The functional requirements of shipboard antenna multicouplers are:

- Impedance matching between the antenna and the transmitter or between the antenna and the receiver
- Efficient coupling between the antenna and the transmitter or between a common antenna and the transmitter
- Isolation of electromagnetic coupling between transmitters
- Rejection of collocated transmitter signals
- Reduction of spurious radiation from receivers
- Filtering of harmonic and spurious transmitter output
- Protection of receiver input circuits from high RF voltages

11.5.2 The VHF/UHF Antenna Coupler CU-1559

The CU-1559/U antenna coupler is a two-port impedance matching device and isolation network that allows duplex operation of two VHF/UHF transceivers. Figure 11.29 shows a block diagram of the CU-1559/U antenna coupler. The CU-1559 may be used to couple two UHF transmitters to one antenna and two VHF/UHF receivers to a second antenna. It can also couple either four transmitters or four receivers to two antennas. The CU-1559/U is normally used with two VHF/UHF transceivers AN/SRC-31 and two antennas.

REFERENCES

[BE–1] Bell, C. R. and R. E. Conley, "Navy Communications Overview," *IEEE Transaction on Communications*, September 1980, p. 1573.

[BL–1] Blake, B., ed., "LAMPS MK III," *Jane's Underwater Warfare Systems 1989–1990*, Jane's Information Group, Inc., Alexandria, Va., 1989, p. 124.

[BO–1] Bowen, D. G. and B. V. Cox, "Tactical Communications to Support Intelligence," *Signal*, June 1988.

[DD–1] *Subsystem Design and Engineering Standard for Tactical Digital Information Link (TADIL) C*, MIL-STD-188-203-3, Department of Defense, Washington, D. C., October 5, 1983.

[DD–2] *Engineering Standards for Meteor Burst Communications*, MIL-STD-188-135, Department of Defense, Washington, D. C.

[DD–3] *Aircraft Internal Time Division Command/Response Multiplex Data Bus*, MIL-STD-1553B, Department of Defense, Washington, D. C., September 8, 1986.

[9]The voltage standing wave ratio (VSWR) is the ratio of maximum voltage to the minimum voltage along the antenna. The VSWR is a measure of the impedance matching between antennas and couplers. The greater the mismatch, the larger the VSWR.

[GR–1] Graf, K. L., "SINCGARS: The New Generation Combat Net Radio System," *Signal*, August 1987, pp. 43–45.

[HA–1] *Hughes Aircraft Company Brochure on Position Location Reporting System (PLRS)*, Hughes Aircraft Company, Fullerton, Calif., 1986.

[HO–1] Howeth, Captain L. S. (Under the Auspices of Bureau of Ships and Office of Naval History), *History of Communications—Electronics in the United States Navy*, U. S. Government Printing Office, Washington, D. C., 1963.

[HS–1] *ITT Reference Data for Radio Engineers*, Sixth Edition, Howard W. Sams & Co., Indianapolis, Ind., pp. 28.16–28.17.

[IS–1] Isaacs, J., W. Robertson, and R. Morrison, "A Cosite Analysis Method for Frequency Hopping Radio System," *1991 IEEE Military Communications Conference*, November 4–7, McLean, Va., 1991.

[IS–2] Isaacs, J., T. McNair, and R. Morrison, "A Program for Analyzing Cosite Interference Between Frequency Hopping Radios," *Proceedings of IEEE Tactical Communications Conference*, Ft. Wayne, Ind., April 28–30, 1992, pp. 119–124.

[JE–1] Jessop, G. R., ed., *VHF/UHF Manual*, Radio Society of Great Britain, Potters Bar, Hartfordshire, EN6 3JE, U.K., 1983.

[KI–1] Kiely, D. G., *Naval Electronic Warfare*, Brassey's Sea Power Series, Vol. V, Pergammon-Brassey's International Defense Publishers, McLean, Va. 22102, Chapter 4, p. 51.

[LA–1] Law, Preston, "AN/SRQ-4 (LAMPS MK III)," *Shipboard Antennas*, Artech House, Inc., Dedham, Mass. 02026, 1983, pp. 502–504.

[MG–1] *Magnavox Brochure on HAVE QUICK/HAVE QUICK II-A Brief Description*," Magnavox, Ft. Wayne, Ind. 46808, March 1986.

[NA–1] *Tactical Data Link Assessment*, Copernicus Project Office, Department of the Navy, Washington, D. C., June 15, 1992.

[NC–1] *Link 16 Communications Planning-Quick Reference Guide*, NRADWARM-92-TRG-001, NCCOSC RDT&E Division, Warminster, Pa., August 1, 1992.

[NC–2] *Communication Support System (CSS) Overview*, NOSC Technical Report, CSS-10001-00, NOSC (now NCCOSC RDT&E Div.), San Diego, Calif., 92152-5185, July 26, 1990.

[NC–3] Naval Computer and Telecommunication Command, *Naval Telecommunication Procedure–Spectrum Management Manual,* NTP6(D), Washington, D. C. 20394–5000, March 1992

[NT–1] National Telecommunications and Information Administration, *Preliminary Spectrum Reallocation Report,* Washington, D. C. 20230, February 1994.

[NT–2] National Telecommunications and Information Administration, *Manual of Regulations and Procedures for Federal Radio Frequency Management,* Washington, D. C. 20230, September 1993.

[PA–1] *Paramax Data Sheet on Common High Bandwidth Data Link-Ship Terminal AN/USQ-123(V)* , Paramax Systems Corporation, Salt Lake City, Utah 84116, 1993.

[PA–2] *Paramax Data Sheet on Modular Interoperable Surface Terminal (MIST)*, Paramax Systems Corporation, Salt Lake City, Utah 84116, April 1989.

[RE–1] Reagan, J., B. Gaspard, and G. Massa, "Technique for Effective Evaluation of Collocation Interference Designs in Government Solicitation," *1991 IEEE Military Communications Conference*, November 4–7, 1991, McLean, Va.

[RI–1] Ricci, F. and D. Schutzer, *U.S. Military Communications—C3I Force Multiplier*, Computer Science Press, Rockville, Md. 20850, 1986.

[SC–1] Schoppe, W. J., "The Navy's Use of Digital Radios," *IEEE Transaction on Communications*, December 1979, pp. 1935–1945.

[SN–1] *Common High Bandwidth Data Link Shipboard Terminal (CHBDL-ST)*, System Specification, SPAWAR-L-841, Space and Naval Warfare Systems Command, Washington, D. C., July 6, 1990.

[SN–2] *Use of Navy Shipboard SINCGARS System As a CSS Resource*, Space and Naval Warfare Systems Command, Washington, D. C., May 18, 1990.

[SN–3] *AN/SRQ-4 Radio Terminal Set*, Technical Manual, EE185-AA-OMI-0120, NAVELEX (now Space and Naval Warfare Systems Command), Washington, D. C., September 1, 1984.

[TO–1] Toone, J. W. and S. Titmas, "Introduction to JTIDS," *Signal*, August 1987, pp. 55–59.

[WA–1] Walsh, E. J., "Navy's Tactical Data Link Advances Fleet Technology," *Signal*, February 1990, pp. 61–65.

[WI–1] Williamson, J., ed., "Modular Interoperable Surface Terminal (MIST)," *Jane's Military Communications 1991–1992*, Jane's Information Group, Inc., Alexandria, Va., 1991, p. 366.

[WI–2] Williamson, J., ed., "AN/URC-93(V) UHF Transceiver," *Jane's Military Communications 1989*, Jane's Information Group, Inc., Alexandria, Va., 1991, p. 267.

[WI–3] Williamson, J., ed., "AN/VRC-12 and AN/VRC-43 to -49 Series Radio Sets," *Jane's Military Communications 1985*, Jane's Information Group, Inc., Alexandria, Va., 1991, p. 119.

[WI–4] Williamson, J., ed., "AN/WSC-3 UHF SATCOM/LOS Transceiver," *Jane's Military Communications 1989*, Jane's Information Group, Inc., Alexandria, Va., 1991, p. 356.

[WI–5] Williamson, J., ed., "AN/GRT-21 and AN/GRT-22 VHF/UHF Transmitter," *Jane's Military Communications 1989*, Jane's Information Group, Inc., Alexandria, Va. 22314-1651, 1991, pp. 172–173.

[WI–6] Williamson, J., ed., "AN/GRR-23 and AN/GRR-24 VHF/UHF Receiver (Models 3101 and 3102)," *Jane's Military Communications 1989*, Jane's Information Group, Inc., Alexandria, Va. 22314-1651, 1991, p. 172.

[WI–7] Williamson, J., ed., "Single Channel Ground and Airborne Radio System VHF (Sincgars-V)," *Jane's Military Communications 1989*, Jane's Information Group, Inc., Alexandria, Va. 22314-1651, 1991, pp. 134–135.

[WI–8] Williamson, J., ed., "AN/ARC-182 VHF/UHF AM/FM Transceiver," *Jane's Military Communications 1989*, Jane's Information Group, Inc., Alexandria, Va. 22314-1651, 1991, p. 311.

[WI–9] Williamson, J., ed., "Joint Tactical Information Distribution System (JTIDS) Class 2 Terminal," *Jane's Military Communications 1989*, Jane's Information Group, Inc., Alexandria, Va. 22314-1651, 1991, p. 474.

12

Satellite Communications (SATCOM)—Fundamentals

Naval shipboard SATCOM provides circuits via the three military SATCOM systems: the UHF Fleet Satellite (FLTSAT), the SHF Defense Satellite Communications System (DSCS), and the EHF Military Strategic and Tactical Relay Satellite (MILSTAR). Each of these systems will be described in separate chapters to follow. The technical fundamentals which are helpful for the understanding of SATCOM in general are given below.

Communications satellites are orbiting relays which receive, amplify, process, and retransmit signals from one point to another point on the surface of the Earth. Communications satellites have been in use since the 1960s for providing wideband global connectivity and long-range circuits of high quality virtually unencumbered by propagation difficulties. A satellite provides coverage within its footprint, that is, the area from which it is visible. Global coverage can be achieved by multiple hopping, that is, using multiple ground-to-ground SATCOM circuits to extend the length of a circuit. Relaying between satellites is another approach.

12.1 SATELLITE ORBITS

Satellites move like independent celestial bodies following the laws of orbital mechanics. As a first-order approximation, a satellite's path is a two-body orbit, with the Earth being the primary mass which is very much larger than the negligible satellite mass; the latter moves in an ellipse around Earth. Key orbital parameters are the eccentricity ε, the inclination i of the orbital plane with respect to the equatorial plane, the angle ψ of the line intersecting the orbital and equatorial planes (ψ counted from the longitude $\phi = 0$ to the epoch of satellite passage into the northward direction), the epoch of the satellite through the perigee (closest approach to Earth), and the orbital period T ($T=2\pi\sqrt{a^3}/\sqrt{\mu}$), where $\mu = gM_e$ (g = gravitational constant, M_e = mass of the Earth, $\mu = 3.98 \cdot 10^{14}[\mathrm{m^3 s^{-2}}]$). The projection of the satellite position on the surface of the Earth (intersection of the line through the satellite and the Earth's center with the surface) is called the sub-satellite point. The farthest point from the Earth's surface is called the apogee. When the semi-major axis, a, and the semi-minor axis, b, of the orbital ellipse are equal and hence the eccentricity ε is zero ($\varepsilon=\sqrt{1-(b/a)^2}$), the orbit is circular.

12.1.1 Geosynchronous Orbits

The geosynchronous orbit is a circular orbit with a 24-hour period. The satellite altitude is 35,600 km (22,300 statute mi or 19,200 nmi). For an inclination $i = 0°$, the satellite appears to be stationary above the equator, and the satellite is said to be in a geostationary orbit. If the inclination $i > 0$, the sub-satellite point describes a figure eight.

Because the Earth is not an ideal sphere with a uniform mass distribution, there are small gravitational forces acting on a geostationary satellite that tend to change its attitude and position in orbit. Position and attitude must be maintained by station-keeping propulsion devices onboard the satellite. The quantity of onboard propellant is limited, hence position can be kept only for a limited number of years.

The footprint of a geostationary satellite is a circular area with a subtended angle of 81°. At the circumference the elevation angle of the satellite is 0°; the satellite is at the horizon. Generally 5° elevation above the horizon is required for satisfactory signal reception. Thus, the coverage is generally considered to be between the 76°N and 76°S latitudes.

A special type of geosynchronous orbit is an inclined elliptical orbit.[1] A 24-hour period orbit with the apogee in the northern latitudes will cover all of the Northern Hemisphere including the polar region and also a significant part of the Southern Hemisphere.

12.1.2 Polar Orbits

An orbit with the inclination $i = 90°$ is a polar orbit. The satellite will pass over the poles and provide coverage for all northern and southern latitudes. For non-24-hour periods, the orbit will precess, that is, the orbital plane will rotate around the Earth's axis.

---

[1]When the inclination is 63 degrees the orbit is stable and does not precess. The Russian communications satellite Molnya uses this type of orbit, which is consequently referred to as "Molnya-orbit."

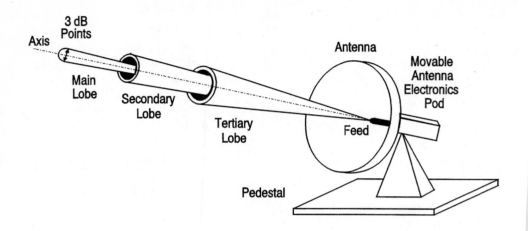

Figure 12.1   High Gain Antenna Pattern

12.2 TRACKING

Generally, SATCOM antennas provide significant gain. The antenna's main lobe, that is, the direction of highest gain, can be viewed as a cone-shaped region around the antenna's axis as shown in Figure 12.1. The figure also shows the secondary and tertiary lobes, regions of progressively less gain, separated from each other by regions of very low gain. The main lobe must be kept directed at the satellite. In order to accomplish that, the antenna must acquire the satellite and then track it.

Only antennas of permanent ground stations aimed at a geostationary satellite can be kept in a fixed orientation, since station-keeping will maintain a geostationary satellite's position to within a few minutes of arc. Satellites in inclined orbits must be constantly tracked. Similarly, shipboard antennas, especially high gain antennas for the SHF DSCS (operating in the X-band) and the EHF MILSTAR (operating in the SHF and EHF bands), will have antenna pedestals fixed to the ship's structure; consequently, the antenna will move with ship's pitch, roll, and yaw. Considering the rolling motion of some ships, a shipboard SATCOM antenna must have the capability to track rapidly ($10° \text{ s}^{-1}$).

In order to accomplish tracking, the received signal strength is rapidly sampled and the antenna orientation is continually adjusted to maximize the signal strength. Either the communications signal or a tracking signal sent by a beacon from the satellite is used. Use of the beacon signal has the advantage of providing a steady carrier signal, while the communications signal may not have a carrier at all and then tracking would require complex electronics.

12.3 PROPAGATION AND NOISE

The signal arriving at a SATCOM receiver is subject to propagation losses; in the receiver, the usable signal is competing with noise.

12.3.1 Propagation

Signals with the frequencies used for SATCOM (UHF, SHF, EHF) propagate along straight-line paths and are virtually unaffected by the ionosphere which bends, reflects, and absorbs HF and lower frequency signals. However, the atmosphere (air and other gases, water vapor, and clouds) will attenuate and absorb radio waves. Figure 12.2 shows the vertical atmospheric absorption and the absorption coefficient. The peak at 22.5 GHz results from water vapor, the 60 GHz line is due to oxygen, the other lines arise from

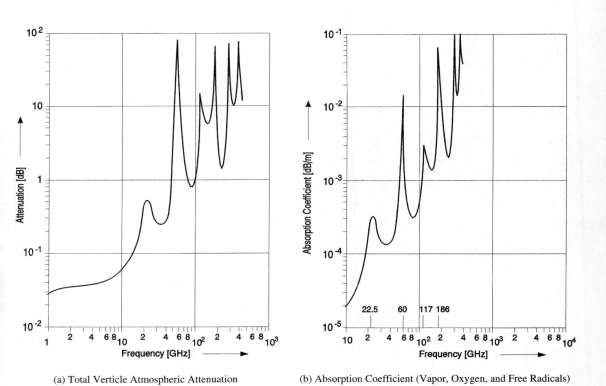

(a) Total Verticle Atmospheric Attenuation   (b) Absorption Coefficient (Vapor, Oxygen, and Free Radicals)

Figure 12.2   Atmospheric Attenuation (Courtesy, Westinghouse Electric Corp.)

other atmospheric gases and from free radicals in the upper atmosphere. Raindrops will also act as scatterers and absorbers of radio waves. A strong cloudburst can significantly reduce SHF signal strengths, and rain may be virtually impenetrable to EHF signals.

12.3.2 Noise

The overall noise is most conveniently expressed as an equivalent noise temperature and is an important parameter in determining a SATCOM link's capacity for signal transmission. The noise temperature is defined as the effective temperature of a thermal noise source at the system input that would produce the same noise at the system output, the system being considered noiseless [SC-1].

The noise power in [W] can then be expressed as:

$$N = GkT_{op}B,$$

where k is Boltzmann's constant ($1.38 \cdot 10^{-23}$ [WK^{-1} Hz^{-1}]), T_{op} the effective noise operating temperature in [K], B the system bandwidth in [Hz], and G the system gain. The noise at the system input is $kT_{op}B$.

The noise at the system input is written in logarithmic form as:

$$10\log N = 10\log k + 10\log T_{op} + 10\log B,$$

or, expressed in [dB]:

$$N[\text{dBW}] = k[\text{dBWK}^{-1}\text{Hz}^{-1}] + T_{op}[\text{dBWK}^{-1}] + B[\text{dBHz}].$$

The notion of the noise temperature facilitates the reflection of the noise to a reference port in a chain or cascade of subsystems. Considering a cascade of two subsystems 1 and 2, reflection of the noise from subsystem 2 to the input of subsystem 1 will add a noise temperature contribution T_{21} from subsystem 2 to the noise temperature T_1 of subsystem 1. If subsystem 1 has a gain G_1, then the noise temperature T_2 (at the input of subsystem 2) will appear as $T_{21} = T_2/G_1$. If subsystem 1 has a loss L_1, the noise temperature of subsystem 2 reflects as $T_{21} = T_2L_1$. If subsystem 1 is distributed and has a loss, for example, an extended feed line, then $T_{21} = T_2(L_1 - 1)/L_1$.[2]

Noise results from several sources: the receiver preamplifier, losses of the transmission line and coupler connecting the preamplifier and the antenna feed, and the background at which the antenna is pointed (sky, atmosphere, stars, sun).

The sky (background) temperature T_{sk} is generally small (<4 K for frequencies above 2 GHz). The atmosphere-induced temperature T_a (from pointing the antenna at the atmosphere) is a function of frequency and antenna elevation. A first-order approximation for an atmosphere of constant density would be given by the expression $T_{at}(L-1)/L$.[3] A

---

[2]This approach is used in calculating the equivalent noise temperature of a receiver as illustrated in Figure 12.5.

[3]L is the loss of the signal in the path through the atmopshere.

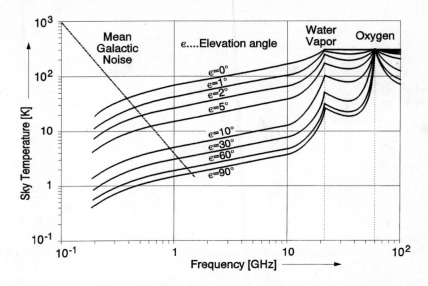

Figure 12.3 Atmosphere Noise Temperature Due to Water Vapor and Oxygen (After D. L. Rice et al., 1967.) (With Galactic Noise Superimposed)

better approach is an evaluation applying a model of the atmosphere with an exponentially decreasing density; a good approximation using this method was developed by Hogg [HO-1] and refined by Rice [RI-1]; it is shown in Figure 12.3.

The noise of the preamplifier is a function of the device used and the frequency. Figure 12.4 shows the noise temperatures of some typical devices. Parametric amplifiers and GaAs semiconductor devices operating at room temperature (or cooled by liquid nitrogen, should cooling result in a significantly better performance) are devices suitable for shipboard applications. They generally contribute significantly to the operating temperature.

The overall operating noise temperature T_{op}, expressed at a reference port, can be

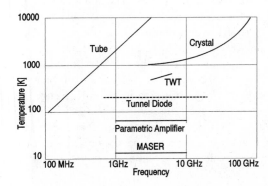

Figure 12.4 Noise Temperature of Front-End Amplifier Devices (Courtesy of Chilton Publications Inc.)

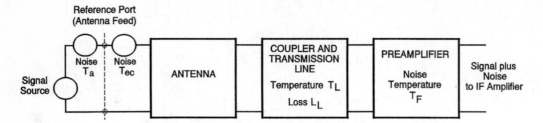

Figure 12.5   Equivalent Circuit for Determining the (Operating) Noise Temperature of a SATCOM Receiver

derived using an equivalent circuit as illustrated in Figure 12.5. In this figure the antenna is shown as receiving the signal plus the antenna noise expressed by the equivalent temperature T_a. At the reference port (the antenna feed) an additional noise source is shown with the equivalent temperature T_{ec}, which comprises the noise contributions from the line connecting the feed to the front-end amplifier and the front-end amplifier itself. Then T_{op} can be expressed as $T_{op} = T_a + T_{ec}$.

The antenna equivalent noise temperature T_a consists of the contribution from the sky (background, T_{sk}/L), the antenna side lobes and back lobes (T_{SL}), and the atmosphere ($T_{at}(L-1)/L$, shown in Figure 12.3). The noise represented by T_{ec} consists of contributions from the front-end preamplifier ($T_F L_L$) and the line losses ($T_L (L_L - 1)/L_L$) reflected to the reference port. Thus, T_{op} can be expressed as:

$$T_{op} = T_{sk}/L + T_{SL} + T_{at}(L-1)/L + T_L(L_L - 1)/L_L + T_F L_L.$$

It should be observed for the evaluation of the above equation that the losses (L, L_L) are numerical factors (not in [dB]), and the temperatures are absolute values in [K].

To illustrate the relative contributions to T_{op} in the above relationship, the noise temperature of a typical shipboard SHF terminal (AN/WSC-6) is calculated. From Figure 12.2 the atmospheric attenuation is about 0.05 dB, that is, $L = 1.0116$. The sky temperature T_{sk} is assumed to be 1 K, thus the background noise $T_{sk}/L \approx 1$ K. The contribution of the atmosphere (taken from Figure 12.3 for $\phi = 30°$) is ≈ 6 K. The antenna back lobe temperature is estimated by approximating the antenna back lobe by a hemispheric pattern (with a space angle of 0.5 sterad) as -15 dB below isotropic gain (corresponding to a factor of 0.03) and pointing at the ground which is at 290 K, hence the temperature of the back lobe is $0.03 \cdot 0.5 \cdot 290 \approx 4.4$ K. The side lobe contribution is estimated by assuming that the (first) side lobe is -10 dB below the main lobe; with the main lobe contributing a noise temperature of 6 K, the side lobe's contribution is ≈ 0.6 K. Hence, T_{SL}, which consists of the back lobe and side lobe temperatures, will be ≈ 5.2 K. The feed line is assumed to have losses of 3 dB ($L_L = 2$) and is at a temperature of 290 K, thus $T_L(L_L - 1)/L_L = 145$ K. The front-end amplifier is assumed to have a noise temperature of 85 K, hence $T_F L_L \approx 170$ K.[4] The total equivalent noise temperature T_{op} is the sum

---

[4]The noise contributions of the amplifier stages following the front-end amplifier, when reflected back to the antenna feed, are reduced by the gain of the front-end amplifier and are negligible.

of all noise contributions (sky, atmosphere, side lobes, line losses and front-end amplifier), which for the presented example becomes 1 K + 6 K + 5.2 K + 145 K + 170 K = 327.2 K ($\approx$ 25.2 dBK). As it can be seen, the line losses and the front-end amplifier contribute the main shares of the noise. Taking an antenna gain (4-ft reflector at 8 GHz) with 37.2 dB, one arrives at $G/T \approx 12$ dBK^{-1}, a result which is characteristic of the AN/WSC-6 in some shipboard applications.

12.4 ANTENNA PARAMETERS

One of the frequently used SATCOM antenna types is the reflector antenna. Generally parabolic reflectors are used (shaped reflectors are also used for special applications) with the feed (the radiating or receiving element) in the focus. The reflector concentrates the radio wave energy into a conical section, the main beam (see Figure 12.1). The half power beamwidth δ (angle where the gain is 3 dB down from the maximum) in degrees is approximately (depending on a good design):

$$\delta \approx 73(\lambda/d),$$

where λ is the operating wavelength and d is the reflector diameter [JA-1]. The antenna gain G is:

$$G = \eta \, (\pi d/\lambda)^2$$

where η is the antenna efficiency. The efficiency depends on the accuracy of the antenna surface. Surface irregularities must be less than $\lambda/16$ (1/16 of the wavelength of the highest radio carrier frequency for which the antenna is designed) in order to avoid noticeable defocusing. Additionally, it is required that $\lambda \ll d$ in order to avoid edge diffractions which would recognizably distort the pattern. Finally, the illumination from the feed must be tapered toward the reflector edge so as to be limited to the antenna surface and thus avoid sizable back lobes that would collect additional unwanted noise in a receiver and would waste energy in a transmitter. With good antenna designs, efficiencies in the range of 50% to 60% can be accomplished.

Figure 12.6 is a useful diagram for determining the gain for parabolic antennas at different diameters and frequencies. Given the diameter and the wavelength (or the operating frequency), one can determine the half-power beamwidth. Then, with a given efficiency one can determine the on-axis gain from the superposed nomogram.

The method to use of the diagram is illustrated using an example of a 4-ft parabolic antenna operating at 8 GHz. First a line is drawn parallel to the 3 GHz line through the 8 GHz point on the frequency scale. This line is intersected with a vertical through the antenna diameter. A horizontal line is drawn through this intersection and traced to the half power beamwidth scale (also labeled in terms of the antenna directivity). Then a straight line is drawn through the half power beamwidth point and a chosen efficiency (50%) and intersected with the gain scale. The result is 37.2 dB, a value typical for the AN/WSC-6 antenna.

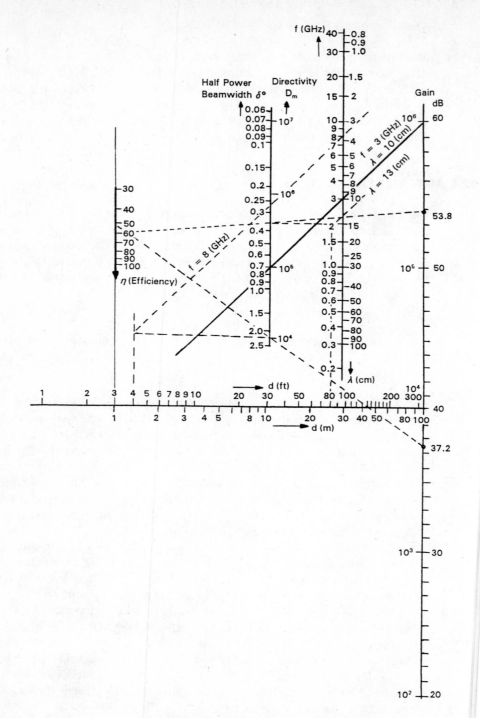

Figure 12.6 Diagram for Parabolic Antennas

238

12.5 THE RANGE EQUATION AND LINK BUDGETS

The range equation is the basis of all link power budgets and performance calculations. It is a relationship among all pertinent link parameters. The first step is to express the received power of the communications link in terms of its range and antenna gains. The received power is then related to the link noise and the required performance.

Owing to the straight-line omnidirectional propagation of electromagnetic signals at the distance R, the transmitted power P_T of an omnidirectional point source is equally distributed over the surface $4\pi R^2$ of the sphere of radius R. The received power P_R of an antenna of effective area A_{eR} is given by:

$$P_R = P_T (A_{eR}/4\pi R^2).$$

If the point source is replaced by an antenna with gain G_T into the direction of the receiving antenna, then:

$$P_R = P_T G_T (A_{eR}/4\pi R^2).$$

Replacing the effective antenna area with the gain ($A_{eR} = \lambda^2 G_R/4\pi$, see [JA-1]) produces:

$$P_R = P_T G_T G_R (\lambda/4\pi R)^2.$$

Expressing this in logarithmic form, one obtains:

$$10\log P_R = 10\log P_T + 10\log G_T + 10\log G_R - 10\log(4\pi R/\lambda)^2.$$

The last term $L_S = 10\log(4\pi R/\lambda)^2$ is called the space loss. The relationship is then customarily expressed in the form:

$$P_R [\text{dBW}] = P_T [\text{dBW}] + G_T [\text{dB}] + G_R [\text{dB}] - L_S [\text{dB}].$$

Given the expression for the received signal power and the expression for the total link noise power ($N[\text{dBW}] = k[\text{dBWK}^{-1}\text{Hz}^{-1}] + T_{op} [\text{dBK}^{-1}] + B[\text{dBHz}]$), the link power budget can be written in the logarithmic form if one takes into account that the required signal-to-noise power ratio $(S/N)_i$ realizable by the link equals the received power less the operating system noise (see section 12.3.2) less a design margin M. Thus:

$$(S/N)_i [\text{dB}] = P_T [\text{dBW}] + G_T [\text{dB}] + G_R [\text{dB}] - L_S [\text{dB}] - k[\text{dBWK}^{-1}\text{Hz}^{-1}]$$
$$- T_{op} [\text{dBK}^{-1}] - B[\text{dBHz}] - M[\text{dB}].$$

A simple rearrangement produces:

$$(S/N)_i [\text{dB}] + B[\text{dBHz}] = (P_T G_T) [\text{dBW}] + (G_R/T_{op}) [\text{dBK}^{-1}] - L_S [\text{dB}]$$
$$- k[\text{dBWK}^{-1}\text{Hz}^{-1}] - M[\text{dB}].$$

The SNR consists of the signal carrier power C divided by N, the noise power, which is the product of noise power density N_0 and the signal bandwidth B ($N = N_0 B$). Thus, the left-hand side of the above relationship, that is, $(S/N) [\text{dB}] + B[\text{dBHz}]$, is the ratio of the carrier power to noise power density ratio in logarithmic form, that is, C/N_0 [dBHz]. The expression $(P_T G_T)$ is called the transmitted effective isotropic radiated power (EIRP), and the expression (G_R/T_{op}) is referred to as the receiver figure of merit. The above expression develops into:

$$(C/N_0) \text{ [dBHz]} = (P_T G_T) \text{ [dBW]} + (G_R/T_{op}) \text{ [dBK}^{-1}] - L_S \text{ [dB]}$$
$$- k\text{[dBWK}^{-1}\text{Hz}^{-1}] - M\text{[dB]}.$$

For digital transmission systems, the detector performance is important in assessing the overall link performance. The detector performance curves are generally plotted with respect to the E_B/N_0, the symbol energy to noise power density ratio (see also Chapter 7). The E_B/N_0 is related to the C/N_0 by the following expression:

$$E_B/N_0 \text{ [dB]} = C/N_0 \text{ [dBHz]} - B_{TR} \text{ [dBHz]} + B_{TR}\tau_{sy} \text{ [dB]}.$$

The first two terms on the right-hand side above (C/N_0 [dBHz] $- B_{TR}$ [dBHz]) are the S/N[dB]; the last term, $B_{TR}\tau_{sy}$ [dB] is the so-called WT product of the signal (see also Chapters 7 and 8), that is, the product of the transmission signal bandwidth B_{TR} and the transmission symbol duration τ_{sy}. The WT product transforms the S/N into the E_B/N_0. One can use the modulation system performance curves (cf. Fig. 7.3) to test whether the achieved E_B/N_0 will be sufficient to meet a required bit error rate. The modulation system performance curves include the processing gain offered by the modulation method and, if incorporated, the coding method used in the communications link.

From the above discussion it follows that the C/N_0 of a link can be used as a basis for calculating its performance. The noise power from different sources can be added. Strictly speaking, this can only be done when the noise from the different sources is not coherent; however, the noise sources in a SATCOM link contribute gaussian noise, and for all practical purposes the noise from different sources is not coherent. Thus, the C/N_0 from two link segments, one ground to satellite (uplink), the other satellite to ground (downlink), can be combined as shown below. (Note that the relationship below does *not* hold for *logarithmic* values but *only* for *numerical* values of C/N_0).

$$1/(C/N_0)_{total} = 1/(C/N_0)_{uplink} + 1/(C/N_0)_{downlink}.$$

An instructive example shows how the performance of a typical DSCS satellite shore-to-ship link and a ship-to-shore link using a terminal like the AN/WSC-6 would be computed. The numerical data is summarized in Tables 12.1 and 12.2.

Tables 12.1 and 12.2 show that the shore-to-ship link requires substantial satellite power. This is due to the fact that the small shipboard antenna and the long waveguide from the antenna to the low-noise front-end amplifier limit the G/T_{op} to 12[dBK^{-1}]. The modem in this example, the MD-1030A(V), uses sophisticated modulation and half-rate coding with a constraint length 7 that will produce a processing gain of 5 dB with respect to a normal PSK modem; a PSK modem would require an E_B/N_0 of 9.6 dB in order to perform at an error rate of 10^{-5}.

A second example illustrates how the performance of a typical FLTSATCOM shore-to-ship link and ship-to-shore link using an AN/WSC-5 shore terminal and an

AN/WSC-3 shipboard terminal would be computed. The data are summarized in Tables 12.3 and 12.4.

The next is an example showing how the performance of a typical INMARSAT shore-to-ship link and ship-to-shore link using a Coast Earth Station (CES) terminal and a Standard A Ship Earth Station (SES) terminal would be computed. The numerical data is summarized in Tables 12.5 and 12.6.

TABLE 12.1 POWER BUDGET FOR A TYPICAL SHORE-TO-SHIP LINK VIA DSCS

UPLINK	
Shore station EIRP (typical)[1]	66.5 dBW
Space loss (at 8.125 GHz)	201.9 dB
Margin	6 dB
k	-228.6 dBWK^{-1}Hz^{-1}
Satellite G_R/T	-15 dBK^{-1}
C/N_0	72.2 dBHz

DOWNLINK	
Satellite EIRP[2]	5.5 dBW
Space loss (at 7.5 GHz)	201.2 dB
Margin	6 dB
k	-228.6 dBWK^{-1}Hz^{-1}
G_R/T (ship terminal)[3]	12 dBK^{-1}
C/N_0 (downlink)	38.9 dBHz

COMBINED	
Total C/N_0 (up and downlinks)	38.8 dBHz
Bandwidth (for a data rate of 1200 bps)[4]	33.8 dBHz
E_B/N_0 (achieved)	5.0 dB
E_B/N_0 (required for an error rate of 10^{-5})[5]	4.6 dB
Additional operating margin available	0.4 dB

[1]Power assigned to this link; it is a fraction of the total power transmitted from the shore station

[2]Assumed to be the power assigned to this link; it is only a fraction of the available satellite repeater power

[3]The value of G_R/T is typical for the AN/WSC-6 terminal

[4]Assuming a modulation that at baseband generates 1 Hz per bit and $B_{TR} = 2400$ Hz

[5]Characteristic for the MD-1030A(V) modem; it includes a coding gain of 5 dB

TABLE 12.2 POWER BUDGET FOR A TYPICAL SHIP-TO-SHORE LINK VIA DSCS

UPLINK	
Ship terminal EIRP (typical)[1]	56 dBW
Space loss (at 8.125 GHz)	201.9 dB
Margin	6 dB
k	-228.6 dBWK^{-1}Hz^{-1}
Satellite G_R/T	-15 dBK^{-1}
C/N_0	61.7 dBHz

DOWNLINK	
Satellite EIRP[2]	-13 dBW
Space loss (at 7.5 GHz)	201.2 dB
Margin	6 dB
k	-228.6 dBWK^{-1}Hz^{-1}
G_R/T (shore terminal)[3]	33 dBK^{-1}
C/N_0 (downlink)	41.4 dBHz

COMBINED	
Total C/N_0 (up and downlinks)	41.3 dBHZ
Bandwidth (for a data rate of 1200 bps)[4]	33.8 dBHz
E_B/N_0 (achieved)	7.5 dB
E_B/N_0 (to obtain an error rate of 10^{-5})[5]	4.6 dB
Additional operating margin available	2.9 dB

[1]Power assigned to this link; it is a fraction of the total transmitter power available at the ship

[2]Assumed to be the power assigned to this link; it is only a fraction of the available satellite repeater power

[3]The value of G_R/T is typical for the AN/GSC-39 terminal

[4]Assuming a modulation that at baseband generates 1 Hz per bit and $B_{TR} = 2400$ Hz

[5]Characteristic for the MD-1030A(V) modem; it includes a coding gain of 5 dB

TABLE 12.3 POWER BUDGET FOR A TYPICAL SHORE-TO-SHIP LINK VIA FLTSATCOM

UPLINK

Shore station EIRP (typical)[1]	20.0 dBW
Space loss	172.1 dB
Margin	6.0 dB
k	-228.6 dBWK^{-1}Hz^{-1}
Satellite G_R/T	-18 dBK^{-1}
Satellite Receiver Gain	6.0 dB
C/N_0	50.5 dBHz

DOWNLINK

EIRP (typical)[2]	22.0 dBW
Space loss	170.4 dB
Margin	6.0 dB
k	-228.6 dBWK^{-1}Hz^{-1}
G_R/T (ship terminal)[3]	11.0 dBK^{-1}
C/N_0[4]	85.2 dBHz

COMBINED

Total C/N_0 (up and downlinks)	50.5 dBHz
Bandwidth (for a data rate of 1200 bps)[5]	33.8 dBHz
E_B/N_0 (achieved)	16.7 dB
E_B/N_0 (required for an error of 10^{-5})[6]	9.5 dB
Additional operating margin available	7.2 dB

[1]Typical AN/WSC-5 EIRP: 100 W per carrier (20 dBW)

[2]This is assumed to be the power assigned to this particular link

[3]The ship terminal is the AN/WSC-3

[4]Satellite regeneration gain = 6 dB

[5]Assuming a modulation that at baseband generates 1 Hz per bit and B_{TR} = 2400 Hz

[6]PSK modulation

TABLE 12.4 POWER BUDGET FOR A TYPICAL SHIP-TO-SHORE LINK VIA FLTSATCOM

UPLINK

Ship terminal EIRP (typical)[1]	14.7 dBW
Space loss	172.1 dB
Margin	6.0 dB
k	-228.6 dBWK^{-1}Hz^{-1}
Satellite G_R/T	-20 dBK^{-1}
Satellite Receiver Gain	6.0 dB
C/N_0	51.2 dBHz

DOWNLINK

EIRP (typical)[2]	22.0 dBW
Space loss	170.4 dB
Margin	6.0 dB
k	-228.6 dBWK^{-1}Hz^{-1}
G_R/T (shore terminal)[3]	12.0 dBK^{-1}
C/N_0[4]	86.2 dBHz

COMBINED

Total C/N_0 (up and downlinks)	51.3 dBHz
Bandwidth (for a data rate of 1200 bps)[5]	33.8 dBHz
E_B/N_0 (achieved)	17.7 dB
E_B/N_0 (required for an error of 10^{-5})[6]	9.5 dB
Additional operating margin available	8.0 dB

[1]Typical AN/WSC-3 EIRP: 30 W per carrier (14.7 dBW)

[2]This is assumed to be the power assigned to this particular link

[3]The shore terminal is the AN/WSC-5

[4]Satellite regeneration gain = 6 dB

[5]Assuming a modulation that at baseband generates 1 Hz per bit and B_{TR} = 2400 Hz

[6]PSK modulation

TABLE 12.5   POWER BUDGET FOR A TYPICAL SHORE-TO-SHIP LINK VIA INMARSAT

UPLINK	
CES terminal EIRP	
(C-Band, typical)[1]	65.0 dBW
Space loss	196.2 dB
Margin	5.0 dB
k	-228.6 dBWK^{-1}Hz^{-1}
Satellite G_R/T	-19.6 dBK^{-1}
Satellite Receiver Gain	5.0 dB
C/N_0	77.8 dBHz

SATELLITE	
EIRP (L-Band, typical)[2]	17.0 dBW
Space loss	187.75 dB
Margin	5.0 dB
k	-228.6 dBWK^{-1}Hz^{-1}
G_R/T (ship terminal)[3]	-4 dBK^{-1}
C/N_0	49.0 dBHz

COMBINED	
Total C/N_0 (up and downlinks)	48.9 dBHz
Bandwidth (for a data rate	
of 1200 bps)[4]	33.8 dBHz
E_B/N_0 (achieved)	15.1 dB
E_B/N_0 (required for an error	
of 10^{-5})[5]	9.5 dB
Additional operating margin	
available	5.6 dB

[1]Typical Coastal Earth Station terminal telephone channel EIRP 65 dBW

[2]This is assumed to be the power assigned to this particular link; 27 dBW for 10 channels (or 17 dBW per channel)

[3]The ship terminal is Standard A Ship Earth Station (SES) terminal

[4]Assuming a modulation that at baseband generates 1 Hz per bit and $B_{TR} = 2400$ Hz

[5]With PSK modulation

TABLE 12.6   POWER BUDGET FOR A TYPICAL SHIP-TO-SHORE LINK VIA INMARSAT

UPLINK	
Ship terminal EIRP	
(L-Band, typical)[1]	26.0 dBW
Space loss	188.31 dB
Margin	5.0 dB
k	-228.6 dBWK^{-1}Hz^{-1}
Satellite G_R/T	-17.0 dBK^{-1}
Satellite Receiver Gain	6.0 dB
C/N_0	50.29 dBHz

SATELLITE	
EIRP (C-Band, typical)[2]	18.0 dBW
Space loss	196.2 dB
Margin	5.0 dB
k	-228.6 dBWK^{-1}Hz^{-1}
G_R/T (shore terminal)[3]	32.0 dBK^{-1}
C/N_0	77.4 dBHz

COMBINED	
Total C/N_0 (up and downlinks)	50.3 dBHz
Bandwidth (for a data rate	
of 1200 bps)[4]	33.8 dBHz
E_B/N_0 (achieved)	16.5 dB
E_B/N_0 (required for an error	
of 10^{-5})[5]	9.5 dB
Additional operating margin	
available	7.0 dB

[1]Typical Standard A SES EIRP: 36 dBW per 10 channels (or 26 dBW per channel)

[2]This is assumed to be the power assigned to this particular link;

[3]The ship terminal is Standard A Ship Earth Station (SES) terminal

[4]Assuming a modulation that at baseband generates 1 Hz per bit and $B_{TR} = 2400$ Hz

[5]With PSK modulation

REFERENCES

[SC–1] Schwartz, M., *Information Transmission, Modulation and Noise*, McGraw-Hill, New York, N.Y., 1970.

[HO–1] Hogg, D. D., "Effective Antenna Temperature Due to Oxygen and Water Vapor in the Atmosphere," *Journal Applied Physics*, Vol. 30, No. 9, Sept. 1959, p. 1417.

[JA–1] Jasik, H., *Antenna Engineering Handbook*, McGraw-Hill, New York, N.Y., 1961, pp. 2–15.

[RI–1] Rice, D. L. et al., "Transmission Loss Prediction for Tropospheric Loss Communication Circuits," *Technical Note 101*, National Bureau of Standards, U.S. Government Printing Office, January 1967.

13

UHF Satellite Communications

The Navy has a fleetwide UHF SATCOM (FLTSATCOM) capability. An early-on capability was provided by the Gapfiller, a Navy UHF transponder added to the MARISAT satellite. The Gapfiller gave the Navy the opportunity to develop the expertise to operate its own FLTSAT and the full system as soon as it became available.

The FLTSATCOM satellite also provides AFSAT service for the Air Force, but the FLTSAT and AFSAT services use completely independent repeaters in the FLTSAT. The satellites are placed into 24-hour equatorial orbits and provide near global coverage between approximately 76° N and 76° S (with no coverage of the polar regions).[1]

The design lifetime of the FLTSAT satellite is 5 years. As the Gapfiller satellites were phased out, they were replaced by leased satellites (LEASAT) which are designed for launch by the space shuttle. The combination of FLTSAT and LEASAT provide the fleet with reliable around-the-globe UHF communications service.

13.1 FLTSATCOM SATELLITES

The two types of satellites which provide the FLTSATCOM service, the FLTSAT and the LEASAT, are functionally similar. There are certain differences, and each of the satellites is described below.

[1]SATCOM coverage with equatorial satellites can be accomplished as far north as Thule, Greenland (approximately 80°N).

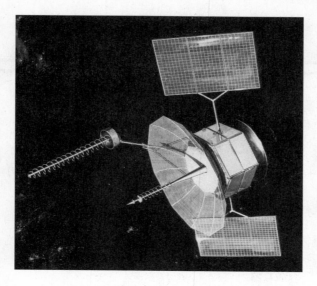

Figure 13.1   The FLTSATCOM Satellite
(Courtesy of TRW Inc.)

13.1.1 The Fleet Satellite (FLTSAT)

Figure 13.1 shows the FLTSATCOM satellite [WI-1]. The satellite is built by TRW and weighs about 1860 kg at launch. It contains two modules, the payload with communications equipment and a spacecraft module with the power supplies and the solar arrays. The communications equipment consists of UHF, SHF, and S-band repeaters, transponders, and antennas.

The spacecraft is 3-axis stabilized, and the antennas are kept pointing at the earth. A clock drive keeps the solar panels oriented for receiving the optimum solar flux. The solar array is designed to provide about 1400 W after 5 years.[2] In addition, a set of NiCd batteries supplies power for peak loads on demand. The batteries supply their power to a bus to which converters are connected, delivering regulated power to all the load modules.

The FLTSATCOM satellite carries four antennas:

- A 4.8m parabolic UHF reflector with a bifilar helix feed for transmission
- An 18-turn helical UHF receiver antenna
- An SHF horn, body-mounted and positioned behind a transparent section of the UHF paraboloidal reflector
- A conical spiral S-band tracking and telemetry antenna.

In addition, certain FLTSATCOM satellites carry an EHF package, the FLTSAT-COM EHF package (FEP). This equipment package provides an initial experimental EHF capability. Eventually, an EHF satellite capability will be provided by the MILSTAR satellite (see Chapter 15).

---

[2]The output of solar cells diminishes during their lifetime because high-energy particles (e.g., solar wind, Van Allen Belt) reduce the cells' semiconductor threshold, which decreases the overall output with time.

Four equatorial satellites provide the FLTSAT coverage; the typical coverage pattern is shown in Figure 13.2.

The FLTSAT frequency plans are given in Table 13.1. The frequencies of the uplinks and downlinks are arranged in three frequency plans in order to avoid interference in areas where coverage overlaps (see Figure 13.2). The FLTSAT characteristics are summarized in Table 13.2.

The ten 25-kHz channels are dedicated for use by the Navy; each 25-kHz channel has a separate transmitter. The transmitter powers are given in Table 13.2. The transmitter power is fixed for the normal mode of operations. A typical link power budget calculation for the FLTSAT is given in Chapter 12. Channel 1 carries primarily Fleet Broadcast transmissions. The uplink for the Fleet Broadcast is in the SHF band, the downlink is in the UHF band. Hence, Channel 1 requires processing onboard satellite.

13.1.2 The LEASAT

Figure 13.3 shows the LEASAT satellite [WI-2]; the LEASAT satellites have been in service since the mid 1980s. The LEASAT satellites have certain features in common with the DSCS-II satellite (see Chapter 13). The satellite is spin-stabilized, with its axis parallel to the Earth axis, and its body rotates at 30 rpm. It has a despun section on top that carries the Earth pointing antennas. The satellite Earth orbit insertion weight (leaving the

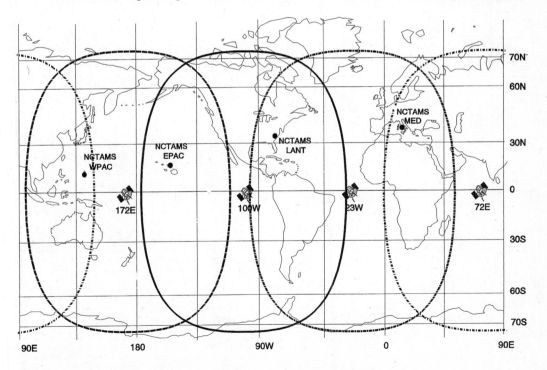

Figure 13.2   FLTSATCOM Coverage

TABLE 13.1 FLTSATCOM TRANSMIT AND RECEIVE FREQUENCIES

Channel	Plan	Downlink Frequency (MHz)	Uplink Frequency (MHz)	Nominal Bandwidth (kHz)	Channel	Plan	Downlink Frequency (MHz)	Uplink Frequency (MHz)	Nominal Bandwidth (kHz)
1	A	250.45	SHF	25	12	A	243.955	317.055	5
	B	250.55	SHF	25		B	244.055	317.155	5
	C	250.65	SHF	25		C	244.145	317.245	5
2	A	251.95	292.95	25	13	A	243.960	317.060	5
	B	252.05	293.05	25		B	244.060	317.160	5
	C	252.15	293.15	25		C	244.160	317.260	5
3	A	252.65	294.65	25	14	A	243.965	317.065	5
	B	253.75	294.75	25		B	244.060	317.165	5
	C	253.85	295.85	25		C	244.165	317.265	5
4	A	255.35	296.35	25	15	A	243.970	317.070	5
	B	255.45	296.45	25		B	244.070	317.170	5
	C	255.55	296.55	25		C	244.170	317.270	5
5	A	256.95	297.95	25	16	A	243.975	317.075	5
	B	257.05	298.05	25		B	244.075	317.175	5
	C	257.15	298.15	25		C	244.175	317.275	5
6	A	258.45	299.45	25	17	A	243.980	317.080	5
	B	258.55	299.55	25		B	244.080	317.180	5
	C	258.65	299.65	25		C	244.180	317.280	5
7	A	265.35	306.35	25	18	A	243.985	317.085	5
	B	265.45	306.45	25		B	244.085	317.185	5
	C	265.55	306.55	25		C	244.185	317.285	5
8	A	266.85	307.85	25	19	A	243.990	317.090	5
	B	266.95	307.95	25		B	244.090	317.190	5
	C	267.05	308.05	25		C	244.190	317.290	5
9	A	268.25	309.25	25	20	A	243.995	317.095	5
	B	268.35	309.35	25		B	244.095	317.195	5
	C	268.45	309.45	25		C	244.195	317.295	5
10	A	269.75	310.75	25	21	A	244.000	317.100	5
	B	269.85	310.85	25		B	244.100	317.200	5
	C	269.95	310.95	25		C	244.200	317.300	5
11	A	243.945	317.045	5	22	A	244.010	317.110	5
	B	244.045	317.145	5		B	244.110	317.210	5
	C	244.145	317.245	5		C	244.210	317.310	5
					23	A	260.660	294.200	500
						B	261.700	295.300	500
						C	262.300	295.900	500

TABLE 13.2   FLTSATCOM SATELLITE CHARACTERISTICS

Channels	Ten 25-kHz channels
Transmitters	Two channels with an EIRP of 28 dBW Six channels with an EIRP of 26 dBW
Receiver G/T	$-16.7\ \text{dBK}^{-1}$
UHF antenna coverage	19° (earth coverage)

space shuttle) is approximately 7000 kg. On-orbit weight is approximately 1300 kg. The solar panel, wrapped around the cylindrical body, is designed to provide 1.2 kW after a 7-year life in orbit.

The despun part of the satellite contains the antennas, electronics, and communications equipment. The LEASAT satellite has one 500-kHz, seven 25-kHz, and five 5-kHz repeater channels. The 25-kHz channels are dedicated to Navy use. The Fleet Broadcast uses an SHF uplink and one 25-kHz UHF downlink. The other six 25-kHz channels support the FLTSATCOM system with or without the demand assigned multiple access (DAMA) equipment.

Figure 13.3   The LEASAT Satellite (Courtesy of Hughes Space and Communications)

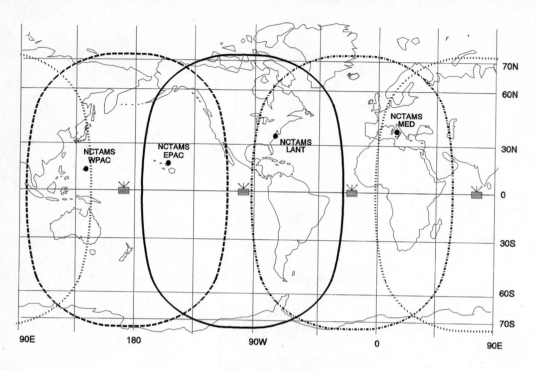

Figure 13.4   LEASAT Coverage

Four satellites in equatorial orbits provide global coverage (except for the polar regions). The coverage is shown in Figure 13.4.

The LEASAT characteristics are summarized in Table 13.3. Although the satellite has fewer repeaters than the FLTSAT, the better utilization through DAMA will provide the required service.

The LEASAT operating frequencies of the uplinks and downlinks are organized in four frequency plans (*W, X, Y,* and *Z*) as summarized in Table 13.4. Interference in areas covered by more than one satellite is avoided by using this frequency plan.

TABLE 13.3   LEASAT CHARACTERISTICS

Repeater Bandwidth (kHz)	500[1]	25	25	5
Number of Repeaters	1	1	6	5
Transmitter Power (dBW)	28	26	26	16.5
G/T (dB/K)	−18	−20	−18	−18

[1]This channel supports the Fleet Broadcast.

TABLE 13.4 LEASAT TRANSMIT AND RECEIVE FREQUENCIES

Channel	Plan	Downlink Frequency (MHz)	Uplink Frequency (MHz)	Nominal Bandwidth (kHz)	Channel	Plan	Downlink Frequency (MHz)	Uplink Frequency (MHz)	Nominal Bandwidth (kHz)
1	W	250.35	SHF	25	8	W	265.25	306.25	25
	X	250.45	SHF	25		X	265.34	306.25	25
	Y	250.55	SHF	25		Y	265.45	306.25	25
	Z	250.65	SHF	25		Z	265.55	306.25	25
2	W	263.80	297.40	500	9	W	243.855	316.955	5
	X	260.60	294.20	500		X	243.955	317.055	5
	Y	261.70	295.30	500		Y	244.055	317.155	5
	Z	262.30	295.90	500		Z	244.155	317.255	5
3	W	251.85	292.85	25	10	W	243.860	316.960	5
	X	251.95	292.95	25		X	243.960	317.060	5
	Y	252.05	293.05	25		Y	244.060	317.160	5
	Z	252.15	293.15	25		Z	244.160	317.260	5
4	W	253.55	294.55	25	11	W	243.875	316.975	5
	X	253.65	294.65	25		X	243.975	317.075	5
	Y	253.75	294.75	25		Y	244.075	317.175	5
	Z	253.85	294.85	25		Z	244.175	317.275	5
5	W	255.25	296.25	25	12	W	243.900	317.000	5
	X	255.35	296.35	25		X	244.000	317.100	5
	Y	255.45	296.45	25		Y	244.100	317.200	5
	Z	255.55	296.55	25		Z	244.200	317.300	5
6	W	256.85	297.85	25	13	W	243.910	317.010	5
	X	256.95	297.95	25		X	244.010	317.110	5
	Y	257.05	298.05	25		Y	244.110	317.210	5
	Z	257.15	298.15	25		Z	244.210	317.310	5
7	W	258.35	299.35	25					
	X	258.45	299.45	25					
	Y	258.55	299.55	25					
	Z	258.65	299.65	25					

13.2 UHF SATELLITE SUBSYSTEMS

The Navy UHF SATCOM system comprises the following information transmission subsystems, sometimes also called FLTSATCOM circuits or links:

- Fleet Satellite Broadcast (FLTBROADCAST)
- Common User Digital Information Exchange Subsystems (CUDIXS)/Naval Modular Automated Communications Subsystem (NAVMACS)
- Officer-in-Tactical Command Information Exchange Subsystem (OTCIXS)
- Submarine Satellite Information Exchange Subsystem (SSIXS)
- Tactical Intelligence Subsystem (TACINTEL)
- Tactical Data Information Exchange Subsystem (TADIXS)
- Secure Voice Subsystem (SECVOX)
- Teletypewriter Subsystem.

Each of these subsystems supports a dedicated network that is a common resource of all network user members. A network member can access the net much in the same manner as using a telephone, that is, when a circuit in the net is free, then service is rendered. To increase the capability of the common resource, the DAMA subsystem was added. The FLTSATCOM system also contains a control subsystem, a quality monitoring subsystem (QMS), and a telemetry, tracking, and command (TT&C) subsystem.

The FLTSAT subsystems or circuits provide a significant part of the communications for the Navy's Command, Control, Communications, and Intelligence $(C^3I)^3$ data. Figure 13.5 illustrates the use of FLTSATCOM circuits in the Navy's C^3I data transfer by showing an example of how the Navy's FLTSATCOM is employed in tactical warfare. The FLTSAT provides information transfer from shore nodes to afloat platforms and among platforms and thus supports the four pillars of the Navy C^3I system which consists of (1) shore nodes, (2) afloat nodes, (3) shore communications systems, and (4) communication between shore nodes and afloat nodes.

Figure 13.5 illustrates the C^3 I data flow. Two external sensor systems (national and theater) provide the tactical intelligence to the Ocean Surveillance Information System (OSIS) located at the Fleet Ocean Surveillance Information Center (FOSIC) or Fleet Ocean Surveillance Information Facility (FOSIF). The data may be carried by various Defence Communications Systems such as the Defense Data Network (DDN), AUTODIN, or the FLTSATCOM TACINTEL channel. The Integrated Undersea Surveillance System (IUSS) also provides ASW information via SATCOM and dedicated landlines to the OSIS. Most of ocean surveillance information products are transferred to the Battle Group Commander via the FLTSATCOM TADIXS-A channel. Additional Special Intelligence (SI) messages are transferred to the Battle Group Commander via FLTSATCOM TACINTEL channel. The fleetwide tactical information, which is generated by the

---

[3]In the documentation either the designation C3I or C^3I can be found.

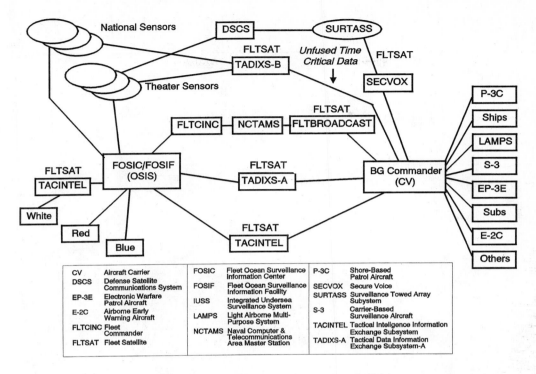

Figure 13.5   FLTSATCOM Circuits in the Navy's C^{3}I Information Transfer

FLTCINC and processed at an associated Naval Computer and Telecommunications Area Master Station (NCTAMS), is transferred via a FLTSATCOM FLTBROADCAST channel. In support of anti-air warfare (AAW), FLTSATCOM TADIXS-B is used for transferring time-critical over-the-horizon targeting (OTH-T) data. TADIXS-B is one-way broadcast channel to afloat platforms.

In the interest of completeness, additional organic Battle Group (BG) communications that do not use the FLTSATCOM are shown in Figure 13.5. These organic BG assets are: Maritime patrol aircraft (P-3C), Light Airborne Multi-Purpose System (LAMPS) helicopter, escort ships, surveillance aircraft (S-3), airborne Electronic Warfare (EW) aircraft (EP-3E), direct support submarines, airborne EW aircraft (E-2C), and others. These organic assets transfer either data or voice via HF or UHF LOS radios such as Link 11, Link 16, and Link 4A.

The FLTSATCOM circuits play a significant role in the information transfer on a Navy platform. To put the magnitude of the information transfer into perspective and show the scope of the FLTSAT contribution, Figure 13.6 illustrates an aircraft carrier (CV) C^{3}I system and the shipboard communications assets associated with it. A typical carrier shipboard C3I system consists of the following systems:

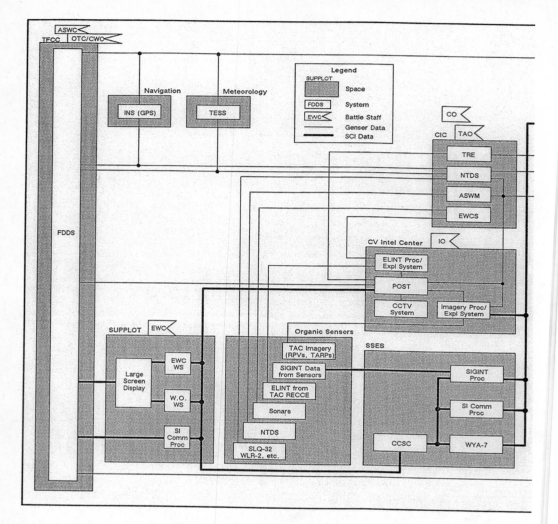

Figure 13.6 FLTSATCOM Circuits on a Navy Platform

- Exterior Communications System (EXCOMM System)
- Main Communications Center
- Combat Information Center (CIC)
- Intelligence Center (CVIC)
- Special Signal Exploitation Space (SSES)
- Organic Sensors
- Tactical Flag Command Center (TFCC) and Supporting Plot (SUPPLOT)
- Navigation and Meteorology Subsystem

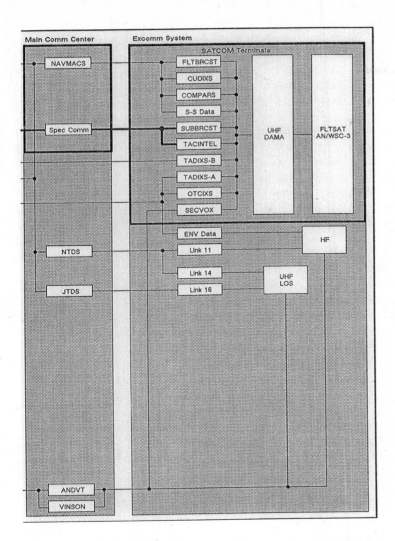

The carrier C^{3}I system is the most complex example of a C^{3}I system in combatant ships. Figure 13.6 describes such a system and shows the interfacing of its shipboard telecommunications assets. Other combatant ships may have scaled down versions of the shown example with appropriately scaled down communications support. The functions of the subsystems shown in Figure 13.6 are described below.

The EXCOMM system provides the RF circuits and/or nets for the platform; this includes UHF FLTSAT, UHF LOS, HF, and other RF systems. The essential communications circuits, which include FLTBROADCAST, CUDIXS, Naval Communications Processing and Routing System (NAVCOMPARS), Ship-to-Shore Data, TACINTEL, TADIXS-A, TADIXS-B, OTCIXS, SECVOX, Environmental Data,

Link 11, Link 14, Link 4A and Link 16, are interfaced to the equipment in the main communications center. The main communications center provides services to various combat systems in the platform. Equipment in the main communications center providing processing support are: the NAVMACS computer, special communications processors, the NTDS receiver, JTIDS equipment, the Advanced Narrowband Digital Voice Terminal (ANDVT) and VINSON secure voice terminals, and associated cryptographic devices.

The Combat Information Center (CIC) acquires, processes, and distributes tactical information to the Commanding Officer (CO), the Tactical Action Officer (TAO), and other warfare commanders and coordinators. Major systems in CIC include: Tactical Receive Equipment (TRE), Naval Tactical Data System (NTDS), Anti-Submarine Warfare Module (ASWM), and Electronic Warfare Control System (EWCS).

The Carrier (CV) Intelligence Center (IC) provides the Intelligence Officer (IO) and other warfare officers onboard the platform with tactical intelligence information. Major systems in a CVIC are: the Electronic Intelligence (ELINT) Processor and Exploitation System, the Joint Operational Tactical System (JOTS), Closed Circuit Television (CCTV), and the Imagery Processing and Exploitation System.

The Special Signal Exploitation Space (SSES) provides intelligence support functions to warfare commanders and coordinators. Major systems in the SESS are: the Signal Intelligence (SIGINT) Processor, the Special Intelligence (SI) Communications Processor, the Combat Cryptological Support Console (CCSC), and the AN/UYA-7 terminal.

In addition to external sensor inputs to the IC and CIC, a CV also has a set of its own organic sensors. These organic sensors include: Tactical Aircraft (TAC) imagery (from UAVs), SIGINT data from sensors, ELINT from TAC reconnaissance (RECCE), data from the AN/SLQ-32 EW countermeasure set and AN/WLR-2 ECCM radars, sonar, and NTDS information.

The Supporting Plot (SUPPLOT) provides various displays on tactical situations. The Electronic Warfare Coordinator (EWC) is stationed in the SUPPLOT. Included in the SUPPLOT are a large screen display, the EW Work Station (WS), and the Watch Officer (WO) WS and the SI Communications Processor.

The Task Force Command Center (TFCC) is a command support system for the Officer-in-Tactical Command (OTC), and the Combined Warfare Commander (CWC) and ASW Commander in the TFCC are supported by the Fleet Data Display System (FDDS).

The carrier C^3I system also includes a navigation and meteorology system. Two complementary navigation systems are used in the platform: an Inertial Navigation System (INS) and the Global Position System (GPS), a precision satellite navigation system providing location and time reference. The weather and environmental data are received and distributed internally by the Tactical Environmental Support System (TESS).

Table 13.5 summarizes the key characteristics of the FLTSATCOM subsystems in terms of information transmission, function, and data processing used. Figure 13.7 gives a pictorial overview of the FLTSATCOM subsystems, shows the end-to-end subsystems configurations, and identifies the equipment used. The sections to follow contain detailed descriptions of the subsystems.

TABLE 13.5 FLTSATCOM INFORMATION TRANSMISSION SYSTEMS

Subsystem	Description	Data Rate	Message Type and Format	Processor (Shipboard)	Function
Fleet Broadcast	Ship receive only; 15 75-bps TTY channels formatted into one 1200 bps data stream	1200 bps TDM, containing 75-bps TTY channel	Message traffic using JINTACCS, USMTF, JANAP 128(I), and ACP 127 formats	NAVMACS, ICS	Broadcast Record Message to Fleet
CUDIXS/ NAVMACS	Ship/shore and shore/ship GENSER data; two networks each 50 primary and 10 special users	2400 bps, half duplex	Message traffic using JINTACCS, USMTF, and JANAP 128(I) formats	NAVMACS	Common user GENSER data and record traffic
OTCIXS	Bidirectional formatted data for TDP and TTY messages on controlled and time shared basis	2400 bps	Message traffic and data using JINTACCS, USMTF, and OTH-T Gold formats	ON-143(V)6, TDP, TTY Terminal	OTH tactical and targeting data for battle groups
SSIXS	Shore/ship group broadcast of record TTY information and ship/shore messages	2400 bps and 4800 bps half duplex	Data and message traffic using OTH-T Gold, JINTACCS, and USMTF formats	Special communications processor	Periodic broadcasts of messages complementing VLF communications links from BCAs to submarines
TACINTEL	Shore/ship and ship/shore tactical data; network can support up to 23 users	2400 bps, half duplex (1200 and 4800 bps optional)	Message traffic using JINTACCS, and USMTF formats	Special communications processor	Special intelligence tactical data
TADIXS	Shore/ship data broadcast	2400-bps half duplex	Data, also messages using the OTH-T Gold format	TADIXS-A processor, TRE and TDP	Tactical data for targeting and surveillance data
SECVOX	Shore/ship data broadcast	2400-bps push-to-talk half duplex	2400-bps encoded data stream	CV-3951V voice digitizer	Secure voice network, extension to AUTOSEVCOM
ORESTES TTY	Ship/shore and shore/ship TTY	75 bps TTY	Various formats		BLOS TTY tactical and report back, backup for CUDIXS and TACINTEL

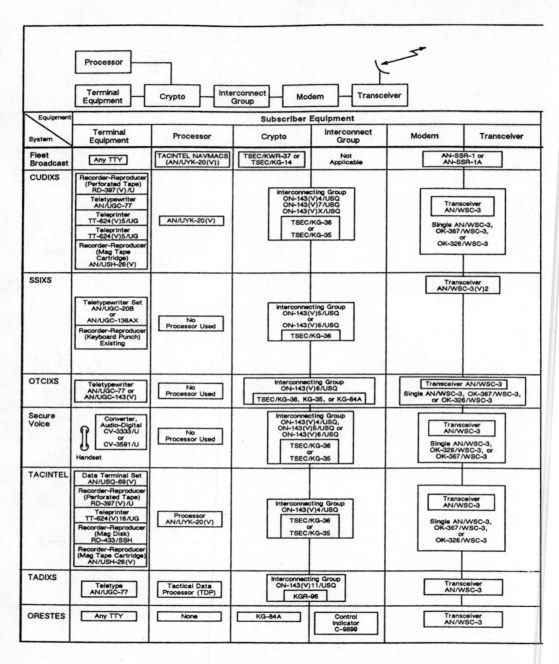

Figure 13.7 FLTSATCOM End-to-End Subsystem Configurations

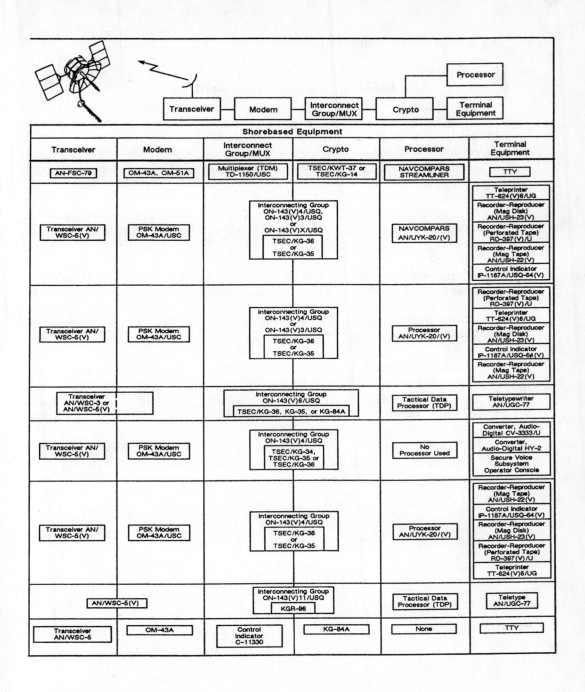

Transceiver	Modem	Interconnect Group/MUX	Crypto	Processor	Terminal Equipment

Shorebased Equipment

Transceiver	Modem	Interconnect Group/MUX	Crypto	Processor	Terminal Equipment
AN-FSC-79	OM-43A, OM-51A	Multiplexer (TDM) TD-1150/USC	TSEC/KWT-37 or TSEC/KG-14	NAVCOMPARS STREAMLINER	TTY
Transceiver AN/WSC-5(V)	PSK Modem OM-43A/USC	Interconnecting Group ON-143(V)4/USQ, ON-143(V)3/USQ or ON-143(V)X/USQ TSEC/KG-36 or TSEC/KG-35		NAVCOMPARS AN/UYK-20/(V)	Teleprinter TT-624(V)6/UG Recorder-Reproducer (Mag Disk) AN/USH-23(V) Recorder-Reproducer (Perforated Tape) RD-397(V)/U Recorder-Reproducer (Mag Tape) AN/USH-22(V) Control Indicator IP-1187A/USQ-64(V)
Transceiver AN/WSC-5(V)	PSK Modem OM-43A/USC	Interconnecting Group ON-143(V)4/USQ or ON-143(V)3/USQ TSEC/KG-36 or TSEC/KG-35		Processor AN/UYK-20/(V)	Recorder-Reproducer (Perforated Tape) RD-397(V)/U Teleprinter TT-624(V)6/UG Recorder-Reproducer (Mag Disk) AN/USH-23(V) Control Indicator IP-1187A/USQ-64(V) Recorder-Reproducer (Mag Tape) AN/USH-22(V)
Transceiver AN/WSC-3 or AN/WSC-5(V)		Interconnecting Group ON-143(V)6/USQ TSEC/KG-36, KG-35, or KG-84A		Tactical Data Processor (TDP)	Teletypewriter AN/UGC-77
Transceiver AN/WSC-5(V)	PSK Modem OM-43A/USC	Interconnecting Group ON-143(V)4/USQ TSEC/KG-34, TSEC/KG-35 or TSEC/KG-36		No Processor Used	Converter, Audio-Digital CV-3333/U Converter, Audio-Digital HY-2 Secure Voice Subsystem Operator Console
Transceiver AN/WSC-5(V)	PSK Modem OM-43A/USC	Interconnecting Group ON-143(V)4/USQ TSEC/KG-36 or TSEC/KG-35		Processor AN/UYK-20/(V)	Recorder-Reproducer (Mag Tape) AN/USH-22(V) Control Indicator IP-1187A/USQ-64(V) Recorder-Reproducer (Mag Disk) AN/USH-23(V) Recorder-Reproducer (Perforated Tape) RD-397(V)/U Teleprinter TT-624(V)6/UG
AN/WSC-5(V)		Interconnecting Group ON-143(V)11/USQ KGR-96		Tactical Data Processor (TDP)	Teletype AN/UGC-77
Transceiver AN/WSC-5	OM-43A	Control Indicator C-11330	KG-84A	None	TTY

13.2.1 Fleet Satellite Broadcast Subsystem

The FLTBROADCAST supports the transmission of broadcast message traffic. The FLTBROADCAST normally operates on channel 1. It has 15 subchannels, each operating at 75 bps, with the subchannels integrated into a time-division multiplexed (TDM) data stream of 1200 bps. The primary transmission from the shore station to the satellite uses SHF transmission via the AN/FCS-79 terminal. Direct-sequence spread-spectrum transmission is used on the SHF uplink to provide jamming protection. A second channel, operating on UHF via the AN/WSC-5 transceiver for both the uplink and downlink, provides a backup capability. This backup channel is normally assigned to secure voice transmission.

Several RF uplink modulation techniques (FSK, PSK, and spread-spectrum) can be used on the FLTBROADCAST channel and since two downlink channels are available, several operating modes are commonly used, as summarized in Table 13.6.

Shipboard reception. The shipboard subscribers receive the UHF broadcast signal that is then demodulated and demultiplexed. The individual data streams are sent to optional Naval Automated Communications System (NAVMACS) and the Tactical Intelligence (TACINTEL) processors for screening and then to teletypewriters. Weather data and General Service (GENSER) data are sent directly to teletypewriter equipment.

Message traffic. The FLTBROADCAST message traffic is channelized and queued prior to transmission from the shore site. The processors performing this function are the Naval Communications Processing and Routing System (NAVCOMPARS) for GENSER data and the STREAMLINER for special intelligence (SI) traffic. In addition, weather data are transmitted that have been entered via a teletypewriter or a recorder. The TDM consists of eleven 75-bps GENSER subchannels, two weather data subchannels, and two SI subchannels. Subchannel 16 carries synchronization signals. Messages in the FLTBROADCAST use formats that adhere to the Joint Interoperability of Tactical Com-

TABLE 13.6 FLTBROADCAST OPERATING MODES

Mode	Uplink	Uplink Modulation	Downlink
1 (Primary)	SHF	Spread-spectrum	Ch. 1
2	SHF	Spread-spectrum	Ch. 2
3	SHF	Narrowband	Ch. 1
4	SHF	Narrowband	Ch. 2
5	SHF	Narrowband (PSK)	Ch. 1
6	SHF	Narrowband (PSK)	Ch. 2
7 (Backup)	UHF (Ch. 2)	Narrowband	Ch. 2

mand and Control System (JINTACCS) standards, the U.S. Message Text Format (USMTF) [JC-1], the Joint Army, Navy, Air Force Publication (JANAP) 128(I) [JC-2], and the Allied Communications Policies (ACP) 121 [NA-1] and 127 [NA-2].

Figure 13.8 shows a block diagram of the FLTBROADCAST subsystem that also identifies the equipment used.

13.2.2 Common User Digital Exchange Subsystem/Naval Modular Automated Communications Subsystem (CUDIXS/NAVMACS)

This subsystem contains two major components: (a) the CUDIXS, a shore-based processor suite that provides message traffic processing and transmission control, and (b) the NAVMACS, a shipboard subscriber terminal and processing system to transmit, receive, and process the messages. The NAVMACS is devised to interoperate with the CUDIXS. If the shipboard system includes a message processing and distribution system (MPDS), the NAVMACS is designed to operate interactively with the MPDS.

Figure 13.9 shows the shipboard configuration of the subsystem (NAVMACS segment). Note that the HF data receivers can be patched into the NAVMACS terminal and the cryptographic equipment; thus the processor and the end-equipment can service both the NAVMACS and the HF transmit/receive equipment.

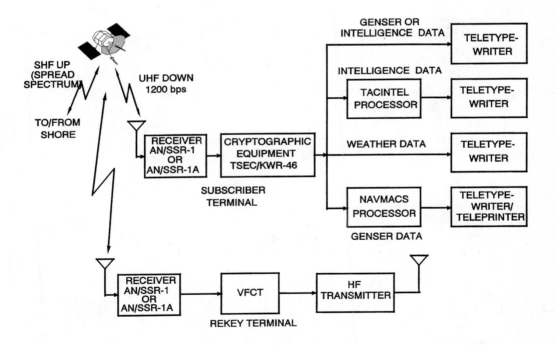

Figure 13.8   Fleet Broadcast Shipboard Configuration

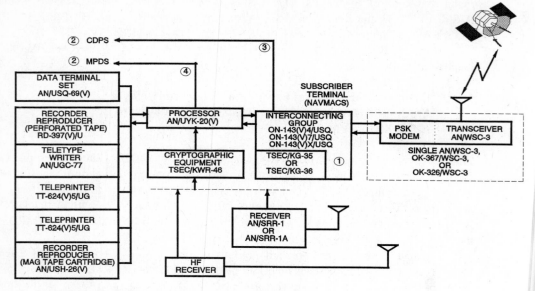

① The ON-143(V)/USQ and ON-143(V)7/USQ will transition to the ON-143(V)9/USQ

② CDPS and MPDS interface only with the few existing ON-143(V)7/USQ, or, in the future, with the ON-143(V)9USQ

③ The AN/UYK-20(V)-based NAVMACS system interfaces only with the existing ON-143(V)4/USQ and the future ON-143(V)9/USQ

④ The ON-143(V)9/USQ interfaces with either the MPDS or CDPS according to platform requirements

MPDS = Message Processing and Distribution System   CDPD = Communications and Data Processing System

Figure 13.9   NAVMACS, the Shipboard Segment of the CUDIXS/NAVMACS Subsystem

The CUDIXS/NAVMACS is a functional replacement for the ORESTES[4] TTY network. It provides improved operational message throughput rates, increased traffic volume, and better link reliability. The CUDIXS supports high-volume users requiring two-way traffic flow (special users) and subscribers having a one-way ship-to-shore traffic message flow on the same link. The ship-to-ship traffic for primary users is carried via the FLTBROADCAST.

The CUDIXS link control supports a network of 50 primary and 10 special users. Usually two CUDIXS/NAVMACS networks operate within one satellite footprint. To provide global coverage, CUDIXS controllers are available in each FLTSAT footprint. All Naval Computer and Telecommunications Area Master Stations (NCTAMS) have CUDIXS installations.

Link control.   CUDIXS/NAVMACS operates over one of the 25-kHz-wide FLTSATCOM channels. Each channel operates at a rate of 2400 bps as a half-duplex link using differential PSK (DPSK). When DAMA is used, CUDIXS/NAVMACS operates in a dedicated DAMA time slot.

[4]ORESTES is a designation for a TTY circuit.

The RF channel is accessed using a polling protocol. One CUDIXS terminal per network in each footprint exercises control by transmitting a sequence order list to all subscribers and the NAVMACS terminals operate in a reactive mode. The sequence order list specifies order, net cycle time slice, and duration of transmission of each net member. Transmission of one message may require more than one cycle to transmit. The subscribers decide when they can transmit. The processors, NAVMACS in the shipboard terminal case, will prepare the messages by queuing and generating blocks that will fit into the allotted time slices. Each block contains a header, which typically carries precedence level, sequence of transmission, and number of blocks. The header is also used to establish the next sequence order list. The protocol includes an acknowledgment (ACK) in the transmission block to provide confirmation of data block reception.

The sequence order list is dynamic, that is, the number of elements in the list varies. It depends on the number of members in the net, the number of blocks authorized for a transmission, and the number of time slices that can be accessed. Hence, the net cycle length varies. Typical cycle times are in the range of 2.5 to 3.5 minutes.

Message traffic input.   The Naval Communication Processing and Routing System (NAVCOMPARS) is the interface for the CUDIXS/NAVMACS message traffic at the shore facilities. Local traffic can be entered via teletypewriters or, from the other location, via AUTODIN. The NAVCOMPARS routes traffic arriving from the fleet to its destinations, and messages can be printed locally or passed on via AUTODIN.

Operations.   Each subscriber to a CUDIXS net is assigned an identification (ID) number that is recognized by the CUDIXS and NAVMACS processors. The ID number facilitates message screening. The NAVMACS processor forwards only the messages that are addressed to the subscriber number to the end equipment, that is, teletypewriter, printer, or recorder. Other traffic is discarded by the processor.

13.2.3 Officer-in-Tactical Command Information Exchange Subsystem (OTCIXS)

The OTCIXS supports full duplex communications to Battle Groups (BGs) for command and control (C^2) and over-the-horizon targeting (OTH-T). The OTCIXS circuit carries TTY message traffic and formatted messages for tactical data processor (TDP) operation. The TDP capability supports primarily the exchange of surveillance formatted track data and targeting data among the platforms in a BG.

Figure 13.10 shows the OTCIXS installation for surface ships, Figure 13.11 for submarines. The configurations also show the secure voice equipment that can be connected to the controller ON-143(V)6/USQ interconnection group.

Platforms of a BG with OTCIXS circuits are command ships, Tomahawk-equipped surface ships and submarines, Frigates (FFs), and Guided Missile Frigates (FFGs) not equipped with Tomahawk. Shore sites also netted with the BG via OTCIXS are Fleet Ocean Surveillance Information Centers (FOSICs) or Fleet Ocean Surveillance Information Facilities (FOSIFs) and Submarine Operating Authority (SUBOPAUTH) Shore Tar-

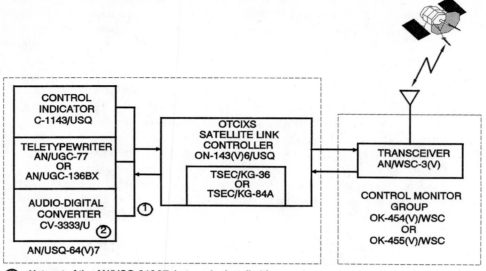

① Not part of the AN/USQ-64(V)7, but can be installed for
   FLTSATCOM Secure Voice Operations
② The CV-3591 is the new generation of this equipment

Figure 13.10   OTCIXS Surface Ship Configuration

geting Terminals (STTs). The FOSICs/FOSIFs have only a receiving capability for TDP data to provide input for data fusion. The STTs have a two-way TTY capability for coordination with the BG.

Operation.   The OTCIXS terminals operate over an automatically controlled time-shared net. The ON-143(V)6/USQ interconnection group performs the link control. This equipment also performs automatic crypto synchronization and controls the storing and forwarding of incoming TTY and TDP data traffic.

One of the ON-143(V)6/USQ controls must be configured as the OTCIXS net control station (NCS) and provide subscriber access to the net. Since the OTCIXS net has only two transmission precedence levels (FLASH or IMMEDIATE), time slots are reserved for the FLASH transmissions, and 20 subsequent slots are allocated for IMMEDIATE precedence. To gain access, the net must be momentarily idle (no traffic). All subscribers with waiting outgoing traffic send a net request either into a reserved FLASH slot or selecting randomly one of the 20 subsequent slots. The first subscriber to gain access will be designated to transmit data. Subscribers which did not gain access cancel their transmission requests until the net is idle again. Transmissions are repeated three times to enhance transmission reliability. When a subscriber with FLASH traffic request needs to access the net, the FLASH traffic will pre-empt redundant retransmissions of IMMEDIATE precedence traffic.

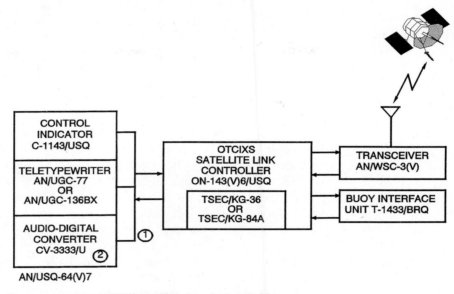

① Not part of the AN/USQ-64(V)7, but can be installed for
   FLTSATCOM Secure Voice Operations

② The CV-3591 is the new generation of this equipment

Figure 13.11   OTCIXS Submarine Configuration

Transmission and reception. Outgoing traffic is entered into the ON-143(V)6/USQ interconnection group. Data and TTY may be entered simultaneously, and TTY can be in a free format. With the TTY, a control indicator declaring precedence and the destination ID must be entered from the TDP console.

Received traffic is checked in the ON-143(V)6/USQ for accuracy by comparing the repetitive transmission. No acknowledgment is returned; this form of reception facilitates operation in an emission control (EMCON) condition.

13.2.4 Submarine Satellite Information Exchange Subsystem (SSIXS)

The SSIXS provides for communications between submarines and shore-based Submarine Broadcast Authorities (BCAs). SSIXS links operate at 2400 and 4800 bps in a half-duplex mode over the single channel AN/WSC-3(V)2 terminal.

The SSIXS complements the existing VLF and MF/HF broadcast circuits from BCAs to submarines. One 25-kHz channel on each FLTSAT satellite is allotted to the SSIXS. Up to 120 submarines may participate in one SSIXS net. Each SSIXS net member has a unique identification (ID) number. SSIXS networks may span more than one satellite footprint to serve other BCA regions. Two BCAs may share a single SSIXS channel by offsetting their group broadcast times in an operationally acceptable manner.

Figure 13.12 shows the submarine configuration of the SSIXS terminal. The relationship to the OTCIXS terminal (see Figure 13.11) is readily apparent. The AN/WSC-3(V)2 transceiver, PSK modem, and interconnect processor ON-143(V)6/USQ are shared with the OTCIXS, SSIXS, and SECVOX subsystems.

Operation.   At the NCTAMS, the SSIXS shares satellite access to the RF terminal equipment with the other UHF SATCOM subsystems. Almost all SSN 688 class submarines are equipped with dual UHF SATCOM installations to allow simultaneous operations on more than one UHF SATCOM circuit.

Two modes of operation are available on SSIXS nets: (a) reception of messages transmitted at scheduled intervals during so-called group broadcast and (b) transmission from a submarine to the BCA at times other than the group broadcasts. The latter may include query for messages. When the BCA receives a transmission from a submarine, it acknowledges the transmission and sends the messages in queue for the submarine. Message transmission may be repeated to enhance communications reliability. Having two modes of operations gives the submarine commander the choice to be passive or communicate actively.

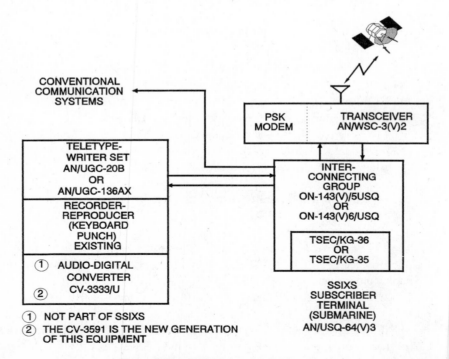

Figure 13.12   SSIXS Submarine Configuration

Message transmission/reception.   Onboard the submarine, the messages are entered via the teletypewriter or the tape reader equipment. The SSN 688 class submarines (SSN) are equipped with the data link control system (DLCS) and have the capability to prepare inputs and obtain outputs via the sensor interface for OTH-T messages.

On reception, the net member ID number plays an important role. One function of the ID number is to determine the number of receptions of the transmissions to the subscriber. Also, a submarine sending to the shore must include the ID number to receive an ACK and requested messages. Third, the ID number is used onboard the submarine to screen the broadcast message traffic. The SSIXS processor discards all messages not directed to the subscriber.

SI SSIXS.   Besides the GENSER messages, dedicated SI communications (SI SSIXS) are used. Each subscriber link uses individual crypto devices to provide security.

13.2.5 Tactical Intelligence Subsystem (TACINTEL)

The TACINTEL subsystem supports the transmission of special intelligence information. This subsystem has an automated message processing system, and is used for transfer of information via satellite in a controlled environment. A 25-kHz channel on each satellite is allotted to TACINTEL. The subsystem is structured similar to the OTCIXS. Up to 23 subscribers can be served in each satellite footprint. The TACINTEL channel operates in a half-duplex mode, using DPSK modulation, nominally at a rate of 2400 bps, with 1200 bps and 4800 bps as options. Figure 13.13 shows the shipboard configuration of the TACINTEL subsystem.

Operation.   The TACINTEL operations protocol resembles the OTCIXS protocol. A polling scheme is used to support the 23-subscriber net. The protocol is computer-controlled and resides in the TACINTEL processor. One TACINTEL station processor serves as controller and maintains net control by transmitting a sequence order list to the subscribers; the sequence order list specifies order and duration of subscriber transmission. The subscriber's processor puts the outgoing message traffic into a queue, and the messages are transmitted according to the sequence order list. Inactive subscribers may rejoin the net and become active by sending a request to the net control during a special time slot reserved for gaining net access. When the controller receives a request, it modifies the sequence order list and gives the requesting subscriber active status during the next sequence order list transmission.

Each TACINTEL subscriber has an ID number that is recognized by the subscriber processor. This ID number is used to screen incoming traffic. Only messages addressed to the subscriber are forwarded to the end-equipment (teletypewriter, recorder, printer), and the other traffic is discarded.

13.2.6 Tactical Data Information Exchange Subsystem (TADIXS)

The TADIXS supports broadcast transmission of TDP data from shore sites to the fleet. There is a TADIXS-A, sending fused data and TADIXS-B, sending raw data. TADIXS-A and TADIXS-B share a common channel, therefore, TADIXS-A is interrupted when

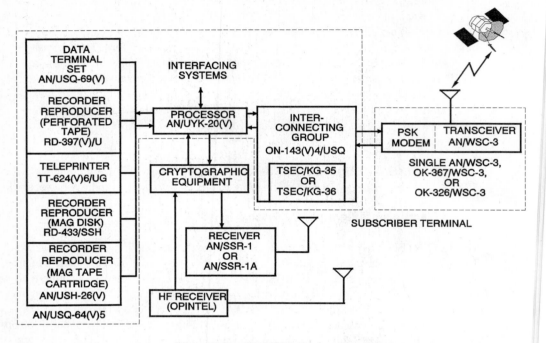

Figure 13.13   TACINTEL Shipboard Configuration

TADIXS-B is active. The TADIXS-B uses a dedicated processor, the Tactical Receive Equipment (TRE) that uses the ON-143(V)11/USQ. The TRE is a computerized message system that demodulates, decodes, decrypts, and processes the TADIXS-B transmissions. The information transmitted is data or messages in the Over-the-Horizon (OTH) Rainform Gold format.[5] Figure 13.14 shows the TADIXS-A shipboard receiver configuration; it uses an ON-143(V)6/USQ interconnection group. The TRE screens duplicate messages and filters them according to user-specified parameters. The TRE can connect to multiple TDPs; one TDP can connect via a point-to-point Military Standard MIL-STD-188-114 interface [DD-1], from others via a MIL-STD-1553B bus [DD-2].

13.2.7 Secure Voice Subsystem (SECVOX)

The SECVOX subsystem supports shore-to-ship, ship-to-ship, and ship-to-shore voice transmission. The narrowband voice (3-kHz channel) is digitized using a vocoder, producing a 2400-bps data stream that is subsequently encrypted. The SECVOX subsystem uses push-to-talk half-duplex transmission over a 2400-bps FLTSATCOM channel on each of the four satellites. A secure voice channel controller (operator) at the NCTAMS

---

[5]Rainform is the designation of a message format; the message formats are named according to colors, hence the terminology "Rainbow Format" or Rainform.

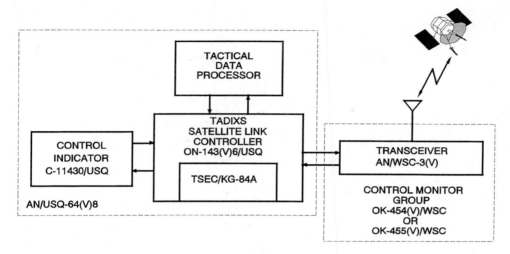

Figure 13.14   TADIXS Shipboard Configuration

or Naval Communications Station (NAVCOMMSTA) responsible for the satellite maintains channel control. SECVOX channels on the satellites are dedicated; when DAMA is used, a dedicated time slot is available.

Figure 13.15 shows the shipboard SECVOX configuration. The shipboard configu-

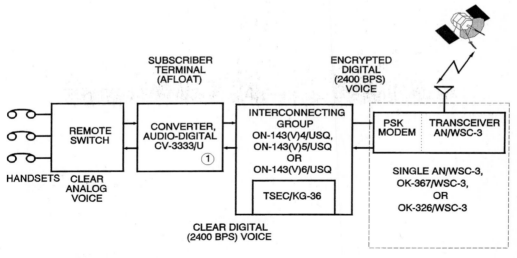

① THE CV-3591 IS THE NEW GENERATION
OF THIS EQUIPMENT

Figure 13.15   SECVOX Shipboard Configuration

ration depends on the type of platform where the installation is located. Two types of platforms are distinguished:

- Large ships, with two or more AN/WSC-3 terminals that can operate more than one FLTSATCOM link at one time.
- Small ships and submarines, with simple AN/WSC-3, where equipment is shared among NAVMACS (or SSIXS) and SECVOX.

SECVOX operation.   SECVOX channel access can be gained in two ways: (a) directly accessing a SECVOX channel that is not in use, and (b) coordinating with the SECVOX operator by sending a request for channel access to the controller if the SECVOX channel is busy.

Small ships (or submarines), which have only one AN/WSC-3 transceiver, use the NAVMACS/CUDIXS (or SSIXS) network to send the request. The voice transmission request is forwarded from the receiving shore station to the SECVOX controller. The controller coordinates the voice transmission by assigning a voice channel and contacting the party with whom the SECVOX link is to be established.

During voice transmission, the small ship (or submarine) drops out from the NAVMACS/CUDIXS (or SSIXS) net. Message traffic can still be received in the FLTBROADCAST net (which uses the AN/SRR-1 receiver and has dedicated crypto equipment; see Figure 13.8). After the SECVOX transmission is terminated, the small ship (or submarine) rejoins the NAVMACS/CUDIXS (or SSIXS) net.

A large ship, for example, an Aircraft Carrier (CV/CVN) or Guided Missile Cruiser (CG/CGN), also initiates a SECVOX request via NAVMACS/CUDIXS. However, the large ship, where two or more AN/WSC-3 terminals plus sufficient crypto and baseband units are available, remains an active member of the NAVMACS/CUDIXS net.

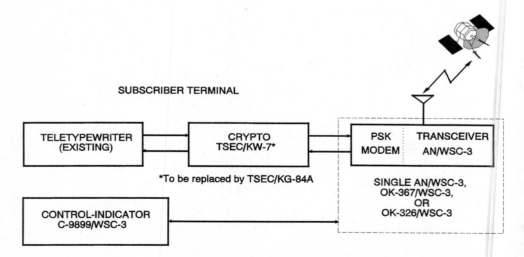

Figure 13.16   Teletypewriter Shipboard Terminal Configuration

13.2.8 Teletypewriter (TTY) Subsystem (ORESTES)

The ORESTES subsystem uses the FLTSATCOM to provide a reliable beyond-line-of-sight (BLOS) Navy TTY capability via satellite. The ORESTES subsystem provides full-period and on-call tactical and report-back TTY circuits that serve as backup for the other networks (e.g., NAVMACS/CUDIXS and TACINTEL). Figure 13.16 shows the shipboard configuration of an ORESTES terminal.

13.3 EQUIPMENT

In this section, the shipboard equipment used for the UHF FLTSATCOM is discussed. Included are transmitters, receivers, antennas, TTY, and voice terminals.

13.3.1 Receivers/Transceivers

There are two types of shipboard UHF FLTSATCOM receivers currently in use. One is the stand-alone AN/SSR-1 (and the AN/SSR-1A variant), and the other is the receiving section of the AN/WSC-3 transceiver.

AN/SSR-1 and AN/SSR-1A.   The AN/SSR-1 receiver is installed aboard most surface ships and supports the FLTBROADCAST message traffic reception. The receiver can be set for FM or PSK demodulation. The AN/SSR-1A receiver can drive high-level or low-level TTY equipment; the AN/SSR-1 interfaces to high-level TTY terminals only.

The AN/SSR-1 receiver consists of antennas, amplifier-converters (both shown in Figure 13.20), the MD-900 combiner-demodulator, shown in Figure 13.17, and the MD-1063 demultiplexer, shown in Figure 13.18. The equipment fits into a standard 19-inch rack of 222 mm (8.75 in) height and 635 mm (25 in) depth. The amplifier-converter AM-6534/SSR-1 converts the RF signals from the frequency region of 240 to 340 MHz to an intermediate frequency (IF). The combiner MD-900 demodulator generates a 1200-bps baseband output from the received RF. The TD-1063 demultiplexer accepts the 1200-bps

Figure 13.17   MD-900 Combiner-Demodulator (AN/SSR-1 Receiver Equipment) (Courtesy Motorola Government Systems and Technology Group)

Figure 13.18   TD-1063 Demultiplexer
(AN/SSR-1 Receiver Equipment) (Courtesy
Motorola Government Systems and
Technology Group)

data stream and demodulates it into the 15 75-bps channels that are connected to the
NAVMACS, TACINTEL, or TADIXS processors or teletypewriters.

AN/WSC-3.   Many variants of the AN/WSC-3 are used for receiving
FLTSATCOM channels and UHF LOS transmissions. Table 13.7 presents an overview.

The AN/WSC-3 is the principal equipment version, capable of supporting UHF
LOS communications and SATCOM. The AN/WSC-3 transceiver has two built-in demodulators for DPSK data rates from 75 to 9600 bps, FSK at 75 bps and AM or FM voice
modulation. The transmitter segment generates 30 W RF output for AM and 100 W for
FM, FSK, and DPSK. It can support AM narrowband and wideband as well as FM narrowband operations, AM voice and SECVOX, PSK data at 75, 300, 1200, 2400, 4800,
and 9600 bps, and FSK at 75 bps. It can be tuned to 7000 channels at a 25-kHz spacing.
The variants are modifications of the AN/WSC-3, with the capabilities modified as shown
in Table 13.7. The table shows which variants can support SATCOM, their audio and
control interfaces, and other key operating capabilities. One variant, the AN/WSC-3(V) 9
can be tuned to 34,995 channels at a 5-kHz spacing.

The AN/WSC-3 is used in different configurations. Small ships and submarines use
a single transceiver. Large ships usually have two or more transceivers installed. Special
racks are provided for the AN/WSC-3 installation. The two-transceiver configuration is
designated OK-367/WSC-3, and the four-transceiver configurations are designated
OK-326/WSC-3 and OK-326A/WSC-3.

A typical representative of the AN/WSC-3 family, the AN/WSC-3(V)6, is shown in
Figure 13.19.

13.3.2 Antennas

AS-2815/SSR-1   This antenna is used for receiving the Fleet Broadcast and is
combined with the AN/SSR-1 receiver. Figure 13.20 shows the antennas; on the bottom
of the figure, the AM-6534/SSR-1 amplifier-converter can be seen. Figure 13.21 shows
the AN/SSR-1 antenna on the topside of a ship. The antenna consists of two loops and
two parasitic dipoles. The loops are fed in phase quadrature and are right-hand circularly

TABLE 13.7 VERSIONS OF THE AN/WSC-3 TRANSCEIVER

Designation	SATCOM CAPABLE	DAMA	VINSON (FM wideband)	LINK 11	STANDARD AUDIO IF	SAS (3)	PARALLEL REMOTE CONTROL	SERIAL REMOTE CONTROL	MIL-STD-1553 BUS REMOTE CONTROL	5-kHz CHANNEL SPACING	IRR	400 Hz POWER	ECCM MODE	MTSC (1)	SAFS (2)
AN/WSC-3	X						X								
AN/WSC-3A	X	X					X								
AN/WSC-3A(V)2 (DAMA IDM)	X	X	X	X	X		X								
AN/WSC-3A(V)3 (DAMA IDM)	X	X	X	X		X	X								
AN/WSC-3(V)1	X						X								
AN/WSC-3(V)2	X		X	X	X		X								
AN/WSC-3(V)3	X		X	X		X	X								
AN/WSC-3(V)6	X		X	X	X		X								
AN/WSC-3(V)7			X	X		X	X								
AN/WSC-3(V)8 (Computer interface)			X	X			X	X							
AN/WSC-3(V)9 (AF)	X		X	X			X	X		X					
AN/WSC-3(V)10 (Have Quick)			X	X	X		X						X		
AN/WSC-3(V)11 (Have Quick)			X	X		X	X						X		
AN/WSC-3(V)12 (Trident)	X		X	X		X	X				X				
AN/WSC-3(V)13 (PHM)	X		X	X	X	X	X					X			
AN/WSC-3(V)14 (LHD)			X	X		X			X						
AN/WSC-3(V)15	X	X	X	X	X	X	X								
AN/WSC-3(V)16 (SAFS)	X		X	X	X		X	X							X
AN/WSC-3(V)17	X	X	X	X	X	X	X							X	
AN/WSC-3(V)18 (Trident)	X		X	X		X	X				X			X	
AN/WSC-3(V)19	X	X	X	X	X	X	X	X						X	X

(1) Modified transmit signal transient and preamble characteristics
(2) SSN AFSATCOM interface
(3) Single audio system

Figure 13.19   AN/WSC-3(V)6 Transmitter/Receiver Unit (Courtesy E-Systems)

polarized. A set of four antennas is used to provide full hemispherical coverage and space diversity reception.

AS-2410/WSC-1(V) and AS-3018/WSC-1(V).

These antennas are part of the antenna groups OE-82A/WSC-1(V) and OE 82B/WSC-1(V). These antenna groups interface with AN/WSC-3 installations in shipboard installations (an antenna group consists of the antenna itself and additional equipment for tracking, control, sensing, switching, and preamplification). Figure 13.22 shows the AS-3018/WSC-1(V). The AS-3018/WSC-1(V) is an improved version of the AS-2410/WSC-1(V) and has the same appearance as the AS-2410/WSC-1(V). The antenna consists of an array of four crossed dipoles mounted above a ground plane. Elements are fed in phase quadrature and generate right-hand circularly polarized waves. The antenna is mounted on a pedestal that permits 360° azimuthal and 2° to 110° elevational rotation. The antenna is used for simultaneous

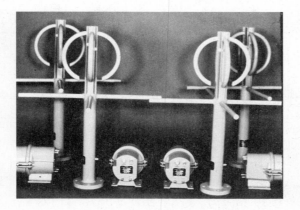

Figure 13.20   AN/SSR-1 Antenna and AM-6534/SSR-1 Receiver Equipment (Courtesy Motorola Government Systems and Technology Group)

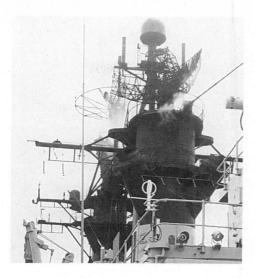

Figure 13.21 AN/SSR-1 Antenna on the topside of a Navy Vessel (Courtesy Motorola Government Systems and Technology Group)

transmission and reception. Automatic tracking keeps the antenna oriented toward the satellite during the ship's motion. To eliminate blocking by the ship's superstructure, two antennas are generally used with automatic switching between the antennas, should the actively operating one become blocked. The dimensions of the antenna are $1.25 \times 1.25 \times 0.7$ m ($50 \times 50 \times 28$ in.) (H × W × D), the weight is 147 kg (325 lbs). It operates in the frequency region of 240 to 318 MHz (maximal VSWR 1.5:1), has 150 W CW output power (250 W pep), is right-hand circularly polarized, and has a minimum gain of 12 dB.

AS-3018A/WSC-1(V).   This antenna, shown in Figure 13.23 [LA-1], electrically equivalent to the AS-3018/WSC-1(V) and used for the same purpose, is physically totally different. It is a component of the OE-82C/WSC-1(V) antenna group. The antenna consists of a single crossed dipole element recessed within a barrel-shaped drum struc-

Figure 13.22 AS3018/WSC-1(V) Antenna (Courtesy US Navy, Naval Sea Systems Command)

Figure 13.23   AS3018A/WSC-1(V) Antenna (Courtesy U.S. Navy, Naval Sea Systems Command)

ture. The back plane of the drum serves as reflector and the rim in front of the crossed dipole functions as director. This antenna has the same RF characteristics as the AS-3018/WSC-1(V) (same operating frequency region, VSWR, polarization, and gain) but has a higher power rating (400 W CW, 800 W pep), its size is 1.37×0.85 m (54×33 in.) (Diameter $\times$ Depth), and it weighs 133 kg (295 lbs). The antenna mount permits 360° azimuthal and 0° to 90° elevational orientation, and the antenna tracks the satellite automatically. Generally, two AS-3018A/WSC-1(V) antennas are used on a ship to eliminate superstructure blocking.

13.3.3 Secure Voice (SECVOX) Equipment

For naval communications, voice transmissions must be secure, that is, they have to be encrypted. To this end, voice must be digitized. In order to use the existing (narrowband) SATCOM capabilities, the digitized voice must be kept to 2400 bps.

    If normal voice, occupying a bandwidth of 300 to 3000 Hz, is sampled, coded, and digitized using PCM, the resulting data stream would be somewhere from 50 to 100 kbps in order to accommodate the needed number of samples and levels to provide reasonable fidelity. In order to achieve a low data rate of 2400 bps, data compression is needed. This can be realized by using voice processing such as performed by a VOCODER, or by using linear predictive coding (LPC-10), code-excited linear predictive coding (CELP), or continuously variable slope delta modulation (CVSD) (see Chapter 6, Table 6.2).

    The VOCODER and digitizer approach is used in narrowband secure voice terminal equipment such as the audio-digital converter CV-3333/U (see Figures 13.15 and 13.32). A more advanced and less complex means of compressing voice uses linear prediction coding (LPC). The LPC-10 encoding and decoding process is used in the current generation equipment such the CV-3591 ANDVT [also known under the nomenclature of AN/USC-43(V)]. The use of the LPC-10 process provides an improved voice fidelity.

    The AN/USC-43(V), shown in Figure 13.24, is capable of processing voice and signaling information. It operates in half-duplex mode only; two units are needed for full-duplex operation. The output of the AN/USC-43(V) is a 2400-bps data stream.

Figure 13.24 AN/USC-43(V) (CV-3591 ANDVT) (Courtesy ITT Aerospace/Communications)

Figure 13.25 AN/UGC-143A(V) Teleprinter (Courtesy NAI Technologies, Inc.)

The AN/USC-43(V) consists of four components: the CV-3591(P)/U converter unit (or basic terminal unit, BTU), a cryptography adapter, the KYV-5, the C-11006/U converter-modem control unit, and the J-3953/U interface unit. The BTU provides voice processing, coding, and modulation. It has a built-in modem for LOS applications; for SATCOM applications an external modem is used. The KYV-5 is an applique or plug-in unit; it is in principle similar to the KG-84 cryptography equipment family. The C-11006/U can control the CV-3591(P)/U so that it operates as a voice processor or modem processor. The J-3953/U interface unit is an ancillary device for tactical ground vehicle use.

13.3.4 Teletypewriter (TTY) Terminal Equipment

The Navy has several generations of TTY equipment in use. The prior generation equipment, still in use in many installations, is the AN/UGC-6 and AN/UGC-48. A newer teletypewriter consisting of a keyboard and a printer is the AN/UGC-77 that is used for 100 words per minute (wpm) TTY traffic. It uses a 5-unit code plus a start and stop bit (start-stop Baudot code). Since the start and stop bits are different in length, this code is also known as a 7.42-unit code. These are the type of TTY terminals using the Baudot code, where each character is on line as soon as it has been typed and where the operator must type at a minimum speed in order to maintain the flow of message transmission.

The Navy is in the process of replacing the old TTY terminals with new equipment.

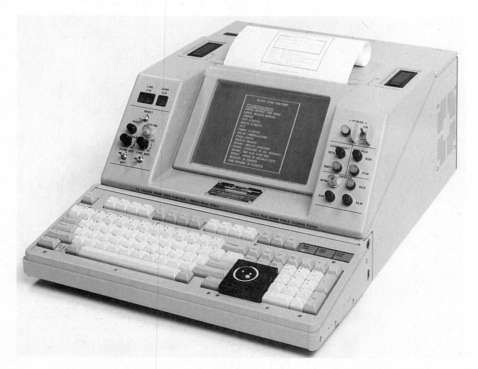

Figure 13.26   AN/UGC-143B(V) Teleprinter (Courtesy NAI Technologies, Inc.)

One device is the Navy Standard Teleprinter (NST) AN/UGC-143A(V) shown in Figure 13.25; Figure 13.26 shows a follow-on version, the AN/UGC-143B(V). The NST includes the facilities for message preparation and editing. Received data can be buffered. All information, transmitted and received, is printed for maintaining a record. The information is transmitted using the American Standard Computer Information Interchange (ASCII) code; various rates of transmission are supported. For encryption, the KG-84 family of equipment can be used.

An advanced keyboard/printer for use on submarines is the AN/UGC-136AX, shown in Figure 13.27. It can transmit and receive TTY messages at data rates from 50 to 2400 bps, and it can operate using Baudot and ASCII codes. The terminal supports message preparation and editing. The AN/UGC-136 has variants, the 136BX and the 136CX. The latter is used in SSN AFSATCOM Interface (SAFS) applications (see Table 13.7).

13.3.5 Interconnection Groups

The interconnection devices of the ON-143(V) USQ family (see Figures 13.9 through 13.15) provide the capability to interface UHF SATCOM links to control processors. They also perform synchronization of cryptographic units and they provide separation of plain and encrypted text. An example of an ON-143 device is also described in Chapter 18.

Figure 13.27   AN/UGC-136AX Teleprinter (Courtesy E-Systems)

A typical member of this equipment family is the ON-143(V) 6. It is very flexible, and can be used in several applications. It is used in OTCIXS, SSIXS, and SECVOX installations (see Figures 13.10, 13.12, and 13.15), it provides support for TTY message traffic, it has the capability to operate in an OTCIXS and SSIXS net, and it also supports SECVOX communications.

13.4 DEMAND ASSIGNED MULTIPLE ACCESS (DAMA) SUBSYSTEM

The Demand Assigned Multiple Access (DAMA) Subsystem was developed to relieve the congestion on the UHF Fleet Satellite. DAMA provides a flexible time division multiplexed (TDM) capability, which accommodates multiple channels in various formats. DAMA allows the sharing of one satellite repeater by several operational links. One station of the FLTSAT network, usually an NCTAMS in the satellite footprint, operates as master control station (MCS). The DAMA multiplexers at these MCSs serve as channel controllers. In order to provide a backup capability, any DAMA equipped platform can be designated as the channel controller [NO-1; NE-1].

13.4.1 The Principle of DAMA

The DAMA multiplexer (TD-1271B/U) accepts inputs from multiple channels at a normal rate, for example, SECVOX at 2400 bps, and generates a burst at either 9.6, 19.2, or 32 kbps for shipboard use. Interleaving of the transmitted bit stream in conjunction with using one-half or three-fourth rate convolutional coding is used to alleviate potential radio frequency interference (RFI). The bursts from various information carrying and auxiliary channels are combined. Three basic frame formats are used; Figure 13.28 shows the structure of the DAMA frame used for shipboard communications (TDMA-1) which combines the channels within a 1.386-s frame. Frame formats TDMA-2 and TDMA-3 consist of 8.32-s frames using 2400-bps bursts combining slots of 75-bps channels and are for aircraft (not for shipboard) use.

In order to accomplish the sharing, channel coordination and clocking are required. For establishing the first element of clocking, the beginning of the frame, the MCS transmits a channel control order wire (CCOW) signal. The preamble of the CCOW information contains control data such as frame format assignments, status information requests, and crypto initialization. The receiving platforms use the CCOW to establish the frame synchronization of the receiving data.

The distance between user platforms (ships) and the satellite, and hence the path delay between satellite and platforms, will be different. Differences of up to 40 ms are possible, depending on the relative positions of the platforms, yet operation using the TDMA-1 format requires an accuracy of 1 ms or better. Thus a second element of clocking is needed, the range between a platform and the satellite. This is accomplished by sending a ranging burst from each platform to the satellite. This burst is returned to the platform and the TD-1271B/U computes the range. Based on this computation, it is determined when a data burst must be transmitted so as to arrive at the satellite at the correct time slot established by the CCOW. Note that the path delay for each platform is differ-

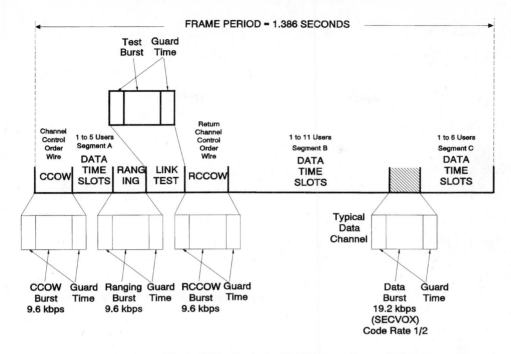

Figure 13.28   Structure of DAMA Frame Format (TDMA-1)

ent, hence the transmission in one channel can coincide with the reception in another channel. Thus, either independent AN/WSC-3 transmit and receive terminals are needed, or a specially configured AN/WSC-3 must be used in order to avoid interference or contention between transmission and reception onboard a platform [NE-2].

The link test is performed in order to measure the satellite circuit quality. The TD-1271B/U sends a short burst from the MCS to the satellite where it is retransmitted. The TDMA-1 frame format is designed such that the link test is transmitted during the CCOW receive slot, and the retransmission is received during the test time slot. Received bursts are compared to the transmitted ones for bit errors, erased symbols, and missed acquisitions.

Each DAMA multiplexer has a return channel control order wire (RCCOW) capability. The RCCOW can be used for sending operator-to-operator messages to channel controllers. A set of 255 prerecorded messages is used for this purpose.

The frame format TDMA-1 has three data segments A, B, and C, as shown in Figure 13.28. Segments A and C can carry 15 different formats, and segment B can carry 16 [NE-1]. An example is presented in Figure 13.29, which illustrates the DAMA operation. The example shows DAMA format 297, which uses format 2 in segment A, format 9 in segment B, and format 7 in segment C. The circuits in each segment are numbered; the example combines the TTY circuit 10066 (second circuit in segment A, format 2), which is one-half rate coded and at a burst rate of 19.2 kbps, with the TACINTEL DAMA circuit number 10296 (seventh circuit in segment B, format 9), which is three-fourth rate coded at a burst rate of 32 kbps, and the DAMA SECVOX circuit number 10242 (eigh-

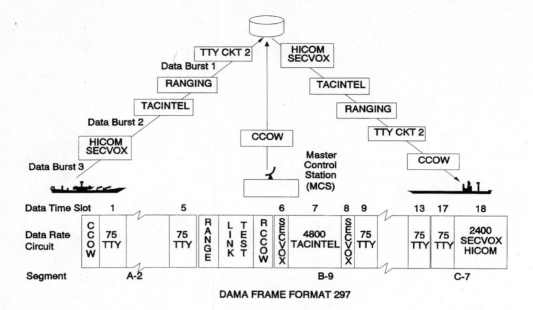

Figure 13.29   Illustration of DAMA Operation

teenth circuit in segment C, format 7), which is three-fourth rate coded at a burst rate of 19.2 kbps. Depending on which framing format is chosen, a maximum of 22 time slots are available in a DAMA frame.

Figure 13.29 also shows the MCS providing network coordination through the CCOW. The figure depicts sharing of TACINTEL, SECVOX, and TTY. The illustration indicates the ranging and portrays the composite data burst transmitted from one platform to another. Multiple platforms can be on the same net and share common circuits. A maximum of four circuits can be shared through DAMA in one FLTSAT transponder using the four inputs of the TD-1271B/U multiplexer.

13.4.2 DAMA Operational Considerations

For DAMA operation, one DAMA MCS is present in each satellite footprint. These MCSs are at NCTAMS LANT, NCTAMS MED, and NCTAMS PAC. These MCSs have multiple DAMA multiplexers TD-1271B/U installed, and each can accommodate up to four circuits. The number of subscriber multiplexers onboard each ship varies according to platform circuit requirements.

Four types of FLTSAT circuits are on DAMA. These are SECVOX, TACINTEL, CUDIXS/NAVMACS, and TTY. A mixed operation mode (a DAMA-equipped platform connected to a non-DAMA-equipped platform) is available; in this mode, the circuit is patched around the DAMA multiplexer.

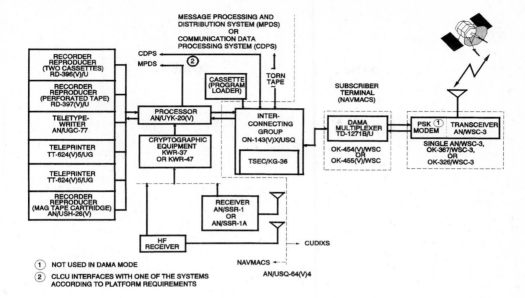

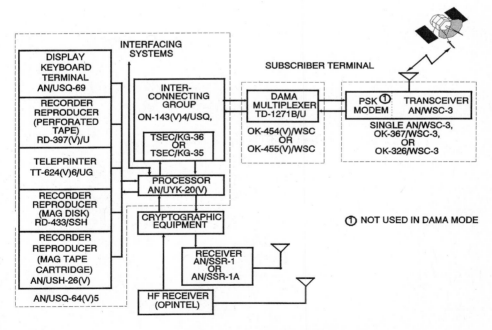

Figure 13.30   CUDIXS/NAVMACS DAMA Configuration

Figure 13.31   TACINTEL DAMA Configuration

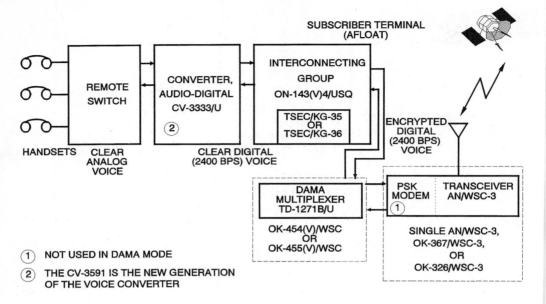

Figure 13.32   SECVOX DAMA Configuration

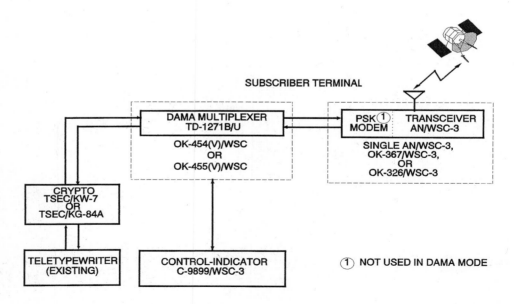

Figure 13.33   Teletypewriter DAMA Configuration

Figures 13.30 to 13.33 show the shipboard configurations of the FLTSAT subsystems that are supported by DAMA. The configurations are in essence the same as shown in section 13.2 for these subsystems, with the addition of the TD-1271B/U DAMA multiplexer.

13.4.3 DAMA Equipment

Ship installations can have either single or double DAMA subscriber terminals. Tables 13.8 and 13.9 summarize the equipment configurations.

Figures 13.34 and 13.35 show DAMA multiplexers. Figure 13.34 shows the TD-1271B/U multiplexer, and Figure 13.35 shows the mini-DAMA, developed for submarine applications.

Figure 13.34   TD-1271B/U DAMA Multiplexer (Courtesy of Motorola, Government Systems Technology Group)

TABLE 13.8 SHIPBOARD EQUIPMENT, SINGLE DAMA

OK-454(V)/WSC Control Monitor Group, Single DAMA	
Electrical cabinet equipment	CY-7970/WSC
Monitor Panel	MX-10342
Switchboard patching, receive/transmit control	SB-4124/WSC
Intermediate frequency patching panel	SB-4125/WSC
Multiplexer	TD-1271B/U

TABLE 13.9 SHIPBOARD EQUIPMENT, DUAL DAMA

OK-454(V)/WSC Control Monitor Group, Dual DAMA	
Electrical cabinet equipment	CY-7971/WSC
Monitor Panel	MX-10342
Switchboard patching, receive/transmit control	SB-4124/WSC
Intermediate frequency patching panel	SB-4125/WSC
Multiplexer	TD-1271B/U

13.5 UHF SATELLITE CONTROL SUBSYSTEM

The successful operation and coordination of a SATCOM system depends on a control system. With coordination, the existing resources can be adjusted to meet operational demands. The control system controls resources, monitors the system performance (satellite degradation), and keeps track of status information. It is the responsibility of the Commander of the Naval Space Command (COMNAVSPACECOM) to manage and control the operation of the Navy's SATCOM assets. The Naval Telecommunications Command

Figure 13.35 Mini-DAMA Multiplexer (Courtesy of Motorola, Government Systems Technology Group)

Operation Center (NTCOC) performs the day-to-day operational control of the SATCOM assets. The NCTAMS, USAF Satellite Operations Center, and the contractor operated control facilities support the NTCOC in performing the required control actions.

FLTSATCOM.   Figure 13.36 shows the FLTSATCOM control. The Navy monitors the operation and status of the assets and performs the operational control. The Air Force controls the satellite through tracking and telemetry.

The controlling NCTAMSs monitor the satellites and collect status data. Each station has a spectrum analyzer that monitors the FLTSAT channels. The data collected include RF signal levels, channel usage rates, and interference. Each location also monitors link quality using a bit error rate (BER) monitor on the Fleet Broadcast channel. A TTY network is used for data transmission among the Naval Space Command, NTCOC, FLEETCINCs, NCTAMSs, and the USAF Operations Center.

LEASAT.   The LEASAT is fully controlled by the Navy. The shore-based AN/FSC-79 terminal serves as FLTBROADCAST station and telemetry and control station for the LEASAT. This control includes satellite altitude, propulsion, electrical power, telemetry, tracking, and command (TT&C). Figure 13.37 shows the LEASAT control.

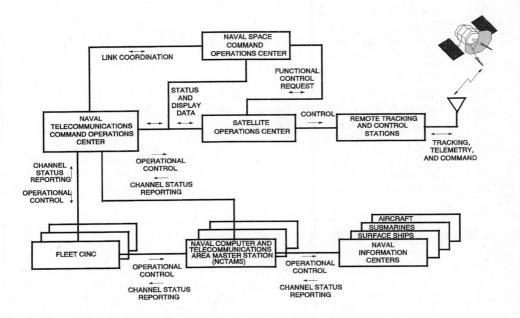

Figure 13.36   FLTSATCOM Operational Control

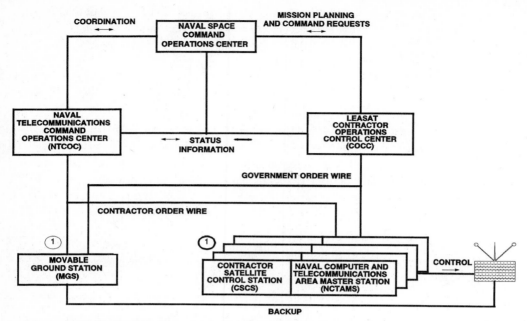

CSCS SITES: HAWAII (EASTPAC), GUAM (WESTPAC), NORFOLK (LANT), NAPLES (MED)
SCS SITES: NCTAMS EASTPAC, NCTAMS WESTPAC, NCTAMS LANT, NCTAMC MED
MGS SITES: GUAM, NORFOLK

SIGNAL FLOW:
FROM COCC TO MGS OR CSCS: GROUND COMMANDS; SPACECRAFT COMMANDS; MESSAGES, REPORTS, LIAISON; VOICE
FROM MSG OR CSCS TO COCC: TELEMETRY; RANGE, AZIMUTH, ELEVATION; GROUND EQUIPMENT STATUS REPORT; VOICE

Figure 13.37 LEASAT Operations Control

For performing the control, the AN/FSC-79 has the capability of tracking and ranging the LEASATs. The TT&C links operate in the S-band.

The system control is, like for the FLTSAT, the responsibility of the Naval Space Command. The NTCOC controls the FLTSAT network and coordinates the actions of the LEASAT Operations Control Center. The control command and telemetry are sent to the satellite from the contractor satellite operations control stations, which are colocated with the NCTAMS (Hawaii, Guam, Norfolk).

13.6 INTERNATIONAL MARITIME SATELLITE (INMARSAT)

The U.S. Navy uses the INMARSAT system to augment tactical shipboard communications; the INMARSAT also supports other services such as the broadcast of Armed Forces Radio Transmission Service (AFRTS). The INMARSAT is an international maritime satellite providing communications services for the maritime shipping industry [LO-1]. Currently, the INMARSAT organization includes 48 member nations. The COMSAT corporation is the U.S. signatory to the INMARSAT Organization. The INMARSAT network includes over 2000 Ship Earth Stations (SESs), 30 Coastal Earth Stations (CESs), and eight spacecraft.

The INMARSAT uses both C- and L-band frequencies [PR-1]. Table 13.10 lists the INMARSAT frequency plan for communications (uplink and downlink), command and ranging, and telemetry beacons. INMARSAT services include telephone, facsimile, slow-scan TV, telex, and data. While both C-band (SHF frequencies) and L-band (high UHF frequencies) are used, the shipboard equipment uses the L-band; hence this SATCOM capability is described in the chapter discussing UHF SATCOM.

The INMARSAT system consists of three subsystems, and they are the space segment, the CES, and the SES, as shown in Figure 13.38. All shore-to-ship and ship-to-shore communications must be routed via a CES which acts as a hub node. It is noted that a SES cannot establish a direct connection to another SES without going through a CES.

For shore-to-ship communications, the CES calls a ship in the ocean when requested by a shore INMARSAT user. Each INMARSAT SES terminal is assigned an identification code, and each ocean region is assigned a region code. Then, the CES automatically assigns a channel for the called ship and establishes a circuit.

The CES is a C-band Earth station with a 27-m parabolic reflector antenna. The shore-to-satellite uplink uses 6.4 GHz. The satellite-to-ship downlink uses 1.5 GHz. The SES is an L-band terminal with a 0.85-m to 1.2-m reflector antenna. There are five standardized INMARSAT SESs: they are the INMARSAT-A, INMARSAT-A64/A56, INMARSAT-C, INMARSAT-M, and INMARSAT-B. The INMARSAT-A SES supports 2400bps voice, facsimile, and telex services. The INMARSAT-A64/A56 service is a high-speed version of the INMARSAT-A services, operated at the data rate of 64/56 kbps. This service operates either in full-duplex or half-duplex mode. It supports communications for research and exploration vessel and imagery transmission. The INMARSAT-C service provides message and data communications at the rate of 600 bps [SU-1]. The INMARSAT-M is an extension of INMARSAT services to land-mobile users, operated at a data rate of 2400 bps. The INMARSAT-B service is the improved version of the INMARSAT-A, which operates at a data rate of 2400 bps [LO-1].

TABLE 13.10   INMARSAT FREQUENCY PLAN

Channel Designation	Frequency (GHz)	Polarization
Communications (Uplink Ship-to-Satellite)	1.6365–1.644	Right-hand Circular
Communications (Uplink Shore-to-Satellite)	6.420–6.424	Right-hand Circular
Communications (Downlink Satellite-to-Ship)	1.535–1.5425	Right-hand Circular
Communications (Downlink Satellite-to-Shore)	4195–4199	Left-hand Circular
Command & Ranging	6.175	Horizontal
Telemetry Beacon 1	3.945	Left-hand Circular
Telemetry Beacon 2	3.9545	Left-hand Circular

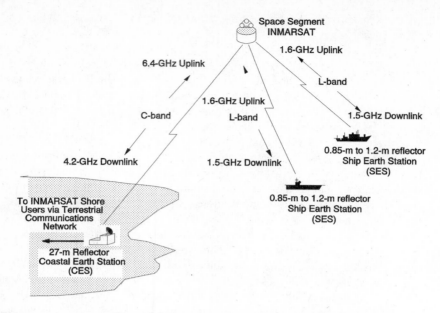

Figure 13.38   Typical INMARSAT Connectivity

For ship-to-shore communications, the SES places a call to a shore INMARSAT user. This connection is established by the SES sending a request to establish a connection to the CES. The CES automatically assigns a channel and returns this information to the SES, including the frequency selection. No operator intervention is required at either the SES or the CES.

The space segment.   The INMARSAT space segment leases transponders from three satellite carriers: MARISAT, INTELSAT, and MARECS. MARISAT is the system that serves the U.S. Navy and merchant marine; the COMSAT Corporation is the MARISAT system manager. The MARECS is spacecraft of the European Space Agency (ESA). The INTELSAT V satellite is owned by the INTELSAT consortium, and is equipped with a Maritime Communications Subsystem (MCS) for INMARSAT applications.

Figure 13.39 shows the MARISAT satellite. Several equatorial satellites provide the INMARSAT coverage in three regions of the International Telecommunications Union (ITU). The three INMARSAT ocean regions are the Atlantic Ocean Region (AOR), the Indian Ocean Region (IOR), and the Pacific Ocean Region (POR). The typical INMARSAT coverage pattern for these three regions is shown in Figure 13.40. A communications circuit operates via five in-orbit primary satellites, as described in Table 13.11. In addition, three satellites are in-orbit to provide spare capability.

Technical characteristics of INMARSAT leased satellites (MARISAT, INTELSAT V and MARECS) are shown in Table 13.12.

Figure 13.39   The MARISAT Satellite (Courtesy of Hughes Space and Communications Company)

TABLE 13.11   THE INMARSAT SATELLITE LOCATIONS

	Atlantic Ocean	Indian Ocean	Pacific Ocean
MARISAT	106.5°W (MARISAT F1)	72.5°E (MARISAT F2)	176.5°E (MARISAT F3)[1]
INTELSAT V	18.5°W (INTELSAT V F6)[1]	63°E (INTELSAT V F5); 60°E (INTELSAT V F7)[1]	None
MARECS	26°W (MARECS A)	None	177.5°E (MARECS B2)

[1]In-orbit spare.

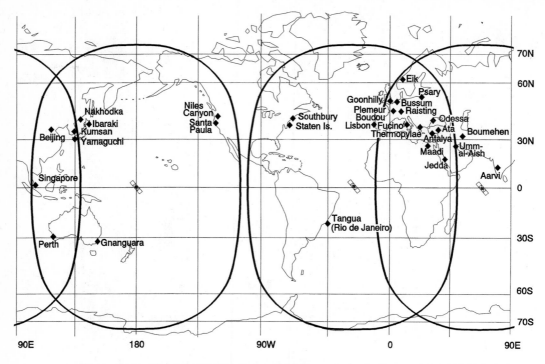

Figure 13.40ʹ The INMARSAT Coverage in Three Ocean Regions and INMARSAT CES Locations

TABLE 13.12 INMARSAT TECHNICAL CHARACTERISTICS

	MARISAT	INTELSAT V with MCS	MARECS A & B
Ship-to-Shore			
• TX Frequency (GHz)	6.420–6.424	6.4175–6.425	6.420–6.425
• RX Frequency (GHz)	1.537–1.541	1.535–1.5425	1.5375–1.5425
• RX G/T (dBK^{-1})	−19.6	−12.1	−15.0
• Sat EIRP (dBW)	27.0	33.0	34.5
• No. of Channels	12	35	12
Shore-to-Ship			
• TX Frequency (GHz)	4.195–4.199	4.195–4.200	4.1945–4.200
• RX Frequency (GHz)	1.6385–1.6425	1.6365–1.644	1.6385–1.644
• RX G/T (dBK^{-1})	−17.0	−13.0	−11.2
• Sat EIRP (dBW)	18.8	16.5	15.4
• No. of Channels	20	90	90

Coastal earth station.   INMARSAT Coastal Earth Stations (CESs) are operated by the INMARSAT member nations. These CESs serve as the hub node for telephone, telex, and data traffic for ship-to-shore and shore-to-ship communications in assigned ocean regions.

Currently there are 30 INMARSAT CESs in operation at locations in three assigned ocean regions as shown in Figure 13.40. Three of these CESs serve the Network Coordination Station (NCS), one for each International Telecommunications Union (ITU) region. The Atlantic Ocean Region (AOR) NCS is located at Southbury, Connecticut; the Indian Ocean Region (IOR) NCS is located at Yamaguchi, Japan; the Pacific Ocean Region NCS is located at Ibaraki, Japan.

Table 13.13 shows the technical characteristics of the INMARSAT CES and SES. An example of the performance and link budget calculations for a shore-to-ship and ship-to-shore communications link using an INMARSAT CES and an INMARSAT SES is given in Chapter 12.

Ship earth stations.   INMARSAT Standard-A SES terminals are equipped with 0.85- to 1.2-meter parabolic antennas. These antennas are mounted on a gyroscopically stabilized platform to prevent ship movements from interfering with its auto-tracking mechanism. These antennas are enclosed in a fiberglass radome to protect them against adverse sea environments [OV-1]. Figure 13.41 shows an INMARSAT-A SES terminal including 1.2-m parabolic antenna. A typical INMARSAT Standard-A SES terminal includes a telephone handset for voice communications and a telex machine for the creation or reception of printed messages.

TABLE 13.13   INMARSAT COASTAL EARTH STATION (CES) AND SHIP EARTH STATION (SES) CHARACTERISTICS

	Uplink	Downlink
Coastal Earth Station		
• Frequency Band	C-Band	C-Band
• Antenna Gain (dB)	54.0 minimum	50.50 minimum
• Polarization	Right-hand circular	Left-hand circular
• EIRP (dBW)	70	N/A
• G/T (dBK^{-1})	N/A	32 dBK^{-1} minimum
• Antenna Aperture (m)	11 to 13	11 to 13
Standard A Ship Earth Station		
• Frequency Band	L-Band	L-Band
• Antenna Gain (dB)	20.5–23 minimum	20.5–23 minimum
• Polarization	Right-hand circular	Left-hand circular
• EIRP (dBW)	36	N/A
• G/T (dBK^{-1})	N/A	−4 dBK^{-1} minimum
• Antenna Diameter (m)	0.85 to 1.2	11 to 13

Figure 13.41   INMARSAT-A Ship Earth Terminal (Courtesy of Magnavox Electronic System Company)

13.7 CURRENTLY PLANNED NAVY UHF SATCOM PROGRAMS

The Navy's UHF SATCOM capability is operationally highly important. In this section, the currently planned programs to upgrade and extend the Navy's UHF SATCOM capabilities are described.

13.7.1 UHF Follow-On Satellite System

The UHF follow-on Satellite Communications System is the replacement system for the Navy FLTSATCOM network. The satellite is shown in Figure 13.42. The FLTSATCOM satellite network was originally intended to have a useful service life of five years. Although in operation the satellites have proved to have more than double the intended design life, the FLTSATs (the first ones were launched in 1978) have already passed their intended design life. With over 1500 UHF user terminals, the Navy is more dependent now upon UHF satellite communications than at any time in the past. UHF satellites provide the lead service for shipboard satellite communications. This dependency on UHF satellites has necessitated that the Navy consider several replacement options for the

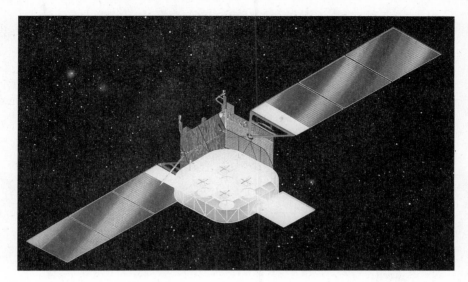

Figure 13.42   The UHF Follow-On Satellite (Courtesy of Hughes Space and Communications Company)

FLTSATCOM. While the MILSTAR joint program for EHF satellite communications and the Navy EHF satellite program are progressing, the Navy requires UHF satellite circuits to support the Navy's everyday administrative and tactical communications. Consequently, the UHF follow-on program is designed to provide future SATCOM service for the Navy. It is designed to provide reliable and redundant tactical satellite communications well into the future.

Eight satellites in equatorial orbits are planned to provide near-global coverage between the latitudes of 71° North to 71° South, covering the CONUS, the Atlantic, Pacific, and Indian Oceans, but not the polar regions. The UHF follow-on constellation will consist of two satellites over each of the four Earth coverage areas; the required capacity for each coverage area is divided between the two spacecraft. In addition there will be one in-orbit spare. The nine UHF follow-on satellites will be completely compatible with existing FLTSATCOM satellites.

The space segment.   Each satellite (see Figure 13.42) contains a communications module, a propulsion module, power supplies, transmit and receive antennas, and a telemetry, tracking, and command (TT&C) subsystem. The spacecraft is three-axis stabilized and weighs approximately 1078 kg without propellants. The spacecraft measures 1.7 m between the ends of the fully deployed solar arrays and has a body that approximates 0.21 m³.

The UHF follow-on satellite communications module contains a UHF and an SHF communications subsystem. The UHF communications subsystem consists of several 25- and 5-kHz channels. Signals received at the satellite are up-converted, amplified to a specified EIRP, and retransmitted. The SHF communications subsystem receives broadband jamming protected uplink signals of the Fleet Broadcast channel; up to three Fleet Broadcast channels can be multiplexed for simultaneous reception. Received signals are processed by the onboard processor MD-942A and are transmitted to the ground via the UHF communications subsystem. The MD-942A also processes the tracking beacon data which are downlinked by the SHF transmitter.

The UHF follow-on satellites carry the following four types of antennas:

- UHF transmit antennas with a nonreflective cup, reflecting disks, and four crossed dipole radiators for the UHF IXS channels
- UHF receive antennas with a planar, four-element patch array for the UHF IXS channels
- Horn-type SHF transmit antennas for beacon data
- Horn-type SHF receive antennas for the Fleet Broadcast channel.

TABLE 13.14   UHF FOLLOW-ON SATELLITE CHARACTERISTICS

Group	Channel characteristics
Group I	• Two 25-kHz bandwidth channels with a variable satellite translation frequency • EIRP of 28 dBW • Jam-resistant SHF uplink • Four Downlink Frequencies: 250.350, 250.250, 250.550, and 250.650 MHz • Four Uplink Frequencies: SHF
Group II	• Nine 25-kHz bandwidth channels with satellite translation frequency of 41 MHz • EIRP: Four channels—28 dBW Five channels—26 dBW • Four Downlink Frequencies per channel: 251.400–269.950 MHz • Four Uplink Frequencies per channel: 292.850–310.950 MHz
Group III	• Eight 25-kHz bandwidth channels with a satellite translation frequency of 33.6 MHz • EIRP of 26 dBW • Four Downlink Frequencies per channel: 260.375–263.925 MHz • Four Uplink Frequencies per channel: 293.975–297.525 MHz
Group IV	• Eight 5-kHz bandwidth channels with a satellite translation frequency of 73.1 MHz • EIRP of 20 dBW • Four Downlink Frequencies per channel: 243.925–244.225 MHz • Four Uplink Frequencies per channel: 317.015–317.325 MHz
Group V	• 13.5-kHz bandwidth channels with a satellite translation frequency of 53.6 MHz • EIRP of 20 dBW • Four Downlink Frequencies per channel: 248.845–249.355 MHz • Four Uplink Frequencies per channel: 302.445–302.955 MHz

The TT&C subsystem provides jamming-protected and secure satellite control while eliminating the need to procure new ground control and tracking facilities. The UHF follow-on satellite system also utilizes the same SHF TT&C signals as do current LEASAT satellites and their ground facilities.

The UHF follow-on satellite characteristics are summarized in Table 13.14. The specific transmit and receive frequencies are shown in Table 13.15. The frequencies of the uplinks and downlinks are arranged in four frequency plans to avoid interference in areas where coverage overlaps. Forty channels are dedicated for use by the Navy; each channel has a separate transmitter. The power for each transmitter is also given in Table 13.14; the transmitter power is fixed for the normal mode of operations. Channels 1 and 2 carry primarily Fleet Broadcast transmissions with SHF uplink and UHF downlink transmission.

The ground segment.   The overview of the UHF follow-on satellite ground system is shown in Figure 13.43. The Consolidated Space Operations Center (CSOC) has the overall control of the UHF follow-on satellite, as delegated by JCS. The normal day-by-day control of the UHF follow-on satellite is delegated to the Navy Satellite Control Station (NSCS) via the mission control station.

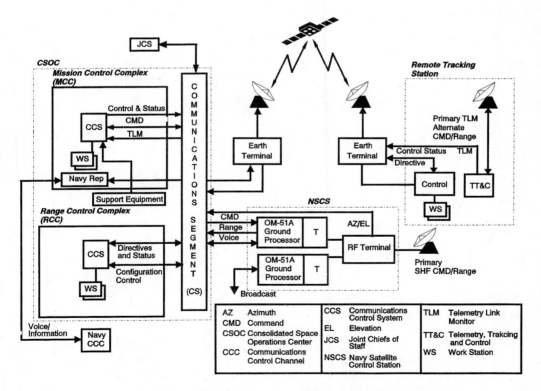

Figure 13.43   UHF Follow-On Satellite Ground Segment Overview

TABLE 13.15   UHF FOLLOW-ON TRANSMIT AND RECEIVE FREQUENCIES

Group	Channel	Plan	Downlink Frequency (MHz)	Uplink Frequency (MHz)	Nominal Bandwidth (kHz)
I	1	N	250.350	SHF	25
		O	250.450	SHF	25
		P	250.550	SHF	25
		Q	250.650	SHF	25
I	2	N′	250.400	SHF	25
		O′	250.500	SHF	25
		P′	250.600	SHF	25
		Q′	250.700	SHF	25
II	3	N	251.850	292.850	25
		O	251.950	292.950	25
		P	252.050	293.050	25
		Q	252.150	293.150	25
II	4	N	253.550	294.550	25
		O	253.650	294.650	25
		P	253.750	294.750	25
		Q	253.850	294.850	25
II	5	N	255.250	296.250	25
		O	255.350	296.350	25
		P	255.450	296.450	25
		Q	255.550	296.550	25
II	6	N	256.850	297.850	25
		O	256.950	297.950	25
		P	257.050	298.050	25
		Q	275.150	298.150	25
II	7	N	258.350	299.350	25
		O	258.450	299.450	25
		P	258.550	299.550	25
		Q	258.650	299.650	25
II	8	N	265.250	306.250	25
		O	265.350	306.350	25
		P	265.450	306.450	25
		Q	265.550	306.550	25
II	9	N	266.750	307.750	25
		O	266.850	307.850	25
		P	266.950	307.950	25
		Q	267.050	308.050	25
II	10	N	268.150	309.150	25
		O	268.250	309.250	25
		P	268.350	309.350	25
		Q	268.450	309.450	25

TABLE 13.15 UHF FOLLOW-ON TRANSMIT AND RECEIVE FREQUENCIES
(*Cont.*)

Group	Channel	Plan	Downlink Frequency (MHz)	Uplink Frequency (MHz)	Nominal Bandwidth (kHz)
II	11	N	269.650	310.650	25
		O	269.750	310.750	25
		P	269.850	310.850	25
		Q	269.950	310.950	25
III	12	N	260.375	293.975	25
		O	260.475	294.175	25
		P	260.425	294.025	25
		Q	260.625	294.225	25
III	13	N	260.475	294.075	25
		O	260.575	294.275	25
		P	260.625	294.125	25
		Q	260.725	294.325	25
III	14	N	261.575	295.175	25
		O	262.075	295.675	25
		P	261.625	295.225	25
		Q	262.125	295.725	25
III	15	N	261.675	295.275	25
		O	262.175	295.775	25
		P	262.725	295.325	25
		Q	262.225	295.825	25
III	16	N	261.775	295.375	25
		O	261.275	295.875	25
		P	261.825	295.425	25
		Q	262.325	295.925	25
III	17	N	261.875	295.475	25
		O	262.375	295.975	25
		P	261.925	295.525	25
		Q	262.425	295.025	25
III	18	N	263.575	297.175	25
		O	263.775	297.375	25
		P	263.225	297.225	25
		Q	263.425	297.425	25
III	19	N	263.675	297.275	25
		O	263.875	297.475	25
		P	263.725	297.325	25
		Q	263.925	297.525	25
IV	20	N	243.915	317.015	5
		O	243.995	317.095	5
		P	244.075	317.175	5
		Q	244.155	317.255	5

TABLE 13.15 UHF FOLLOW-ON TRANSMIT AND RECEIVE FREQUENCIES
(*Cont.*)

Group	Channel	Plan	Downlink Frequency (MHz)	Uplink Frequency (MHz)	Nominal Bandwidth (kHz)
IV	21	N	243.925	317.025	5
		O	243.005	317.105	5
		P	244.085	317.185	5
		Q	244.165	317.265	5
IV	22	N	243.935	317.035	5
		O	244.015	317.115	5
		P	244.095	317.195	5
		Q	244.175	317.275	5
IV	23	N	243.945	317.045	5
		O	244.025	317.125	5
		P	244.105	317.205	5
		Q	244.185	317.285	5
IV	24	N	243.955	317.055	5
		O	244.035	317.135	5
		P	244.115	317.215	5
		Q	244.195	317.295	5
IV	25	N	243.965	317.065	5
		O	244.045	317.145	5
		P	244.125	317.225	5
		Q	244.205	317.305	5
IV	26	N	243.975	317.075	5
		O	244.025	317.125	5
		P	244.235	317.235	5
		Q	244.315	317.315	5
IV	27	N	243.985	295.085	5
		O	244.065	295.165	5
		P	244.145	295.245	5
		Q	244.325	295.325	5
V	28	N	248.845	302.445	5
		O	248.975	302.575	5
		P	319.105	302.705	5
		Q	249.235	302.835	5
V	29	N	248.855	302.455	5
		O	248.985	302.585	5
		P	249.115	302.715	5
		Q	249.245	302.845	5
V	30	N	248.865	302.465	5
		O	248.995	302.595	5
		P	349.125	302.725	5
		Q	249.225	302.855	5

TABLE 13.15 UHF FOLLOW-ON TRANSMIT AND RECEIVE FREQUENCIES (*Cont.*)

Group	Channel	Plan	Downlink Frequency (MHz)	Uplink Frequency (MHz)	Nominal Bandwidth (kHz)
V	31	N	248.875	302.475	5
		O	248.005	302.605	5
		P	249.135	302.735	5
		Q	249.265	302.865	5
V	32	N	248.885	302.485	5
		O	249.015	302.615	5
		P	249.145	302.745	5
		Q	249.275	302.875	5
V	33	N	248.895	302.495	5
		O	249.025	302.625	5
		P	319.155	302.755	5
		Q	249.285	302.885	5
V	34	N	248.905	302.505	5
		O	249.035	302.635	5
		P	249.165	302.765	5
		Q	249.295	302.895	5
V	35	N	248.915	302.515	5
		O	249.045	302.645	5
		P	249.175	302.775	5
		Q	249.305	302.905	5
V	36	N	248.925	302.525	5
		O	249.055	302.655	5
		P	249.185	302.785	5
		Q	249.315	302.915	5
V	37	N	248.935	302.535	5
		O	249.065	302.665	5
		P	249.195	302.795	5
		Q	249.325	302.925	5
V	38	N	248.945	302.545	5
		O	249.075	302.675	5
		P	249.205	302.805	5
		Q	249.335	302.935	5
V	39	N	248.955	302.555	5
		O	249.085	302.685	5
		P	249.215	302.815	5
		Q	249.345	302.945	5
V	40	N	248.955	302.555	5
		O	249.085	302.685	5
		P	249.215	302.815	5
		Q	249.345	302.945	5

The primary command and satellite ranging for the UHF follow-on satellite is performed by NSCS, usually colocated with NCTAMS, using the existing AN/FSC-79 Earth terminal and OM-51A ground processor. The NSCS, which also controls the LEASAT constellation, carries out the primary command and satellite ranging. The TT&C operations are performed by the Remote Tracking Station via the Mission Control Complex (MCC) of the CSOC. The TT&C operations include the Telemetry Link Monitor (TLM) and the control status. The MCC and the Remote Tracking Station are connected to the Air Force Satellite Control Network (AFSCN). The Remote Tracking Station also performs the alternate command and ranging functions for the UHF follow-on satellite in case of NSCS outage.

13.7.2 High-Speed Fleet Broadcast (HSFB)

The Fleet broadcast carries message and record traffic. In the FLTSAT it is supported by one 1200-bps channel that is subdivided into 15 75-bps subchannels. For many requirements 75 bps is too limiting, and the channelization does not provide the flexibility for meeting many current and most future message traffic needs.

The HSFB is designed to provide a significantly improved message and TTY traffic flow from shore-to-ship. The HSFB aggregate data rate is planned to be 9600 bps with a maximum channel throughput of 1200 bps. The aggregate data rate can support up to 12 channels with data rates from 75 to 1200 bps, thus providing flexibility as needed.[6]

Figure 13.44 shows a block diagram of the HSFB shipboard equipment. Both the UHF and the HF HSFB are depicted. The technology chosen for the HSFB uses a VME-bus chassis. The key components, that is, the TDM, encoding and decoding, and the SATCOM modulator and demodulator, are incorporated into the chassis.

Slightly different shipboard equipment configurations for the HSFB are expected to be needed to accommodate the full range of Fleet broadcast reception requirements. The flexibility offered by the VME-bus chassis design and the availability of a wide range of commercial-off-the-shelf (COTS) component cards are expected to support such modifications and provide a spectrum of functionality.

The VME-bus design is consistent with the trend to use nondevelopmental items (NDI) and COTS for future equipment for naval communications. The equipment can be ruggedized for naval applications. The HSFB is designed to interface to current Fleet broadcast equipment and automated equipment such as the NAVMACS II. It is also planned to be consistent with the Navy's Communications Support System (CSS) of the Copernicus Architecture (see Chapter 20). The plan is conceived to employ multiple media such as UHF, EHF (MILSTAR), and HF. At the receiver end, the HSFB will be compatible with the existing cryptographic equipment, that is, the KW-46 and the KG-84 family. The HSFB equipment is currently in the acquisition stage, and its inventory is expected to grow to over 300 units in the next few years.

---

[6]There is also an HF Fleet Broadcast. The HF Fleet Broadcast will be modernized, too. The aggregate rate will be 1200 bps with a flexible channel data rate.

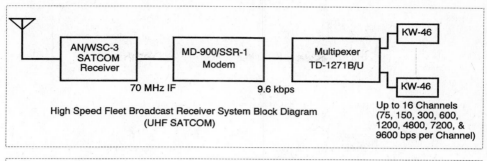

High Speed Fleet Broadcast Receiver System Block Diagram
(UHF SATCOM)

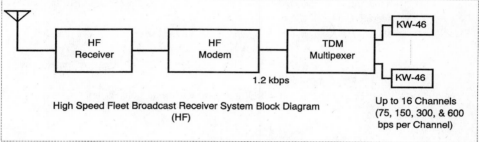

High Speed Fleet Broadcast Receiver System Block Diagram
(HF)

Figure 13.44   High-Speed Fleet Broadcast Block Diagram

13.7.3 New Enhanced Terminals

The current shipboard UHF SATCOM terminals are approaching the end of their design lives. The Navy is currently planning to replace the AN/WSC-3(V) radio family, the TD-1271B/U DAMA multiplexer, the AN/SSR-1 Fleet Broadcast receiver, and the OE-82C/WSC antenna. The equipment will be replaced, where feasible, during the ship improvement program, by NDI using the latest available technology.

The cost of replacing the complete UHF SATCOM equipment on a ship is deemed prohibitive. Consequently, the objective is to use modern equipment and add-on and change-out items. The goals are to reduce weight, improve the bandwidth capability, and reduce antenna costs. In addition, the effectiveness of the SATCOM capabilities are to be increased by expanding the use of DAMA.

REFERENCES

[DD–1] *Electrical Characteristics of Digital Interface Circuits*, Military Standard MIL-STD-188-114A, U.S. Department of Defense, Washington, D.C. 20360, Sept. 30, 1985.

[DD–2] *Aircraft Time Division Command/Response Multiplex Data Bus*, Military Standard MIL-STD-1553B, U.S. Department of Defense, Washington, D.C. 20350, Sept. 8, 1986.

[JC–1] *Description of US Message Text Formatting Program*, JCS PUB 6-04.1 (Formerly JCS PUB 25, Vol I), The Joint Chiefs of Staff, Washington D.C. 20318, Oct. 1, 1989.

[JC–2] *Automatic Digital Network (AUTODIN) Operating Procedures*, JANAP-128(I) JCS Publication, The Joint Chiefs of Staff, Washington D.C. 20318, March 1983.

[LA–1] Law, P. E., *Shipboard Antennas*, 2nd ed., Artech House, Dedham, Mass., 1986.

[LO–1] Long, M., *World Satellite Almanac*, 2nd ed., Howard W. Sams and Co. Indianapolis, Ind., 1989.

[NA–1] *Communications Instructions—General*, ACP-121, NATO Communications Procedures.

[NA–2] *Communications Instructions—Tape Relay*, ACP-127, NATO Communications Procedures.

[NE–1] *UHF DAMA Operator OJT Handbook (Shore)*, Doc. No. FSCS-211-84-2, Naval Electronic System Engineering Activity, St. Inigoes, Md., 20684.

[NE–2] "Contention in the DAMA System," *DAMA Newsletter*, Naval Electronics System Engineering Center, Vallejo, Calif., April 1986.

[NO–1] *Navy UHF Satellite Communication System Description*, Doc. No. FSCS-200-83-1, 31 December 1991, Naval Ocean System Center, San Diego, Calif., 92152.

[OV–1] "An Ocean Voice Buyers's Guide—INMARSAT Terminal Equipment—A, C, and M," *Ocean Voice*, April 1993.

[PR–1] Pratt, T. and C. Bostian, *Satellite Communications*, John Wiley & Sons, New York, N.Y., 1986.

[SU–1] Subramanian, R., and O. Ichiyoshi, "Development of INMARSAT-M Mobile Earth Station for Land Mobile Application," *AIAA Conference on SATCOM-92*, paper number AIAA-92-1816-CP.

[WI–1] Williamson, J., ed., "FLTSATCOM Fleet Satellite Communications System," *Jane's Military Communications, 12th Ed., 1991–1992*, Jane's Defence Data, pp. 361–362, 1991, Jane's Information Group, Sentinel House, 163 Brighton Road, Coulsdon Surrey, CR5 2NH, U.K. (In the U.S.: Jane's Information Group, 1340 Braddock Pl, Suite 300, Alexandria, Va. 22314).

[WI–2] Williamson, J., ed., "LEASAT Communications Satellite System," *Jane's Military Communications, 12th Ed., 1991–1992*, Jane's Defence Data, p. 363, 1991, Jane's Information Group, Sentinel House, 163 Brighton Road, Coulsdon Surrey, CR5 2NH, U.K. (In the U.S.: Jane's Information Group, 1340 Braddock Pl, Suite 300, Alexandria, Va. 22314).

14

SHF Satellite Communications

The Defense Communications Satellite System (DSCS) satellites provide a means for military SHF satellite communications (SATCOM). The DSCS supports primarily strategic long-haul communications and some tactical communications (as compared to the UHF satellites, which support primarily tactical communications). The DSCS is a tri-service resource, which is administered by the Defense Information Systems Agency (DISA). Resource allocation is validated by the Joint Chiefs of Staff (JCS).

The use of the SHF X-band frequency region (7.25–8.4 GHz), where the DSCS operates, offers certain advantages for SATCOM operations. They are:

- Substantial bandwidths are available for allocation, which enables the operation of several 50-, 60-, and 85 MHz repeaters on the DSCS satellite.
- High data rates can be supported.
- Propagation is very stable and propagation-related outages (e.g., fading or scintillation) are minimal.
- High-gain antennas with narrow beamwidth can be used, which reduces the probability of interception and provides advantages against jamming.
- Band spreading can be used to provide jamming-resistant communications.[1]
- Interoperability among the services is possible.

[1]Also known as anti-jamming (AJ) communications.

The DSCS assets are shared by many users, that is, military departments (Army, Navy, and Air Force) and defense-related agencies (e.g., the National Security Agency—NSA, the Central Intelligence Agency—CIA, the Drug Enforcement Agency—DEA, and others). Communications requirements are supported for the following:

- Presidential missions
- Contingency extensions of the Defense Communication System (DCS)
- Navy ship-shore communications
- Diplomatic Telecommunications Service (DTS)
- Airborne command post
- Ground Mobile Forces (GMF)
- Wideband data transmission
- Interarea trunking
- AUTOVON/AUTODIN trunking
- AUTOSEVOCOM
- Electronics Counter Countermeasures (ECCM) C^3I[2]
- NATO.[3]

The diversity of users and types of terminals, (large, small, fixed, transportable, land-based, seaborne, airborne) requires careful load balancing and control of the DSCS operation. This will be discussed in section 14.4.

The Navy is a user of the DSCS; DSCS ship terminals are installed on aircraft carriers (CV and CVN) and on Surveillance Towed Array Subsystem (SURTASS) ships. DSCS communications to these ships are maintained via Navy DSCS shore terminals. The shore terminals are located at [ND-1]:[4]

- Northwest Virginia, supporting NCTAMS LANT, serving the Atlantic Ocean
- Wahaiawa, Hawaii, supporting NCTAMS EPAC, serving the Eastern Pacific Ocean
- Finegayan, Guam, supporting NCTAMS WPAC, serving the Western Pacific Ocean
- Lago di Patria, Italy, supporting NCTAMS MED and NAVEUR, serving the Mediterranean Sea and the Indian Ocean.

---

[2]The abbreviation C^3 stands for Command, Control, and Communications; it is referred to as C-cubed. In the literature one may find the abbreviation C^3 or C3.

[3]NATO maritime SATCOM (e.g., Skynet, Syracuse) uses only SHF; hence, U.S. ships that interoperate with NATO forces, e.g., COMSECONDFLT flagships, require an SHF terminal. This is one of the reasons why LCC-19/20 and CVNs are equipped with SHF AN/WSC-6 SATCOM terminals.

[4]The Navy has also acquired special-purpose, smaller commercial terminals with 9-m (30-ft) antennas for support of operations at sites to which it is difficult to provide wideband land-based communications from the main shore terminals.

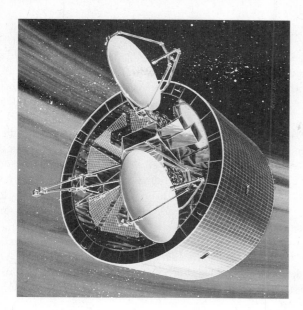

Figure 14.1   DSCS II Satellite (Courtesy TRW Inc.)

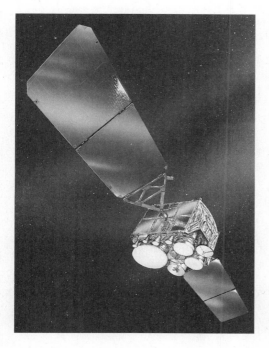

Figure 14.2   DSCS III Satellite (Courtesy of Martin Marietta Astro Space)

14.1 SATELLITES

The DSCS is supported by two types of satellites [WI-1], the TRW-built DSCS II [TR-1], shown in Figure 14.1, and the Martin Marietta-built DSCS III [MA-1], shown in Figure 14.2. Each of the DSCS II satellites has reached or exceeded its design life and they are replaced by DSCS III satellites, which are described in more detail below. The DSCS III satellites are first lifted into a low altitude orbit. The satellites are then boosted into a 24-hour equatorial orbit using the Integrated Apogee Booster System (IABS). Figure 14.3 shows the footprints of the DSCS satellites. There are eight DSCS satellites in the geostationary orbit. They serve the Pacific area (EPAC, WPAC), the Atlantic area (ELANT, WLANT), and the Indian Ocean (IO).

The main characteristics of the DSCS III are summarized in Table 14.1. Figure 14.4 shows the DSCS III satellite transponder channel configuration. It illustrates the flexibility of the available channels, repeaters, amplifiers, and antennas, which can be connected by switching elements to adapt the satellite to area communications needs. There are six transponders with bandwidths from 50 to 85 MHz with power amplifiers of 10 and 40 W. One transponder is hard-limiting for support of direct sequence spread-spectrum AJ communications. This transponder performs onboard processing and is protected against jamming. Repeater operation can be controlled from the ground. Figure 14.4 also shows that the satellite contains a UHF single-channel transponder and a UHF antenna which can be

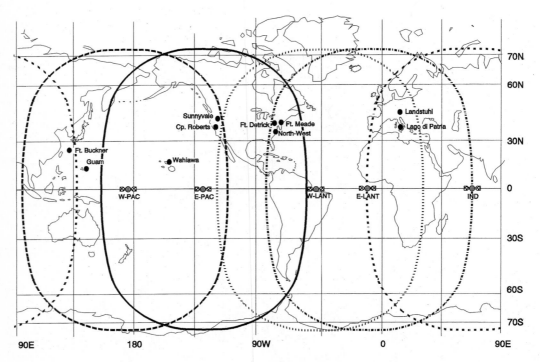

Figure 14.3   DSCS Satellite Footprints

TABLE 14.1 SUMMARY OF DSCS III CHARACTERISTICS

Communications X-band transponders			Satellite (spacecraft)	
Effective Isotropic Radiated Power (EIRP)			Spacecraft Control	
Channel	Earth Coverage	Narrow Coverage	Beam Pointing Accuracy: 0.2° Circular Error Station Keeping Accuracy: ±0.1°	
CH 1 (MBA)	29.0 dBW	40.0 dBW		
CH 2 (MBA)	29.0 dBW	40.0 dBW	Attitude Control Accuracy	
CH 3 (MBA)	23.0 dBW	34.0 dBW	Roll	0.09°
CH 4 (MBA)	23.0 dBW	34.0 dBW	Pitch	0.09°
CH 3 (EC[1] Horn)	25.0 dBW	–	Yaw	0.2°
CH 4 (EC Horn)	25.0 dBW	–	Size	
CH 5 (EC Horn)	25.0 dBW	–	Length (w.o. Panels)	2.07 m (81.5 in)
CH 6 (EC Horn)	25.0 dBW	–	Length (w. Panels)	11.63 m (458 in)
CH 1,2 (Dish)	44.0 dBW		Width	1.93 m (76 in)
CH 4 (Dish)	37.5 dBW		Depth	1.95 m (77 in)
G/T (Earth Coverage)			Weight (Approximately, Dry)	
EC Horn	-14 dBK^{-1}		900 kg (2000 lbs)	
MBA	-16 dBK^{-1}		Tracking Beacons	
RF Power Output			Frequencies	7600 and 7605 MHz
CH 1,2	40 W		EIRP	13 dBW
CH 3,4,5,6	10 W		Design Life	
Bandwidth			10 Years	
CH 2	75 MHz		Overall Reliability	
CH 3,4	85 MHz		< 0.7 at 7 Years	
CH 5	60 MHz		Power	
CH 1,6	50 MHz		1240 W (Arrays, Start of Life) 980 W (Arrays, after 10 Years)	
Gain Control Range				
39 dB				
Antenna Earth Coverage (Gain)				
MBA	15.0 dB			
EC Horn	17.0 dB			
Antenna Narrow Beam Coverage				
MBA	26 dB			
Gimballed Dish	30.2 dB			

[1]EC = Earth coverage

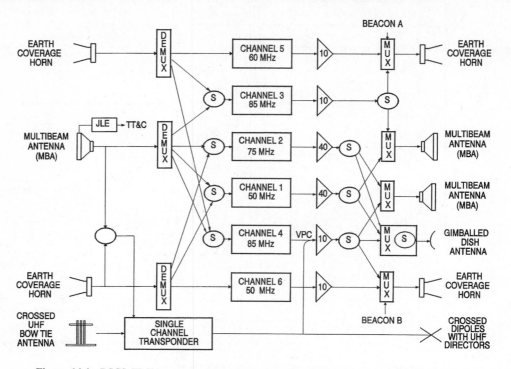

Figure 14.4   DSCS III Transponder Channel Configuration (S indicates a switch) (Courtesy of Martin Marietta Astro Space)

used as required. The single-channel transponder (SCT) serves as the communications means for Emergency Action Messages (EAMs) and the Single Integrated Operation Plan (SIOP). This transponder receives signals either via UHF or SHF from ground or airborne command post terminals. The satellite transmits the SCT signal on UHF for global coverage. The DSCS III satellite contains two beacons, one with full Earth coverage (EC), the other with narrow coverage for acquisition and tracking of the satellite from mobile platforms. A typical SHF SATCOM link budget analysis is given in Chapter 12.

      The DSCS III satellite has several antennas. Low-gain horns provide Earth coverage, high-gain, multibeam antennas (MBAs) provide contoured patterns for selective coverage with increased gain and selective nulling to counteract jamming. The DSCS III MBAs are waveguide lens antennas, one with 61 beams for receiving, two with 19 beams for transmission. The MBAs are high-efficiency antennas (60%). Each MBA has a beam-forming network for the rapid formation of accurate selective coverage patterns. Antennas can be connected to the transponders by ground command from the DSCS satellite control facilities to create configurations as required to satisfy operational needs.

      Figure 14.5 and Table 14.2 show the DSCS III satellite frequency plan. The graphic presentation of Figure 14.5 is augmented by Table 14.2 that includes the frequency allocation of the single-channel UHF transponder and also shows which antenna can be connected to each of the transponders.

TABLE 14.2 DSCS FREQUENCY PLAN (COURTESY OF MARTIN MARIETTA ASTRO SPACE)

Spectrum	SHF	from 7900 to 8400 MHz Receive
		from 7250 to 7750 MHz Transmit
	UHF (SCT)	from 300 to 400 MHz Receive
		from 225 to 260 MHz Transmit
	SHF (SCT)	Channel 1
	SHF (TT&C)	Channels 1, 5 Receive
		7600.0, 7604.70558 MHz Transmit
	S-Band (TT&C)	SGLS[1] Channels 12, 16
		1807.764, 1823.779 MHz Receive
		2257.5, 2277.5 MHz Transmit
Channels 1–5	725 MHz Up-Down Translation	
Chanel 6	200 MHz Up-Down Translation	
Guard Bands	25 MHz, 15 MHz	

RECEIVE PLAN

Channel	Antenna Multibeam	Earth coverage horn	UHF bow tie
Channel 1	X	X	–
Channel 2	X	X	–
Channel 3	X	X	–
Channel 4	X	X	–
Channel 5	–	X	–
Channel 6	–	X	–
SCT	X	X	X

TRANSMIT PLAN

Channel	Antenna Multibeam	Earth coverage	Gimballed dish	UHF cross dipole
Channel 1 (40 W)	X	–	X	–
Channel 2 (40 W)	X	–	X	–
Channel 3 (10 W)	X	X	–	–
Channel 4 (10 W)	X	X	X	–
Channel 5 (10 W)	–	X	–	–
Channel 6 (10 W)	–	X	–	–
SCT	X	–	X	X

[1]The Satellite Ground Link System (SGLS) is a DSCS III-provided TT&C Link for IABS

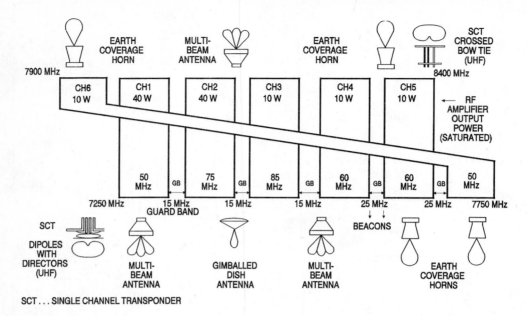

Figure 14.5   DSCS III Frequency Plan (Courtesy of Martin Marietta Astro Space)

14.2 SHIPBOARD TERMINAL CONFIGURATION

Figure 14.6 shows the elements of a shipboard DSCS terminal configuration. The antenna control, radio transmitter, and receiver and the modems are located in a terminal room. A patch panel permits patching of the IF signal to the modem(s). The figure shows two antennas; unless the antenna is at the topmost location of the platform, two antennas are needed in order to eliminate signal blocking by the ship's superstructure.

The digital data stream to or from the modem is connected to digital equipment in the radio room. Typically there is multiplexing and demultiplexing equipment connected to end equipment such as TTY, data, and SECVOX terminals. The cryptographic equipment is generally in a separate security vault.

14.3 EQUIPMENT

The AN/WSC-6 is the standard SHF shipboard SATCOM terminal [NO-1]. Figure 14.7 shows the transmitter that consists of a high-power klystron amplifier, capable of producing an output power up to 8 kW (39 dBW); with the antenna gain (37.2 dB for the 1.2-m reflector) the EIRP is approximately 76 dBW. The electronics cabinet contains the up and down converters from IF to the X-band region. The terminal also includes a cesium or rubidium frequency standard for supplying precise clocking for the modems and up/down converters.

The low noise amplifier (LNA) is a solid-state device that is generally placed as

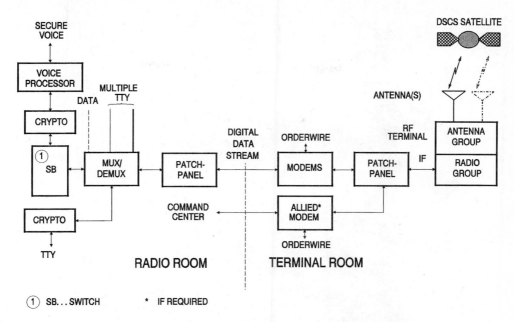

Figure 14.6   Typical DSCS Ship Configuration

closely to the antenna as feasible to reduce the noise contribution of the waveguide connecting the antenna feed to the LNA. The LNA provides preamplification of the X-band signal and has an equivalent noise temperature of approximately 90 K.

The AS-3399/WSC antenna is a 1.2-m (4-ft) reflector antenna with a Cassegrainian feed. It is part of the OE-279/WSC-6 antenna group for the AN/WSC-6 SATCOM terminal. The antenna and its major components are shown in Figure 14.8. The antenna unit includes reflector, feed, waveguide assembly, drive system gyroscope, and radome.

The AN/WSC-6 antenna pattern is illustrated in Figure 14.9. The AN/WSC-6 SATCOM terminal antenna is kept locked on the satellite automatically by a monopulse tracking system. The tracking compensates for ship and satellite motion.

The AN/WSC-6 antenna mounting is very rigid. The system is required to acquire and track in sea state six and winds up to 185 km/h (100 knots). Hence, limiting the flexibility of the antenna support is an important factor in achieving the overall AN/WSC-6 operating capability. The radome will protect the antenna in high winds and prevent icing.

The placement of the low noise amplifier (LNA), which weighs about 90 kg (200 lbs), is limited to a maximum distance of 15 m ($\approx$50 ft) from the antenna. The shorter the waveguide, the better the G/T; however, due to the LNA's weight it must be mounted on deck; if possible, it should be located below deck to minimize exposure to the elements.

The AN/WSC-6 EIRP of about 76 dBW (40 MW); the main beam is a source of high-intensity RF energy that can cause severe electromagnetic interference (EMI) and personnel radiation hazard (RADHAZ) problems. In addition, a hazard to ordnance on nearby ships and aircraft exists up to 100 m (300 ft).

The antenna (including the radome) has a diameter of 1.83 m (72 in.) and is 2.20 m

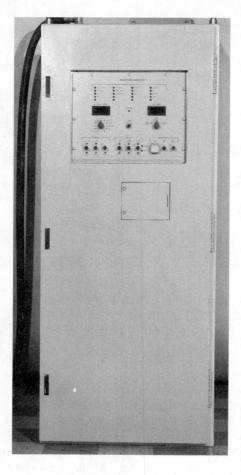

Figure 14.7 AN/WSC-6 High Power Amplifier (Courtesy of the Raytheon Corp.)

(87 in.) high. It weighs 263 kg (580 lbs). The transmitter is right-hand circularly polarized, the receiver left-hand circularly polarized.

A modem used for shipboard SHF communications is the MD-1030/A(V), shown in Figure 14.10. The MD-1030/A(V) is a digital data modem using differentially encoded binary phase-shift keying (DBPSK) in order to resolve the detection ambiguity associated with BPSK modulation.[5] The modem also includes an error correction device in the form of a 1/2-rate encoder and a constraint-length seven decoder. The error correction feature

---

[5]The detector of a BPSK modem transforms the PSK phase reversals into 1-0 bits; however, there is no way to assure that a transmitted 1 is also detected as a 1. Differential encoding, such as nonreturn to zero mark (NRZM) encoding, will interpret each transition occurring at clocking time as a 1, no transition is interpreted as 0. This resolves the ambiguity, but slightly reduces the modem's error rate performance. Note that a precise clock is necessary for correct interpretation of the data. The MD-1030/A(V) does recover the clock from the received data and synchronizes it with a 5-MHz external reference.

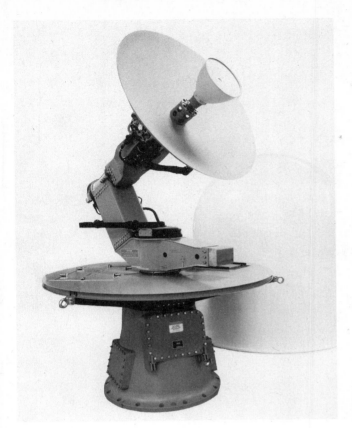

Figure 14.8   AN/WSC-6 Antenna
(AS-3399/WSC) (Courtesy of Electrospace
Systems Inc.)

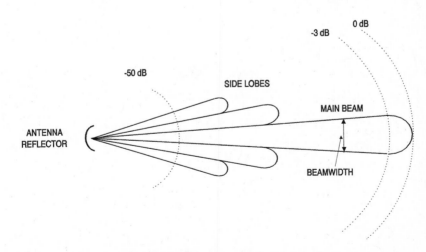

Figure 14.9   AN/WSC-6 Antenna Pattern

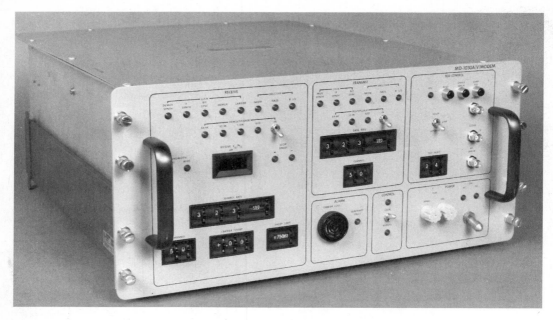

Figure 14.10   MD-1030/A(V) Modem (Courtesy Harris Corp.)

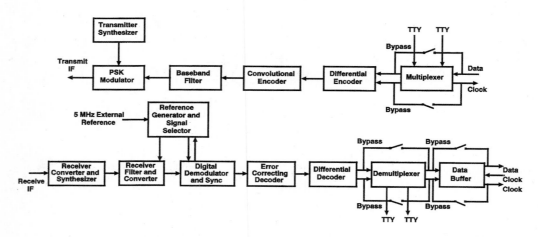

Figure 14.11   Functional Block Diagram of the MD-1030/A(V) Modem

provides a coding gain of 5 dB.[6] Thus, the MD-1030/A(V) will operate at an E_B/N_0 of 4.6 dB with the same error rate performance as a normal BPSK modem would at 9.6 dB.

Figure 14.11 shows a functional block diagram of the MD-1030/A(V) modem. The upper part shows the modulator, the lower part the demodulator. The elements performing the modulation (DBPSK) and the error correction are readily apparent. In addition, the modem also contains a multiplexer and demultiplexer, thus providing the capability to add two TTY channels to the data. This is important considering overall operations, where an order wire circuit between the ship and the SATCOM receiver facility and a command and control (C^2)[7] circuit are needed besides the data circuit. A bypass is provided for applications where the modem is used only for data transmission.

The multiplexer can operate in different modes to accommodate a set of multiple channels (3, 7, or 17), so that TTY transmission from several sources can be selected for two TTY channels at each ship. Using this feature, several ships can operate in a net and have individual TTY channels to meet various operational needs.

In addition, the modem contains a unique buffer to adjust for small timing differences between a shipboard and a shore-based clock in order to assure synchronous data operation at either link end. One needs to consider the fact that a signal transmitted from a moving ship will exhibit a Doppler shift, hence the data (and the clocking derived from it) will be different from the reference clock.[8] The buffer reads the data from the multiplexer at the clock rate derived from the received signal and reads them out at the rate provided by the local reference clock. The buffer is designed to limit the periods between the loss of synchronism between the received data and a local reference to at most once in 24 hours. The buffer can be bypassed for applications where it would not serve a useful purpose.

The modem data rate is selectable; the multiplexer data rates of 600 bps, 32.3, or 64.6 kbps can be chosen. Without multiplexer data rates from 75 bps to 64 kbps can be selected (128 kbps without 1/2-rate coding). The modulated DBPSK output is at an IF of 70 MHz, and the IF bandwidth is 120 kHz. The modem is tunable, that is, the IF can be varied from 69 to 71 MHz in increments of 120-kHz channels. Thus multiple modems can operate in a 2-MHz RF bandwidth in an FDM arrangement.

14.3.1 QUICKSAT SHF SATCOM Terminals

The QUICKSAT terminal is a shipboard SHF SATCOM terminal that is a modified version of the AN/WSC-6 and the Army Ground Mobile Forces (GMF) terminal. Figure 14.12 shows a block diagram of the QUICKSAT system, including the shipboard system,

---

[6]The link budget example for an SHF link in Chapter 12 assumes a modem with the operational characteristics of the MD-1030/A(V).

[7]In the literature one may find the abbreviation C^2 or C2 for Command and Control; it is generally referred to as C-square.

[8]Even without a Doppler shift resulting from (1) the relative motion of the ship with respect to the shore terminal and (2) small movements of the satellite needed for station keeping, clocking signals derived from precise cesium or rubidium reference sources will differ on the ship and the shore. Loss of synchronism will always occur, and processes depending on precise clocking and data synchronization, such as encryption, will occasionally lose lock. This can adversely impact operations and needs to be controlled.

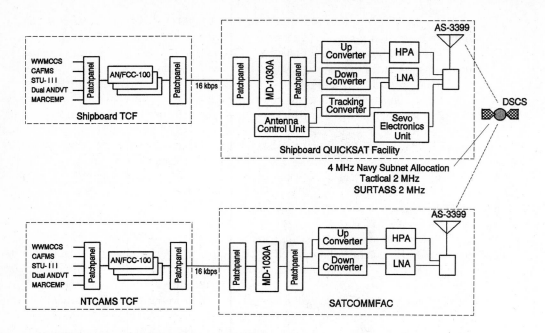

Figure 14.12   Typical QUICKSAT Configuration

the DSCS satellite, and the shore connectivity [NA-1]. The coverage provided by the QUICKSAT is the same as currently available through the DSCS. The maximum data rate of the QUICKSAT is 128 kbps. As shown in Figure 14.12, the QUICKSAT operates a separate 2-MHz subnet of the DSCS; this type of operation is similar to the SURTASS mode of operation.

The major new item is the shipboard terminal. In the QUICKSAT terminal design, the RF subsystem is new. The AS-3399 antenna and its associated control unit, the tracking converter, and the servo electronic unit are the same as in the AN/WSC-6 terminal. The modem is the MD-1030/A(V), currently used in other AN/WSC-6 terminals, for example, SURTASS.

14.4 DSCS CONTROL

The DSCS satellite control is shown in Figure 14.13. The controlling authority is the DCA Operation Control Center (DCAOC). Area control (LANT, PAC) is coordinated by two Area Coordination Operation Centers (ACOCs). The control of the individual satellites is performed by DSCS ground facilities. A primary and an alternate control facility are designated for each satellite. The control is also coordinated with the Air Force Satellite Control Facility. Each DSCS user is accessible from the control facilities via order wires.

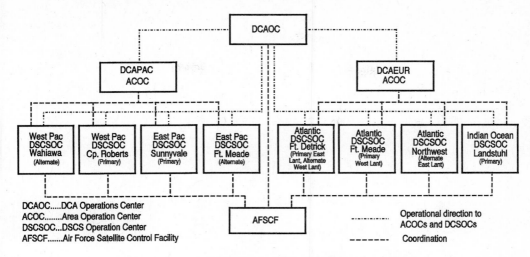

Figure 14.13   DSCS Operational Control

The control facilities monitor the spectrum and power of all transmitted signals. Since power balancing of all signals is essential for satisfactory operation, alerts are automatically generated when the power of a DSCS user either exceeds or falls below predetermined thresholds. The DSCSOC then notifies the user via the order wire and requests readjustment of the increased power level.

Small nets may operate a DSCS subnet control. Examples are QUICKSAT and SURTASS, where the DSCSOC monitors the total allocated band, and a subnet control monitors each emanation from and to the ships on the subnet control set.

14.5 AN EXAMPLE OF NAVY DSCS SHIP COMMUNICATIONS

The Navy has implemented a system using the DSCS to communicate data and record traffic between a group of ships and shore stations [MU-1]. The system serves the Atlantic, Pacific, and Indian Ocean regions and connects shipboard processors with processors at shore-based stations. The system is used for data, record traffic, and order wire message transmission. The supported communication circuits are summarized in Table 14.3.

Figure 14.14 shows the system connectivity. Several ships transmit via the DSCS

TABLE 14.3   COMMUNICATIONS CAPABILITIES

Ship-to-Shore	Shore-to-Ship
32-kbps data 75-bps C^2 TTY and record traffic 75-bps order wire TTY	75-bps data 75-bps C^2 TTY and record traffic (shared broadcast) 75-bps order wire TTY (shared)

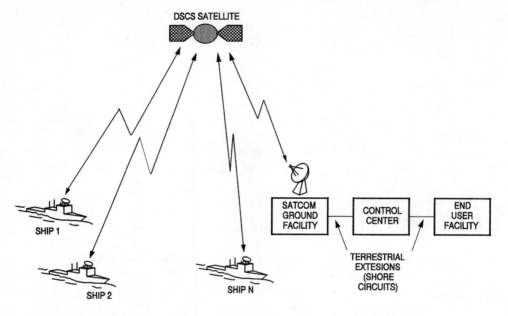

Figure 14.14   Connectivity of a Ship-Shore Data Transmission System via DSCS

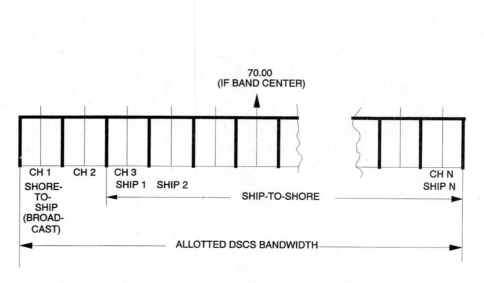

Figure 14.15   FDM Channel Arrangement

using a frequency-division multiplexing (FDM) schema. The overall system has one bandwidth allocation in the DSCS, and each ship uses a prescribed slot in that bandwidth for the ship-to-shore transmission as illustrated in Figure 14.15. The figure shows a single slot for the shore-to-ship transmission, which uses a time-division multiplexing (TDM) approach. The TDM was chosen because for the low data rate links to the ship (75 bps), the link power budget is not driven by the SNR needed to have an acceptable error rate for 75 bps, but by the need to keep the modem's phase-locked loop in lock. (The power needed to keep the phase-locked loop operating corresponds to the power needed to operate a 600-bps link with an error rate of 10^{-5}). By combining the links to all ships into a single broadcast TDM, the system saves power on the shore-to-ship link. This is important since the broadcast shore-to-ship link operates with the relatively low G/T of the AN/WSC-6 (10 to 12dBK^{-1}), and has a higher power requirement than the multiple 32-kbps ship-to-shore links.

Figure 14.16 shows a block diagram of the shipboard equipment. The AN/WSC-6 terminal connects to the MD-1030/A modem and uses the incorporated multiplexer and demultiplexer to combine data and TTY circuits. On the transmit side (ship-to-shore), the 32-kbps data circuit is combined with two 75-bps TTY circuits; one is the C^2 record traffic circuit, the other the DSCS order wire. On the receiver side (shore-to-ship), the demultiplexer is set to receive three 75-bps circuits (out of a group of 7 or 17 circuits). One is the shore-to-ship data channel, the other two are the C^2 record traffic circuit and the DSCS order wire circuit.

At the shore facility, the MD-1030/A modem's buffer equalizes clocking differences between the ship clock and the shore clock. The buffer adjusts for link instabilities caused by the Doppler shift due to the ship's motion and satellite drift, which limits the

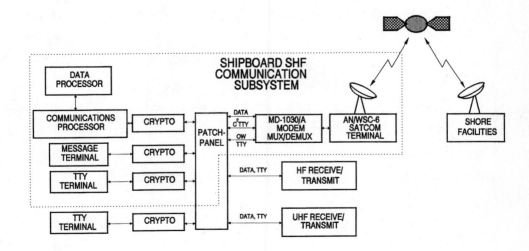

Figure 14.16 Ship System Block Diagram

timing accuracy of the received signals to 10^{-7}. This accuracy is quite sufficient for the low data rate shore-to-ship circuits, but it is inadequate for the high data rate ship-to-shore circuits.

Figure 14.16 also shows interfaces for HF, UHF LOS, and UHF SATCOM circuits. These circuits serve other purposes and can also be used as low data rate backup for the SHF ship-to-shore circuits.

This system uses a separate subnet on the DSCS. Subnet control equipment is installed at the shore sites where the receivers are located. This equipment monitors the allocated spectrum (shown in Figure 14.14) and uses the DSCS order wire to communicate with the ship, should an operational transmitter power adjustment be needed.

REFERENCES

[MA–1] *DSCS III Data Sheet*, Martin Marietta Astro Space, P.O. Box 800, Princeton, N.J. 08543.

[MU–1] Muehldorf, E. I., P. C. Hildre and R. C. Hobart, "Global Ship-Shore Satellite Data Transmission System for Connecting Digital Data Processors," *Record of the AIAA 10th Communication Satellite System Conference*, March 19–22, 1984, Orlando, Fla., pp. 380–386.

[NA–1] *Tactical SHF Terminal Concept of Operations*, NAVSPACECOM, Dahlgren, Va., June. 30, 1990.

[NO–1] *Navy SHF Satellite Communications Systems Description*, Naval Ocean System Center, San Diego, Calif., 92152, Doc. No. NSHFC 301, November 1983.

[TR–1] *DSCS II Satellite Data Sheet*, TRW Inc., 1 Space Park, Redondo Beach, Calif., 90503.

[WI–1] Williamson, J., ed., "Defense Satellite Communication System," *Jane's Military Communications, 12th Ed., 1991–1992*, Jane's Defence Data, 1991, Jane's Information Group, Sentinel House, 163 Brighton Road, Coulsdon, Surrey, CR5 2NH, U.K. (In the U.S.: Jane's Information Group, 1340 Braddock Pl, Suite 300, Alexandria, Va. 22314), pp. 358–360.

15

EHF Satellite Communications

The extremely high frequency (EHF) band offers a number of advantages for satellite communications. They are: (1) wide operating bandwidths[1] (several GHz) that are available and can be easily used, (2) antennas that create narrow shaped and spot beams that concentrate the energy on limited regions for high performance, and (3) small antennas, for example, a 1.8-m (6-ft) reflector can produce a 54-dB gain at 30 GHz. These advantages are counterweighed by certain disadvantages: (1) high propagation losses (see Figure 12.3) where rain may block out communications completely, (2) very high precision mechanical engineering required for waveguide ducts and antennas for the small EHF wavelengths (around 1 cm), and (3) less powerful transmitter devices than at lower frequencies. Compromises are necessary, but the advantages outweigh the disadvantages and EHF SATCOM technology is rapidly developing.

For military communications, the large available bandwidths can be utilized to provide substantial anti-jamming (AJ) protection. The larger EHF uplink bandwidths (2 GHz) when compared to SHF (0.5 GHz) give EHF an advantage in AJ processing gain. In a jamming environment the smaller EHF antenna beam widths also offer a significant advantage in AJ capabilities [VA-1]. In addition, man-made electromagnetic disturbances

---

[1]Note that at 30 GHz a bandwidth of 300 MHz is only 1% of the carrier frequency and the equipment (receivers, transmitters, waveguides, and antennas) can easily accommodate such bandwidth without special tuning devices.

(e.g., scintillations resulting from high altitude nuclear bursts) are not as harmful as at lower frequencies and can be counteracted more easily. Thus, military communications are moving ahead with the EHF Military Strategic Satellite Relay (MILSTAR) [RI-1; WI-1; DE-1], and the Navy participates in this development through the Navy EHF satellite program (NESP).

The use of the MILSTAR technology will offer small portable EHF terminals compatible with existing and planned satellite payloads (e.g., the Navy FLTSAT EHF Package or FEP). Through the MILSTAR, military communications evolve toward a more integrated architecture based on new operational concepts using applications of rapidly evolving satellite technology. The use of small terminals facilitates rapid mobility to crisis and conflict areas, for example, the Persian Gulf, providing direct reliable tactical and strategic communications to the U.S. military command structure.

The MILSTAR supports the modernization of the U.S. command and communications for the control of strategic and tactical forces. It will accomplish worldwide coverage by a constellation of geostationary and inclined orbit satellites with satellite crosslinking [RI-1].

Physical survivability is an important capability of the MILSTAR. It can operate in low-intensity conflict theaters supporting conventional warfare. By appropriately hardening the satellite, payload, and ground terminals, the MILSTAR EHF communication system can also support a nuclear conflict. Narrowly shaped spot beams allow it to operate in a low probability of detection (LPD) mode.

The MILSTAR has transponders with two data rate capabilities: a low data rate (LDR, 2400 bps) and a medium data rate (MDR, 1.544 Mbps). The transponders are interconnected and use complex onboard signal processing for spectrum spreading. LDR and MDR data bursts are interspersed in the MILSTAR; the MDR channel bandwidth is spread less than the LDR channel and is thus not as highly protected.

The MILSTAR onboard processing improves the AJ performance as compared to a repeating satellite; this is shown in Figure 15.1. The degree of AJ protection offered by

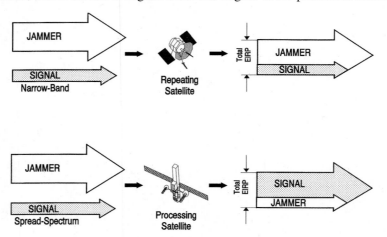

Figure 15.1   MILSTAR Onboard Processing

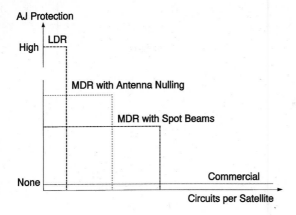

Figure 15.2   MILSTAR AJ Performance

the MILSTAR is shown in Figure 15.2. Commercial satellites have no protection but can support many channels. The MILSTAR LDR has a high AJ protection, but can accommodate only a limited number of channels. The MDR, with antenna nulling, can sustain more channels, but with less protection. With spot beams the coverage of the number of channels increases, but the protection decreases.

The LDR throughput is limited to several hundred kbps with a maximum channel data rate of 2.4 kbps and 15 users at one time [RI-1]. It is intended to serve a limited set of users. User prioritization is essential for MILSTAR operations. The MDR transponder has a maximum throughput of about 40 Mbps, with user channel data rates from 4.8 kbps to 1.544 Mbps (T-1 wideband data channel).

Operational discipline is essential for successful MILSTAR operation. The system timing is highly important and requires atomic standard time clocks that will ensure synchronization of the network's time-division multiple access communications. For rapid access and synchronization of users joining the MILSTAR network, the satellite provides synchronization probing; a coarse probing followed by a fine probing will ensure quick synchronization. A set of complex protocols between satellite and terminals is used to provide the required communications connectivity as well as accomplish robust connectivity in varying threat environments. A part of these protocols is also executed in the satellite payload.

Anti-jamming protection is provided by a complex system of frequency hopping. Both slow and fast hopping can be used on MILSTAR links. In addition, a variety of narrowband steerable antennas are available on the satellite to provide pencil beams to the area of communications and low gain into the direction of a jammer. These antennas can support forces in a limited operating area such as naval task forces.

For naval communications, the characteristics of the MILSTAR high-gain antennas and internal processing permit approaches for design of communications links different from the UHF and SHF satellites; this is illustrated in Figure 15.3. The MILSTAR receiver can be treated as having an inherent high G/T^2 and high EIRP for transmission.

---

[2]By comparison the UHF and SHF satellites have receivers and antennas with a G/T ranging from -11 dBK^{-1} to -20 dBK^{-1}.

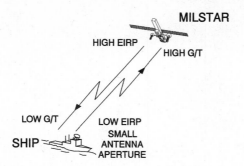

Figure 15.3   MILSTAR Operating Characteristics

Consequently, even with small shipboard antennas, reliable operation can be accomplished. This is especially true for submarines where a reflector of about 15 cm (6 in) in diameter can be used for a MILSTAR link. Figure 15.4 shows the 3-dB beamwidth and antenna gain as function of frequency for three sizes of antenna reflectors; in the area of interest (20 GHz to 40 GHz), a significant gain and a narrow beam can be accomplished by small reflectors. It should be observed that for a 15-cm reflector, the values shown at frequencies below 20 GHz are hypothetical; for the formulas to be approximately applicable, the reflector diameter needs to exceed 10 times the wavelength. As the beamwidth decreases, so does the coverage spot on the surface of the Earth; a 1° beamwidth corresponds to a spot of about 640 km in diameter for direct overhead illumination.

Typical baseband capabilities for naval applications are summarized in Table 15.1. Table 15.1 is intended to provide a general overview rather than providing exact applica-

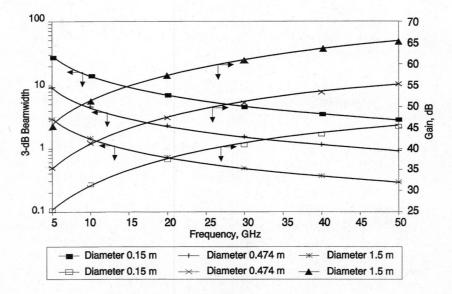

Figure 15.4   Antenna Gain and Beamwidth

TABLE 15.1   NAVY BASEBAND CAPABILITIES—EHF MILSTAR

Terminal type	Circuit type	No. of circuits	Communications service	Data rates (bps)/circuit
Shore	Primary	4	Full duplex, half duplex, receive only, transmit only	75, 150, 300, 600, 1200, 2400, synchronous
	Secondary	4	Full duplex, half duplex, receive only, transmit only	75, 150, 300, synchronous
	Receive only	4		75, 150, 300, 600, 1200, 2400, synchronous
	Submarine report-back receive	1	Receive only	75, 150, 300, 600, 1200, 2400, synchronous
Ship	Primary	4	Full duplex, half duplex, receive only, transmit only	75, 150, 300, 600, 1200, 2400 synchronous
	Secondary	4	Full duplex, half duplex, receive only, transmit only	75, 150, 300, synchronous
	Receive only	4		75, 150, 300, 600, 1200, 2400, synchronous
Submarine	Primary	2	Full duplex, half duplex, receive only, transmit only	75, 150, 300, 600, 1200, 2400, synchronous
	Secondary	2	Full duplex, half duplex, receive only, transmit only	75, 150, 300, synchronous
	Receive only	2		75, 150, 300, 600, 1200, 2400
	Submarine report-back transmit	1	Transmit only	75, 150, 300, 600, 1200, 2400, synchronous

tion data. Note that while the terminal types are listed as shore, ship, and submarine, the same communications processing and terminal equipment is used, that is, the AN/USC-38; it will be discussed in a section to follow. The primary difference in the terminals is the antenna. While shore-based antennas are about 1.8 m (6 ft) in diameter, the antenna for a

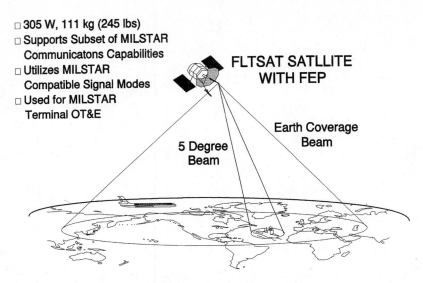

Figure 15.5   FEP Characteristics

surface ship is about 0.9 m (3 ft) in diameter, while the submarine antenna has a diameter of about 15 cm (6 in.).

A forerunner of the EHF satellite capability is the FEP that is hosted on the FLTSATCOM satellite as illustrated in Figure 15.5 [VA-1]. The FEP is an experimental military satellite communications package built by the MIT Lincoln Laboratories. It provides limited coverage but is instrumental in developing the EHF experience for transitioning the Navy to the MILSTAR Program. The NESP provides AJ and LPI communications through the FLTSAT satellites.

15.1 THE MILSTAR SATELLITE

Figure 15.6 shows a picture of the MILSTAR satellite [TR-1]. The satellite has several antennas; they comprise several narrow spot beam antennas, a smaller reflector, an SHF multibeam fast steerable (agile) antenna, an EHF Earth coverage antenna, and five EHF steerable antennas. The steerable antennas are lenses with changeable beam apertures and movable beams. The antennas can form multiple independently steerable beams for implementing connectivity between different regions using independent transmitters and receivers onboard the satellite as illustrated in Figure 15.7. In addition the satellite has crosslinks between two MILSTAR satellites. The satellite also has UHF antennas that support links similar to the FLTSAT links.

The MILSTAR operates at SHF and EHF; the uplink is at 44 GHz, the downlink at 20 GHz [BE-1]. Crosslinking between satellites is implemented at 60 GHz (near the atmosphere's oxygen absorption lines) in order to minimize atmospheric penetration. In ad-

Figure 15.6   MILSTAR Satellite (Courtesy of TRW Inc.)

dition, there is a UHF transponder that supports the FLTBROADCAST and other UHF Navy networks. Within the satellite, the links can be cross-connected to provide a multitude of services.

The data communications connectivity for the LDR capability within the satellite is shown in Figure 15.8 and is summarized in Tables 15.2 and 15.3. Figure 15.8 shows the main elements of the LDR transponder of the MILSTAR payload. The EHF receiver subsystem encompasses the receiver antennas, de-hoppers, and IF amplifiers/receivers. There are nine receiver antennas, each connected to a down converter and frequency de-hopper. The de-hopped signals are connected to the IF amplifier/receivers. Three de-hoppers and IF units are combined into one channel group each; there are three channel groups, A to C. In each channel group, the signals are combined for further processing.

The management subsystem contains an EHF signal processor, message processor (LDR router and TDM multiplexer), and a crosslink processor. It performs message

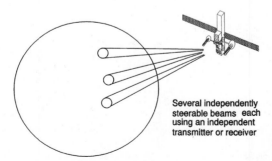

Several independently
steerable beams each
using an independent
transmitter or receiver

Figure 15.7   MILSTAR Independently Steerable Beams

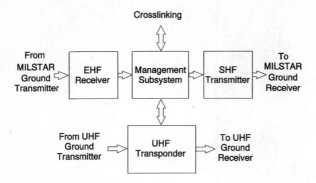

Figure 15.8   MILSTAR LDR Transponder

disassembly, distribution, and reassembly. The messages are then sent to the SHF transmitter subsystem, the crosslinking processor, and the UHF transponder.

The SHF subsystem contains the modulators, power amplifiers, multiple traveling wave tube (TWT) power output stages, the antenna beam selection switch, and the downlink antennas. Only one downlink antenna is actively transmitting. The SHF subsystem generates frequency-hopped DPSK RF signals out of the TDMA signals supplied by the management subsystem and connects the RF signals to the TWT amplifiers. A 25-W TWT generates the SHF downlink signal (there are several spare TWT amplifiers provided in the MILSTAR payload). The primary (active) TWT output is connected to a beam selection switch; there are multiple down transmission beams, and each of the TDM slots may be assigned to a different beam, depending on the location of the receiving terminal [MU-1].[3]

The UHF subsystem contains a FLTBROADCAST receiver and transmitter; in ad-

TABLE 15.2   TYPICAL MILSTAR CONNECTIVITY

Outbound / Inbound	Cross Links	SHF Downlinks			UHF FLTBROADCAST Downlink	UHF Down Link
		High Hop Rate DPSK	High Hop Rate FSK	Low Hop Rate FSK		
EHF High Hop Rate Uplink	X	X	X	X	X	X
EHF Low Hop Rate Uplink	X	X	Limited	X	None	X
UHF Uplink	X	X	Limited	X	None	X
Crosslink	Link 1 to Link 2	X	Limited	X	X	X

[3]The concept of beam switching for satellite communications is described in reference [MU-1].

TABLE 15.3   MILSTAR LDR CAPABILITY SUMMARY

Overall
  – Assured AJ protection
  – TTY, narrowband data and voice
  – Channel rates up to 2400 bps
  – Satellite processing for 200+ users
  – Priority based multiple access

EHF Uplink
  – Nine uplink antennas
  – Three channel groups (small, medium, large)
  – Three channels per group
  – De-hopping and down conversion to IF (high hop rate DPSK, low hop rate FSK)

SHF Downlink
  – 25 W power output TWT
  – Modulation and RF hopping (high hop rate DPSK and low hop rate FSK)
  – TDM to multiple receivers
  – Earth coverage and multiple spot beam antennas

UHF Transponder
  – FLTBROADCAST channel at 1200 bps
  – AFSATCOM four channels
  – Connection to EHF and SHF and crosslinking

Crosslinking
  – 60 GHz crosslinking to other MILSTAR satellites

dition it contains a four-channel Air Force Satellite Communications (AFSATCOM) receiver and transmitter. It connects to the management subsystem and can interchange data with the EHF receiver, SHF transmitter, and crosslinking subsystems.

In addition to the LDR capability, the MILSTAR also has an MDR transponder. The information flow through this transponder is shown in Figure 15.9; the MDR capability is summarized in Table 15.4.

The MDR transponder consists of an RF subsystem and a digital subsystem. The RF subsystem comprises the EHF receiver, the RF processor, and the SHF transmitter

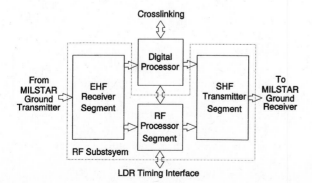

Figure 15.9   MILSTAR MDR Transponder

TABLE 15.4　MILSTAR MDR CAPABILITY SUMMARY

Overall
- – Worldwide coverage
- – Voice, command, and bulk data
- – Channel rates 4.8 kbps to 1.544 Mbps (T-1)
- – Maximum throughput 43+ Mbps
- – Can serve up to 1000 terminals
- – Uses NESP-tested technology

RF Subsystem
- – Eight independently pointable antennas
- – All antennas can transmit and receive
- – Interference control
- – EHF uplink (44 GHz)
- – FDM on uplink, DPSK modulation
- – Frequency hopping uplink AJ protection
- – SHF downlink (20 GHz)
- – 40-W TWT amplifiers
- – Frequency hopping downlink protection

Digital Subsystem
- – Demodulation
- – Signal reassembly and routing
- – TDM to multiple receivers
- – Crosslinking
- – Coordination with LDR data

segments. The RF subsystem performs signal amplification, de-hopping, down conversion, modulation, up conversion, and frequency hopping for retransmission.

The receiver segment consists of eight antennas, two of which are narrow spot beam antennas, and low noise down converters for each antenna. The down-converted signals are combined into channel groups in the RF processor. From there, they are connected to the digital processor for demodulation and processing. The RF processing segment also contains the downlink synthesizer and modulator for the 20-GHz retransmission signals. The modulator receives an input from the LDR transponder in order to properly align the LDR and MDR time slots.

The SHF transmitter segment receives the modulated and FDM slotted signals. It amplifies them through the TWTs and connects them via the beam selection switch to the eight transmit antennas.

The MDR digital subsystem consists of the demodulator, data router, crosslink processor, and resource processor. It demodulates the signals to the individual message streams and reassembles them in a TDM format. The reassembled frames are sent to the RF subsystem for transmission.

The MILSTAR internal connectivity is controlled from the ground. As a result of this internal connectivity, the MILSTAR can support a wide variety of military communications, including inter-service (Navy, Army, Air Force, and Marine Corps) communications.

The principle of the end-to-end data flow is shown in Figure 15.10. On the transmission side, the signal is encoded, modulated, and hopped. In the MILSTAR satellite

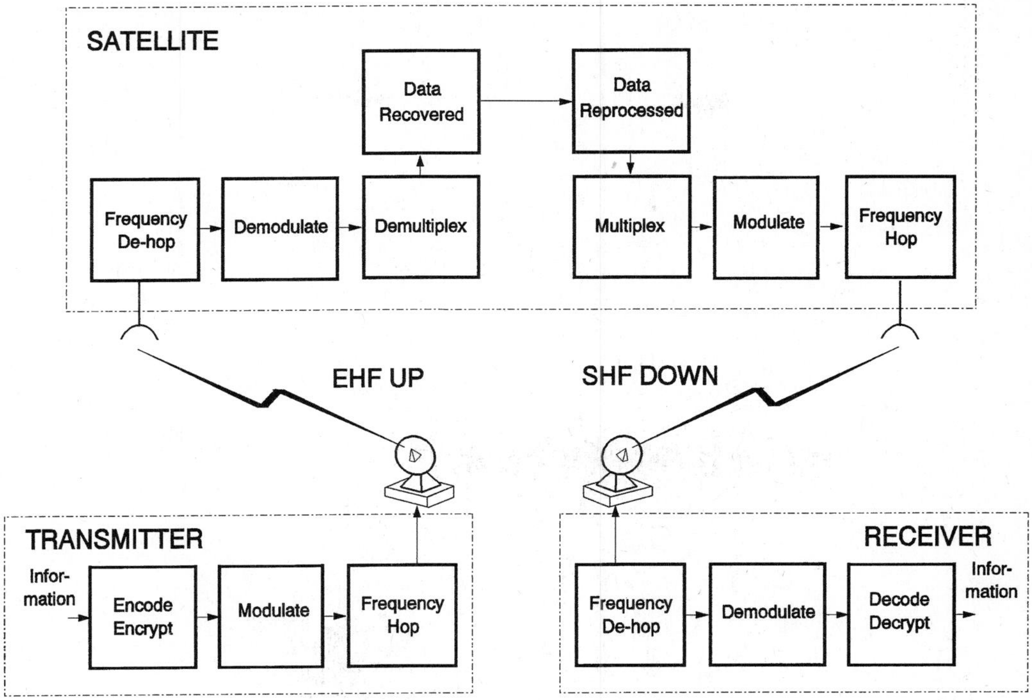

Figure 15.10   MILSTAR End-to-End Information Flow

payload, the signal is de-hopped, demodulated to the data symbol level, and the signals are then ordered according to their destinations. The signals are then assembled into a TDM format, modulated, and hopped. The hopped signals are transmitted to their destinations where they are de-hopped, and restored.

Figure 15.10 shows the path of a single signal in a complex multiplexing schema accommodating multiple transmitters and receivers. A more detailed presentation is given in Figure 15.11. Each of the N transmitters generates an uplink spectrum that fits into a frequency slot of the uplink FDM. In each FDM channel, the data is arranged into groups of three bits, and for each three-bit code a tone is transmitted. Thus the uplink can be described as 8-ary FSK (or multiple frequency-shift keying, MFSK). Inside the FDM, channel frequency hopping is used as AJ protection. A high hopping rate (HHR) at 256 hops per channel may be used.

In the satellite payload, the spectrum is de-hopped and the tones are demodulated; the signals are then demultiplexed. In the router, the data from the N uplinks are allocated to the M downlinks; each of the data elements is now assigned to a time slot. A TDM data stream is generated where each time slot corresponds to one of the P receiver stations.

The TDM data stream is modulated using DPSK. The resulting radio signal is frequency hopped. In addition, the data must be directed to the correct receiver, since the downlink has a single transmitter. Beam control is accomplished by the beam selection

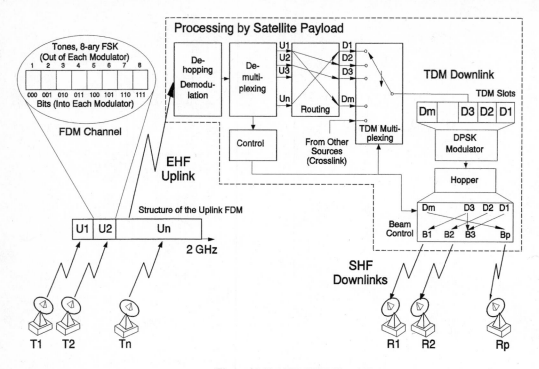

Figure 15.11 MILSTAR Signal Flow

switch that assigns the appropriate time slot to the correct beam; several time slots can be combined in one beam.

The foregoing discussion emphasizes the need for precise clocking and frequency control. It is essential to accomplish the frequency hopping, having a clean FDM uplink, and operating precisely on the TDM downlink.

The MILSTAR communications utilize a set of standard (serial tone) waveforms (MIL-STD-188–110) [DD-1]; STANAG 4253 is the NATO equivalent for the MFSK uplinks and the frequency hopping crosslinks and downlinks. The detailed MILSTAR waveforms are defined in MIL-STD-1582 [AF-1; BE-1]. The purpose of the standardization is to facilitate the interservice communication (Navy, Army, and Air Force) to provide the necessary command and control communications when required.

15.2 SHIPBOARD TERMINALS FOR EHF SATELLITE COMMUNICATIONS

The satellite terminal to be used in naval shipboard applications is the AN/USC-38. This terminal has already demonstrated its operational capabilities in the NESP program. The AN/USC-38 is configured for use on surface ships, submarines, and shore installations [DS-1; WI-2; SC-1].

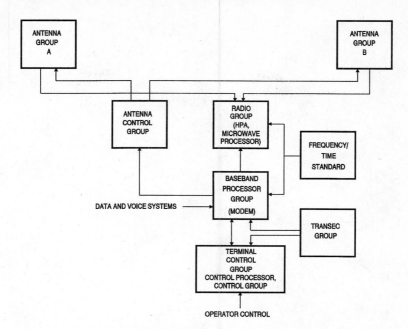

Figure 15.12   Typical MILSTAR Terminal Block Diagram

A typical block diagram of the EHF terminal is given in Figure 15.12. The diagram shows the various components essential for the EHF terminal operation. The Terminal Control Group provides the various control functions, accepts operator control, and displays status to the operator. The information (voice and data) is interfaced to the baseband processor group that contains the modem and all related functions (coding, multiplexing, acquisition, and security control). A frequency and time standard provides the necessary precise timing signals. The radio group includes a microwave processor and High Power Amplifier (HPA). The radio group translates and amplifies the signals for transmission, and separates and tunes the received signals. Two antenna groups are shown; this type of configuration is required should ship superstructure blocking exist. The antenna control group provides the antenna pointing and control signals.

Figure 15.13 shows the AN/USC-38 terminal equipment. The EHF terminal is packaged in three equipment groups. The groups are:

- A high-power amplifier (HPA)
- A communications equipment group (CEG)
- An antenna pedestal group.

Careful design of the AN/USC-38 has been performed to accomplish a high degree of hardware commonality and thus simplify the logistic support requirements.

The HPA contains the oscillators, amplifiers, and up converters to provide the Q-band (44 GHz) high-power signal. Its power component is a cavity coupled TWT.

Communication Equipment Group   **High Power Amplifier**

Figure 15.13   AN/USC-38 Terminal Equipment (NESP) (Courtesy of the Raytheon Company)

Circuitry uses solid state technology where possible to enhance reliability. Liquid cooling is provided to remove the heat of the power amplifier.

The HPA uses 2.6-kW three phase power, 60 Hz, 440 V AC. Its dimensions are: 1.37 m (54 in.) high, 0.49 m (19 in.) wide, and 0.61 m (24 in.) deep. The weight is limited to 279 kg (615 lbs).

The CEG contains several subsystems. They are:

- A power distribution unit and the terminal control unit (for circuit initiation, status monitoring, and fault isolation)
- The terminal control processor (for terminal and network control); it also provides an interface to the support systems
- The modem (containing all typical modem functions, i.e., modulation, coding, multiplexing, acquisition, and TRANSSEC)
- The microwave processor (providing the frequency reference and hop signal generator)
- The antenna position control
- The heat exchanger.

The CEG is packaged in one cabinet with the dimensions 1.83 m (72 in.) high, 0.61 m (24 in.) wide and 0.76 m (30 in.) deep. Its weight is not to exceed 463 kg (1050 lbs). It requires three phase, 60 Hz, 440 V, AC power; the power consumption is 1.6 kW (submarine), 2.3 kW (ship), and 21 kW (shore).

The antenna pedestal group also includes the antenna. For shipboard applications it consists of a three-axis mount using gyros to provide line-of-sight stabilization to compensate for platform motion due to sea conditions. The pedestal group includes the LNA and a down converter. A picture of pedestals and three different MILSTAR antennas is shown in Figure 15.14. The shipboard antenna is a reflector of 0.88 m (35 in) in diameter, with a dual band feed and a rotatable subreflector to provide a conical scan for downlink frequencies. The antenna is enclosed for environmental protection in a radome, a 1.42-m (56-in.) hemisphere that is 1.53 m (60 in.) high. Two antenna groups and a switch are provided for full hemispherical coverage should mounting on the ship's superstructure be blocking. A single antenna can be used should mounting on the ship's superstructure be possible without blocking. The waveguide run from the HPA to an antenna should not exceed 25 m (75 ft). The submarine antenna uses a 0.15 m (6 in.) reflector. It is sufficiently

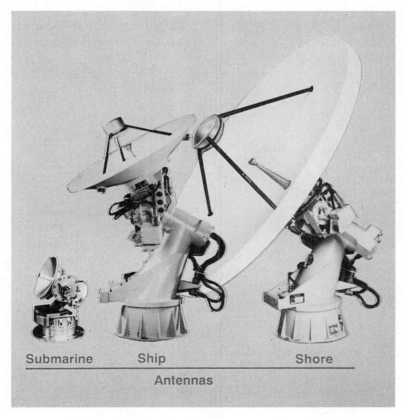

Figure 15.14   MILSTAR Antennas (Courtesy of the Raytheon Company)

small to be combined with the other submarine antennas in the conning tower (see Figure 9.16). The largest of the three antennas shown is the shore station antenna.

15.3 MILSTAR OPERATIONAL CONTROL

The MILSTAR operational control is a computer system. It is illustrated in Figure 15.15. It is distributed to provide system survivability.

The primary element is the MILSTAR Operations Center (MOC) that provides the system management. The MOC controls the MILSTAR Control Elements (MCEs) and provides for operational reporting and management of the satellite and system configuration changes.

Another function of the MOC is mission planning. That includes long-term scheduling, the generation of operating procedures, resource utilization, and contingency planning. An important MOC planning function is the management of the COMSEC and TRANSSEC for the MILSTAR System. In addition, the MOC provides the services of operator training and certification.

The MOC also maintains the MILSTAR Master Net. This is a dedicated full-time net to distribute system status and control data among the MCEs. It utilizes a satellite channel at 300 bps or 75 bps; the satellites transmit the status and telemetry information only, the MCEs receive and transmit. The MCEs report status, satellite data succession information, ephemeris/time information, and cryptographic information. A fixed TDM access scheme is used for the master net.

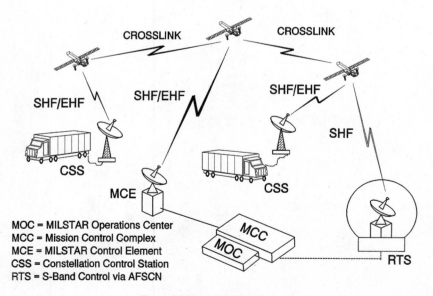

MOC = MILSTAR Operations Center
MCC = Mission Control Complex
MCE = MILSTAR Control Element
CSS = Constellation Control Station
RTS = S-Band Control via AFSCN

Figure 15.15   MILSTAR Control

The actual satellite control is performed by the MCEs. The MCEs are distributed on survivable and enduring host platforms, the Constellation Control Stations (CCS). A set of MCEs with rotating function and responsibility is always active. There are three types of MCEs:

- Designated MCEs (DMCEs) managing the deployed network
- Satellite Control MCEs (SCMCEs), controlling satellites
- Available MCEs (AMCEs), in standby mode, ready to assume active status.

The master net connects the various MCEs and distributes the MOC information. To assure survivability of the MILSTAR net, an MCE has the capability to assume the essential functions of the MOC and thus provide control of the MILSTAR constellation.

The CCS is a transportable configuration sheltered in a van. It contains one or more MCEs, and it is connected to a mobile MILSTAR ground station. It houses MCE operators and MILSTAR operators and can be located where required to provide MCE functions and essential MILSTAR control in contingency situations.

REFERENCES

[AF–1] *Satellite Data Link Standards (SDLS), Uplink and Downlink,* Military Standard MIL-STD-1582, Dept. of the Air Force, Washington, D.C. 20330, Feb. 1, 1989.

[BE–1] Bennett, R. R., "Theater/User Dedicated Communications Satellite System," *Conference Record of the 1990 Military Satellite Conference,* Monterey, Calif., Sept. 30 to Oct. 3, 1990, pp. 12.1.1–12.1.5.

[DD–1] *Interoperability and Performance Standards for Data Modems,* Military Standard MIL-STD-188-110, Dept. of Defense, Washington, D.C. 20350, Sept. 30, 1991.

[DE–1] "MILSTAR Satellite Communications System," *The C³I Handbook, Command, Control, Communications, Intelligence,* Prepared by the Editors of Defense Electronics, EW Communications Inc., Palo Alto, Calif. 94303-4275, 1988, pp. 63–65.

[DS–1] *Navy EHF SATCOM Terminal AN/USC-38 (V),* Data Sheet, the Raytheon Company, Marlboro, Mass.

[MU–1] Muehldorf, E. I., "Switched Beam Communication Satellite," *Digest 1967 IEEE International Conference on Communications,* June 12–14, Minneapolis, Minn., 1967.

[RI–1] Ricci, F. J. and D. Schutzer, *U.S. Military Communications, A C³I Force Multiplier,* Computer Science Press, Rockville, Md. 20853, 1986.

[SC–1] Schultz, J. B., "MILSTAR Progresses Despite High Cost and Technology Risks," *Defense Electronics,* June 1984, pp. 92–97.

[TR–1] *MILSTAR Reference Data,* TRW Inc., 1 Space Park, Redondo Beach, Calif. 09503.

[VA–1] Vaddiparty, S. V., *Satellite System for Tactical MILSATCOM Applications,* Conference Record of the 1990 Military Satellite Conference, Monterey, Calif., Sept. 30 to Oct. 3, 1990, pp. 12.5.1–12.5.11.

[WI–1] Williamson, J., ed., "MILSTAR Satellite Communication System," *Jane's Military Com-*

munications, 12th Ed., 1991–1992, Jane's Defence Data, 1991, Jane's Information Group, Sentinel House, 163 Brighton Road, Coulsdon Surrey, CR5 2NH, U.K. (In the U.S.: Jane's Information Group, 1340 Braddock Pl, Suite 300, Alexandria, Va. 22314), p. 363.

[WI–2] Williamson, J., ed., "AN/USC-38 Terminal," *Jane's Military Communications, 12th Ed., 1991–1992,* Jane's Defence Data, 1991, Jane's Information Group, Sentinel House, 163 Brighton Road, Coulsdon Surrey, CR5 2NH, U.K. (In the U.S.: Jane's Information Group, 1340 Braddock Pl, Suite 300, Alexandria, Va. 22314), p. 364.

16

Shipboard Communications Antennas

The many antennas needed to support the mission of a modern naval ship must be mounted on its topside. There are many kinds of antennas that support shipboard combat systems. The shipboard combat systems include radar, identification, electronic warfare, navigation, underwater, command and decision, weapon control, weapon, telemetry, exterior communications, interior communications, meteorological, and training systems.[1] Exterior communications require only a fraction of the multitude of a ship's antennas. Antennas are required for:

- Radar
- Identification
- Electronic warfare
- Navigation
- Telemetry
- Meteorology
- Exterior communications.

Antenna placement is a difficult problem and there is keen competition among shipboard systems to position the antennas in the most advantageous places. In Chapter 17,

---

[1]These combat systems are described in section 19.1.1, and in Appendix B.

where electromagnetic compatibility is discussed, the problem of antenna placement and the Navy's methodology for placing antennas on a ship's topside are discussed.

Antennas operate in a relatively narrow frequency band. Each antenna supports a specific system. For example, if a UHF air search radar (800 MHz) and a surface search radar (X-band, 9 GHz) are operated on a ship, two antennas are required. If, for reasons of limited space, the X-band antenna cannot search the full space around the ship, two or more X-band radar antennas are required. Most antennas need to be placed in such a way that they provide full hemispherical coverage (e.g., for SATCOM antennas), or 360° planar coverage (e.g., long-distance search radar antennas).

In order to describe the totality of problems in the shipboard antenna area, we first present an overview of the types of non-communications shipboard antennas, listed by frequency range. This will help in gaining a perspective of the antennas competing for space with communications antennas. We then proceed to summarize and discuss the communications antennas.[2]

16.1 SHIPBOARD ANTENNAS (NONCOMMUNICATIONS)

Radar antennas.   The U.S. Navy has a large number of radars in use. These radars support long-range two-dimensional air surveillance, three-dimensional air surveillance, aircraft control, weapons firing, and harbor navigation. Table 16.1 provides a summary of shipborne radar antennas. Details can be found in [LA-1].

Table 16.1 serves mainly as an illustration of the many types of radar antennas that are on US naval ships. Of course, only the radar antennas required by a ship's mission will be placed onboard. However, radar antennas occupy the highest locations available since they need to scan in two or three dimensions. Furthermore, radars are high-powered devices, with 200 kW peak power (e.g., target acquisition radars MK23 and AN/SPN-35) for the more powerful radar systems, and 5 to 10 kW (e.g., AN/SPS-41) for typical navigation radars. Thus, radars are one of the principal sources of EMI on shipboard installations.

Navigation antennas.   Radio navigation has been an important aspect of the use of radio waves in the U.S. Navy. Passive navigation includes methods of direction finding, locating a ship's position using the OMEGA or LORAN systems, and the more modern satellite systems such as the Navigation Satellite (NAVSAT) and the Global Positioning System (GPS), also known as Navigation Satellite Timing and Ranging (NAVSTAR). Tactical Air Navigation (TACAN) is an aircraft navigation system using a constant-duty cycle distance measuring beacon system. The shipboard TACAN unit in conjunction with the airborne unit reduces the transmitted signal to a visual presentation of both azimuth and distance information. Active navigation also uses radars; these were discussed above. Table 6.2 presents a summary of antennas used for U.S. Navy navigation systems.

---

[2]The summary of antennas presented here is based to a large degree on the book by P. E. Law [LA-1].

TABLE 16.1   TABULAR OVERVIEW OF SHIPBOARD RADAR ANTENNAS

Frequency Range (GHz)	Radar Designation	Type of Radar	Type of Antenna
0.2	AN/SPS-29	Long-Range 2-D Air Search	Rotating Folded Dipole Array
0.4	AN/SPS-40	Long-Range 2-D Air Search	Rotating Reflector
0.85–0.94	AN/SPS-49	Long-Range 2-D Air Search	Rotating Reflector
1.25–1.35	AN/SPS-6	Long-Range 2-D Air Search	Rotating Reflector
L-Band	AN/SPS-58	Short-Range 2-D Missile Search	Rotating Reflector
L-Band	MK23	Target Acquisition and Missile Point Defense	Horizontal Horn Array
S-Band	AN/SPY-1	Surveillance, Missile Detection, and Tracking	Phased Array
1.9–3.1	AN/SPS-52	3-D Scanning Surveillance	Rotating Array of Waveguide Slots
2.9–3.1	AN/SPS-48	3-D Air Search	Rotating Planar Array
2.9–3.1	AN/SPA-72	3-D Air Search	Rotating Planar Array
3.4–3.6	AN/SPS-36	Surface Navigation	Rotating Slotted Waveguide
3.5–3.7	AN/SPN-6	Air Control Radar	Rotating Reflector
3.5–3.7	AN/SPN-43	Surveillance and Air Traffic Control	Rotating Reflector
5.4–5.8	AN/SPS-5	Surface Search	Rotating Reflector
5.4–5.8	AN/SPS-10	Surface Search (Ranging and Tracking)	Rotating Reflector
5.5–5.6	AN/SPS-21	Short-Range Surface Tracking	Rotating Parabolic Reflector
C-Band	AN/SPW-2	Guided Missile Fire Control (MK77)	Parabolic Reflector
C-Band	AN/SPG-49	Missile Tracking (Fire Control)	Microwave Lens
C-Band, X-Band	AN/SPG-51	Guided Missile Control	Parabolic Reflector
C-Band	AN/SPQ-5A	Long-Range Guided Missile Control	Combination of Reflectors, Horns, and Waveguide Wedges
C-Band, X-Band	AN/SPG-55	Long-Range Tactical Fire Control	Combination of Reflectors and Horns
8.5–9.6	MK25	Gun-Fire Control System	Parabolic Reflector
8.5–9.6	MK35	Auto-Track Fire Control	Parabolic Reflector
X-Band	AN/SPG-34 (MK34)	Fire Control	Parabolic Reflector

TABLE 16.1   TABULAR OVERVIEW OF SHIPBOARD RADAR ANTENNAS (*Cont.*)

Frequency Range (GHz)	Radar Designation	Type of Radar	Type of Antenna
9–9.16	AN/SPN-35	Landing Approach Air Control	Reflector
X-Band	AN/SPG-53	Gun-Fire Control (MK68)	Parabolic Reflector
X-Band	AN/SPG-60	Gun-Fire Control (MK39)	Parabolic Reflector
X-Band	AN/SPG-9	Fire Control	Reflector
9–10	AN/SPS-55	Surface Search	Rotating Slotted Array
9.3–9.4	AN/SPS-35	Surface Navigation	Slotted Waveguide Array
9.3–9.4	AN/SPS-36	Surface Navigation	Slotted Waveguide Array
9.3–9.4	AN/SPS-41	Surface Navigation	Slotted Waveguide Array
9.3–9.4	AN/SPS-64	Surface Navigation	Slotted Waveguide Array
9.3–9.4	AN/SPS-66	Surface Navigation (Light-Weight Craft)	Slotted Waveguide Array (in Dome)
10–10.2	AN/SPN-44	Navigation (Speed Detection)	Parabolic Reflector
15.4–15.7	AN/SPN-41	Aircraft Approach Control	Parabolic Reflector
Ku-Band	CWIS/Phalanx	Close in Weapons System	Multiple Reflectors in Radome
16.4–16.6	AN/SPG-52	Gun-Fire Control	Parabolic Reflector in Dome
33–33.4 and X-Band	AN/SPN-42	Final Approach Control for Aircraft Landing	Two Parabolic Reflectors, Mounted Jointly

Electronic warfare (EW) antennas.   Electronic warfare is an important branch of military warfare that uses the electromagnetic spectrum to detect the enemy, denies the enemy the use of the spectrum, and operates in spite of an enemy's EW activities. Electronic warfare may be divided into three areas: (1) Electronic Support Measures (ESM), (2) Electronic Countermeasures (ECM), and (3) Electronic Counter-Countermeasures (ECCM). ESM are used to intercept signals (radar, communications, navigation) and to collect data that are strategically and tactically important such as Direction Finding (DF). ECM is based on jamming the enemy's signals. ECCM uses spread-spectrum techniques to accomplish anti-jamming (AJ), Low Probability of Detection (LPD), and Low Probability of Interception (LPI) communications. These communications techniques are discussed elsewhere in this book.

EW antennas for ESM and ECM are summarized in Table 16.3. In order to provide the required omnidirectional coverage required by EW systems to fulfill their mission,

TABLE 16.2   TABULAR OVERVIEW OF NAVIGATION ANTENNAS (PASSIVE SYSTEMS)

Frequency	Navigation Set Designation	Navigation Set	Type of Antenna
10.2 kHz	AN/SRN-12	OMEGA	3-m (10-ft) Whip
10.2–13.6 kHz	AN/SRN-14	OMEGA	3-m (10-ft) Whip
10.2–13.6 kHz	AN/SRN-17	OMEGA	3-m (10-ft) Whip
1.7–2.35 MHz	AN/SRA-42	LORAN	1.8-m (6-ft) Tubular Whip
121.5 MHz	AN/URD-4	Radio Direction Finding (RDF) for Search and Rescue (SAR)	Vertical Dipole in Radome
243.0 MHz	AN/URD-10	RDF for SAR	Fixed Dipole Omni-directional (Electronically Steered)
150 MHz, 400 MHz	AN/SRN-9	NAVSAT (TRANSIT)	Conical Spiral, Monopole
150 MHz, 400 MHz	AN/WRN-5	NAVSAT	Dual-band Monopole
400 MHz	AN/SRN-19	NAVSAT	Cylindrical Monopole, Omni-directional
962–1024 MHz (Transmit) 1025–1150 MHz (Receive)	AN/SRN-6	TACAN (Tactical Air Navigation)	Assembly of Dipole Array (Transmit) and Rotating Fiberglass Cylinder with Embedded Elements (Receive) Encased in a Radome
962–1024 MHz and 1025–1150 MHz	AN/SRN-15	TACAN	Dipole Array and Rotating Cylinder
962–1024 and 1151–1213 MHz (Transmit), 1025–1150 MHz (Receive)	AN/URN-25	TACAN	Multi-element Electronically Scanned Array
1227 MHz, 1575 MHz		GPS (NAVSTAR) Precision Position and Time Reference	Flat Plate or Electronically Steered Array

multiple EW receiving antennas are often distributed around the perimeter of a naval ship. EW transmitting antennas are also placed strategically to cover all aspect angles of the naval ship. Due to the sensitive nature of EW, specific data about the systems such as frequency, range, and power capability are mostly classified.

TABLE 16.3   TABULAR OVERVIEW OF EW ANTENNAS

Frequency	Equipment Designation	System Type	Type of Antenna
LF, MF, HF, and VHF Band (Range Classified)	AN/SRD-19	ESM, DF; Wide-band, Used to Find and Locate Sources of Emission	LF to HF: Several Small Encapsulated Loops; VHF: Several Dipole Arrays
50 MHz to 10.75 GHz	AN/WLR-1	High Sensitivity Early Warning System Using Multiple Antenna (Some Are Listed in the Next Column)	(1) 40–300 MHz: Stub on Fan-Shaped Ground Plane (2) 300–3000 MHz: UHF Cone Over Fan-Shaped Ground Plane (3) 550–2600 MHz: Broadband Omnidirectional, Rotatable for DF (4) 300–7350 MHz: Receiving Broadband Spiral on Cone (5) 1–20 GHz: Rotatable Broadband Multi-feed Shaped Reflector
Classified	AN/SLD-1A	ESM Microwave DF	Four Horn Array, Switched
Classified	AN/SLD-12A	ESM Wideband Micro-wave Directional Finder	Small Reflector in Radome
7–8 GHz	AN/WLR-11A	ESM Wideband Micro-wave Detection and Frequency Analysis	Broadband Omnidirectional Using Biconical Horns with Diplexer for 7–11 GHz and 11–18 GHz
Classified	AN/SLA-12	ECM Microwave System, Complex DF and Jamming (AN/ULQ-6)	Transmit: Horn Assembly Receive: (DF) Horn
Classified	AN/SLQ-32	Advanced EW System, Very Broadband, Divided into three Sub-bands; ESM and ECM: Multiple Antennas	(1) Band 1: Pair of Planar Spirals (2) Band 2 DF: Sectoral Horn Array, Combined into Lens (3) Band 2 Semi-omni: Two 180° Coverage Microwave Antenna in Pillbox Radome (4) Band 3 DF: Sectoral Horn Array (5) Band 3 Semi-omni: Microwave Antenna in Pillbox Radome (6) Band 3 ECM: Sectoral Horns Combined into Lens with Linear Phased Array Feed

IFF Antennas.   Identification, Friend or Foe (IFF) is an important supplement to surveillance radars and approach radars. For example, the AN/SPS-49 antenna (see radar antennas, above) has a special mounting surface for an IFF antenna, and the AN/SPS-40 antenna has as an integral part an IFF antenna. The IFF generally radiates at 1010 to

TABLE 16.4   TABULAR OVERVIEW OF IFF ANTENNAS

Frequency (MHz)	System Designation	System Type	Antenna Type
1030 (Transmit) 1090 (Receive)	AN/UPX-25	IFF, Integral with AN/SPS-48 or Stand-Alone	Pillbox
1010–1030 (Transmit) 1070–1110 (Receive)		Integral Part of AN/SPS-48	Dual Aperture Slotted Array
1030 (Transmit) 1090 (Receive)		Integral Part of AN/SPS-48	Slotted Radiator with Reflector
1030 (Transmit) 1090 (Receive)	AN/UPK-29	IFF Interrogator System	64 Cavity Backed Dipoles, Electronically Steered, in Radome
600–800 (DF) 1030 (Transmit) 1090 (Receive)	AN/UPX-26	IFF, Antenna Integrated with AN/SPS-48 or AN/SPN-43	Rotatable Log-periodic Array

1030 MHz and receives at 1070 to 1110 MHz. Table 16.4 is a summary of typical Naval IFF antennas.

Meteorology and telemetry antennas.   Meteorological information is crucial for naval operations. Hence, U.S. naval ships have antennas for tracking weather satellites. Telemetry is used to receive data from target drones, missiles, and certain warfare systems. Table 16.5 is an overview of typical shipboard meteorology and telemetry antennas.

TABLE 16.5   TABULAR OVER VIEW OF METEOROLOGY AND TELEMETRY ANTENNAS

Frequency (MHz)	System Designation	System Type	Antenna Type
136.5 and 1691	AN/SQM-6D	Polar and Geostationary Weather Satellite Receiver	VHF: Four Dipole Arrays UHF (L-band): Parabolic Reflector
2200–2300	AN/SQM-10	Defense Meteorological Satellite Receiver	1.8-m (6 ft) Parabolic Reflector
1760–1850 and 2200–2290	AN/SKQ-3	Missile Fire and Target Drone Support Telemetry	1.8-m (6 ft) Parabolic Reflector, Multiple Helix Feeds
2200–2290	Carrier ASW Module	Anti-Submarine Warfare Support Telemetry	Array of 32 Fan Dipoles, Can Wrap Around Ship's Mast
Classified	AN/SYR-1	Missile Tracking Telemetry	Phased Array Enclosed in Radome

16.2 OVERVIEW OF SHIPBOARD COMMUNICATIONS ANTENNAS

The U.S. Navy uses a wide variety of communications antennas to cover the principal communications frequency bands, that is, HF, VHF, and UHF. For these frequency bands, many antennas are general-purpose antennas, that is, the antennas can be connected via a matching coupler to the communications equipment that will operate in this band; in addition, transmitter antennas must also be capable of sending the signals at the power level generated by the transmitter. In the SHF and EHF regions, the Navy uses special-purpose antennas for SATCOM; in the VLF and ELF bands, special submarine antennas are used. Table 16.6 provides an overview of the naval communication antennas.

16.3 A BRIEF INTRODUCTION TO SHIPBOARD COMMUNICATIONS ANTENNA FUNDAMENTALS

The previous two sections show how many different types of antennas can be present on a modern naval ship. Besides communications antennas, other antennas will be onboard as needed to fulfill the ship's mission; these may include antennas for radar, navigation, IFF, EW, and meteorology. Hence there will be severe competition for antenna space and there will be a lot of potential EMI sources.[3]

Brief discussion of antenna fundamentals.   At this point only a short recapitulation of antenna fundamentals is given to assist the reader in understanding the material presented here. For details the reader is referred to basic texts, such as [SH-1], or handbooks, such as [JA-1].

Basically, an antenna is a narrowband device. Consider for a moment a whip antenna (such as used for an HF or a VHF transmitter or receiver). A whip, mounted above a ground plane (the ship, the ocean) has an electrical mirror image. Both the antenna and the image act as a dipole, and the dipole acts like a tuned element. When excited, a standing wave will be generated along the dipole, as shown in Figure 16.1. Through the field distribution of the power generated by an antenna, its radiation is generally concentrated to some degree into a limited volume of space. This power distribution is called the antenna pattern. Figure 16.1 also shows the antenna pattern of a thin dipole. At the frequency corresponding to the standing wave, the dipole radiates most efficiently. If the antenna is used at other frequencies, its efficiency is reduced.

For many communications links, it is desirable to concentrate power into one direction and thus produce a directive pattern since many communications links are point-to-point links. The ratio of the power of an antenna into a particular direction to the power transmitted by an ideal omnidirectional antenna is called the antenna gain. The gain is a very important characteristic of an antenna, and providing the best gain is a part of the art of antenna design.

[3]EMI is discussed in a separate chapter that also describes the methodology of how to arrive at a topside antenna arrangement to best meet the ship's mission requirements.

TABLE 16.6 TABULAR OVERVIEW OF SHIPBOARD COMMUNICATION ANTENNAS

Frequency (MHz)	Antenna Designation	Application	Antenna type
VLF/LF	OE-305	Submarine Communications	Towed Buoy
VLF/LF	AN/BRR-6	Submarine Communications	Towed Buoy
VLF/LF	AN/BRR-8	Submarine Communications	Towed Buoy
VLF/LF	AN/BRA-23	Submarine Communications	Mast-Mounted
VLF/LF	AN/BRA-34	Submarine Communications	Mast-Mounted
VLF/LF	OE-176	Submarine Communications	Mast-Mounted
VLF/LF	OE-207	Submarine Communications	Mast-Mounted
VLF/LF	AS-2629B	Submarine Communications	Buoyant Cable
VLF/LF	OE-315	Submarine Communications	Buoyant Cable
2–6	AS-2803/SRC	LHA Amphibious	Twin Fan Wire Rope
2–4	AS-2804/SRC	LHA Amphibious	Single Fan Wire Rope
2–30; Frequency Depends on Coupler or Matching Network	NT-66046	LF/MF/HF Receive, HF Transmit	8.5-m (28-ft) Whip
14–35 Receive 2–30 Transmit	AS-2537/SRC	HF Transmit and Receive	10.7-m (35-ft) Fiberglass Whip
14–35 Receive 2–30 Transmit	AS-2807/SRC	General-Purpose HF Receive Transmit	10.7-m (35-ft) Aluminum Whip
MF/HF Receive 2–30 Transmit	AT-1011/U (AT-1047/U)	Small Craft Transmit and Receive (AN/VRC-35 Radio)	Thin "Fishing Pole" Whip AT-1011 10.7-m (35-ft), AT-1047 5.2-m (17-ft)
2–8	AS-1857	HF Receive	Tuner Whip 1.5-m (5-ft)
2–6	AS-2875/SRC	HF Transmit and Receive; 2–6 MHz 1/4-wavelength mode	Tunable Omnidirectional Helical Monopole 4.9-m (16-ft)
6–30	AS-2876/SRC	HF Transmit and Receive; 6–30 MHz 3/4-wavelength mode	Tunable Omnidirectional Helical Monopole 2.5-m (8-ft)
4–12	AS-2805/SRC	General-Purpose HF Transmit and Receive	10.7-m (35-ft) Broadband Trussed Monopole (Including Matching Network)
10–30	AS-2806/SRC	General-Purpose HF Transmit and Receive	4.1-m (13.5-ft) Trussed Monopole

TABLE 16.6 TABULAR OVERVIEW OF SHIPBOARD COMMUNICATION ANTENNAS (*Cont.*)

Frequency (MHz)	Antenna Designation	Application	Antenna type
4–12 and 10–30	AS-2802/SRC	Broadband General-Purpose HF Transmit and Receive	Discage, 9.8-m (32-ft) High, 5.5-m (18-ft) Diameter
10–30	AS-2865/SRC	Omnidirectional Transmit and Receive (Mount Near Ship's Bow)	Discone, 3.7-m (12-ft) High, 4.3-m (14-ft) Diameter
2–16	MLA-1	HF Transmit and Receive, MLA-1/D 2.25-16.5 MHz, MLA-1/E 1.85-14.5 MHz	Oval 1.5 × 2.1-m (5 × 7-ft) Loop (Tuned) with Small Un-tuned Feed Loop
7.5–30	AS-2874/SRC	Broadband Directional Rotatable Log-period Array (RLPA)	Log-periodic Array, 13.7-m (45-ft) Long, 12.2-m (40-ft) Wide, Requires Sturdy Mast for Support
9–30	AS-2537A Twin Whip	Wideband HF Transmit and Receive, AN/URC-109 ICS-3 on LHD-1 class ships	5.2-m (17-ft) (20) AS-2537A Twin Whip
2–30	Long-wire	HF Transmit and Receive, on LCAC	Long-wire
2–9	LWCA	HF Transmit, AN/URC-109 ICS-3, on DDG-51, LHD-1, and FFG class ships	Lightweight Broadband HF Communications Antenna (LWCA) Twin Fan, 3-wire Type, 18.3-m (60-ft) Long
30–76	AS-1729/VRC	VHF Transmit and Receive	3-m (10-ft) Whip
135	NT-66095	Narrowband (Fixed Frequency VHF Transmit and Receive	Horizontal Dipole
30–76	AS-2231/SRA-60(V)	Broadband VHF Transmit/Receive	Biconical Dipole in Cage, 4-m (13-ft) High, 1.2-m (4-ft) Diameter
30–100	AS-2808/SRC	General-Purpose VHF Transmit and Receive	Broadband Vertical; 2.7-m (9-ft) High, 15-cm (6-in) Diameter
115–162	AS-2809/SRC	General-Purpose VHF Transmit and Receive	Thick Monopole, 1.2-m (4-ft) High, 6.3-cm (2.5-in) Diameter
90–500	AS-2811/SRC	General-Purpose Broadband VHF/UHF Transmit and Receive	Vertical Biconical Dipole, 97-cm (38-in) High, 86-cm (34-in) Diameter
30–76	AS-3226/URC	Broadband General-Purpose VHF Transmit and Receive	3.7-m (12-ft) Cylindrical Pole in Fiberglass Housing

TABLE 16.6 TABULAR OVERVIEW OF SHIPBOARD COMMUNICATION ANTENNAS (*Cont.*)

Frequency (MHz)	Antenna Designation	Application	Antenna type
30–150 and 150–1000	AS-2867/SSR	General-Purpose Broadband VHF/UHF (Receive Only)	Discage with Ground Plane 2.4-m (8-ft) Diameter
30–76	AS-2814/SRC	Broadband Unidirectional Transmit and Receive	Vertical Log-periodic Array, 5.2-m (17-ft) High, 3-m (10-ft) Long
225–400	AS-1735/SRC	General-Purpose UHF transmit and Receive	Dipole Array, Mounted Around the Mast for Omnidirectional Coverage
200–1300	AS-2812/SRC	Broadband Omnidirectional General-Purpose UHF Transmit and Receive	Biconical Dipole, 40-cm (16-in) Diameter, 61-cm (24-in) High
225–400	AS-1018/URC	Broadband Omnidirectional General-Purpose Transmit and Receive	Two Dipole Array (Non-resonant) 1.8-m (6-ft) High, 22-cm (9-in) Diameter
225–400	AS-2810/SRC	Broadband Omnidirectional Low Power General-Purpose Transmit and Receive	Center-fed Dipole, 74-cm (29-in) High, 6-cm (2.5-in) Diameter
225–400	AS-2877/SRC	General-Purpose Broadband Omnidirectional Transmit and Receive	Two Element Dipole Array in Fiberglass Housing, 1.8-m (6-ft) High, 34-cm (13.5-in) Diameter
220–400	AS-390/SRC	Widely Used Lightweight Broadband General-Purpose Omnidirectional Transmit and Receive	Broadband Coaxial Stub with Ground Plane of 8 Downward Directed Elements, 46-cm (18-in) High, 58-cm (23-in) Diameter
225–400	AS-1496/U and AS-1497/U	Phasor-90 UHF Carrier to Aircraft Communications (Phased out, Replaced by AN/SRA-62 Sector Array)	Dual Circularly Polarized Conical Spiral
225–400	AS-2493/SRA-62	High Gain Ship to Air Communications; 4 Arrays (One Fore, One Aft, One Port, One Starboard) to Provide Full Coverage	4 Element Cross Log-periodic Array, 1.4-m (4.5-ft) High, 0.7-m (2.3-ft) Wide, 1.3-m (4.3-ft) Long
240–318	AS-2410/WSC-1, AS-3018/WSC-1	UHF Satellite (FLTSATCOM) Transmit and Receive (for AN/WSC-1 and AN/WSC-3 SATCOM Radio)	4 Crossed Dipole Elements Above Ground Plane, 1.3-m (4.2-ft) High, 1.3-m (4.2-ft) Wide, 0.7-m (2.3-ft) Deep

TABLE 16.6 TABULAR OVERVIEW OF SHIPBOARD COMMUNICATION ANTENNAS (*Cont.*)

Frequency (MHz)	Antenna Designation	Application	Antenna type
240–318	AS-3018A/WSC-1	UHF Satellite (FLTSATCOM) Transmit and Receive (Electrically Equal to AS-3018, But Physically Different)	Single Crossed Dipole Within a Drum Reflector/Director
248–255	AS-2815/SRR-1	FLTSATCOM Receive Antenna for FLTBROADCAST (AN/SRR-1 Receiver)	Dual Loop Plus Dual Parasitic Dipole, Four Antennas Used to Provide 360° Coverage
1400–1600	MX-2400	L-band INMARSAT Coastal Earth Station Antenna	Parabolic Reflector in a Radome 2.8-m (9-ft) Diameter
4000–6000	AS-3274	G-band, Surface Ships, LAMPS III, with AN/SRQ-4	Rotatable Directional Parabolic Section Reflector, 1.4-m (56.88-in) High
4000–6000	AS-3275	G-band, Surface ships, LAMPS III, with AN/SRQ-4	Vertical Dipole, Omnidirectional, 18-cm (7.13-in) High, 10-cm (4-in) Diameter
7250–8400	AS-3399/WSC	SHF SATCOM Antenna, Transmit and Receive for the DSCS SHF Satellite with AN/WSC-6	1.2-m (4-ft) Parabolic Reflector with Cassegrainian Feed in Radome
44,000 (uplink) 20,000 (downlink)		EHF MILSTAR SATCOM Antenna, with AN/USC-38	Parabolic Reflector (Different Sizes for Surface Ships and Submarines)

The far field of a transmitter dipole has only two electromagnetic field components: E_θ and H_ϕ, that is, the electric field component (E_θ) oriented parallel to the dipole and normal to the equatorial plane of a sphere with the dipole as an axis, and the magnetic field component (H_ϕ) oriented in the equatorial plane tangential to the equator of the sphere. Receiving antennas should be aligned with the field for best reception. Thus a dipole should be vertical, a magnetic loop should have its center axis horizontal, and oriented such as to be aligned with the magnetic field generated by the transmitting antenna.

There are several ways to broaden the bandwidth of an antenna. Rather than thin wire dipoles, thick stubs are used, or the antenna is shaped like a spiral, a cone, or a disk. Each shape has its own radiation characteristics. The radiation characteristics are also influenced by the ship's topside structures that may be in the near field of the antenna and shape the antenna pattern.

For higher frequencies, that is, UHF and above, the antenna's near field is limited to a few meters, say 3 m (10 ft) at the most, and reflectors can be used to determine an antenna's radiation pattern. At SHF, for SATCOM applications, parabolic reflector antennas are typically used. They produce a well-defined pattern such as a pencil beam.

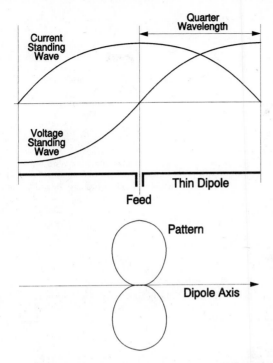

Figure 16.1 Dipole Antenna, Electrical Characteristics

Another important aspect of an antenna's design is impedance matching. The antenna can be modeled as an electrical circuit as shown in Figure 16.2. Impedance matching is particularly important in transmitter antennas; a severe mismatch will cause a reflection back into the RF power source (the transmitter) and can damage the transmitter. It is also important for a receiver antenna to operate at peak efficiency, but an antenna mismatch for a receiver is not destructive.

Antenna couplers.   Since antennas are narrow-band devices and often a transmitter or receiver needs to cover a wide band in the spectrum (HF is a typical example covering 2 to 30 MHz), several antennas are often required to cover the band. The transmitters and receivers are tunable and the devices to couple the antennas to the receivers'

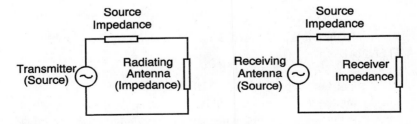

Figure 16.2   Circuit Model of an Antenna

TABLE 16.7   TABULAR OVERVIEW OF SHIPBOARD ANTENNA COUPLERS

Frequency Band	Antenna Coupler Designation	Applications
VLF/LF 10–60 kHz	CU-1441/BRR[1]	Multicoupler, with Mast-mounted, Buoyant Cable, and Whip Antennas
VLF/LF 10–60 kHz	CU-1396/BRA-16	Multicoupler, with Mast-mounted, Buoyant Cable, and Whip Antennas
VLF/LF 10–60 kHz	CU-2364/BRR	Multicoupler, with Towed Buoy, Buoyant Cable, and Towed Buoy Auxiliary Wire
HF 2–30 MHz	OA-9122/SRC	Antenna Coupler Group, with AS-2537/SRC, AN/URC-109 HF Transmitter
HF 2–6 MHz	CU-1772A/SRA-56	Coupler for Connecting to up to Four Transmitter Antennas (Power up to 1 kW)
HF 4–12 MHz	CU-1774/SRA-57	Coupler for Connecting to up to Four Transmitter Antennas (Power up to 1 kW)
HF 10–30 MHz	CU-1776A/SRA-58	Coupler for Connecting to up to Four Transmitter Antennas (Power up to 1 kW)
VHF/UHF 30–400 MHz	CU-1559[2]	Antenna Coupler, with AN/SRC-31 VHF/UHF Transceiver
UHF 225–400 MHz	OA-9123/SRC	Antenna Coupler Group, with AS-1735/SRC, AN/URC-93 UHF Transceiver
VHF 30–76 MHz	TD-1289(V)/URC	Multicoupler, with AS-3226/URC, AN/URC-80 VHF Transceiver

[1]See Chapter 9, Section 9.3.3.
[2]See Chapter 11, section 11.5.2.

multicouplers are used. Multicouplers (and antenna couplers in general) improve imped-ance matching of the antennas to the transmitter or receiver over their frequency range. Couplers may also switch in dummy loads should an antenna have to be disconnected for repair or maintenance. Table 16.7 summarizes antenna couplers.

16.4 DESCRIPTION OF SOME SHIPBOARD COMMUNICATIONS ANTENNAS

Throughout the chapters on HF, VHF, and SATCOM, antennas for the respective fre-quency regions are presented. In this section we provide additional descriptions and pic-tures in order to have a more complete representation of shipboard communications an-tennas.

Monopoles.   A monopole antenna is basically a thin end-fed wire. Together with its electrical image, it forms a dipole. Monopoles are often used as HF antennas, where the wavelength (10 m or longer) is such that the structure of the ship and the surrounding water act as a ground plane. For shorter wavelengths (VHF and UHF), artificial ground planes are provided. An example is the AS-390 VHF antenna where the antenna monopole stub is vertical and the ground plane is formed by eight wire stubs, inclined downward by 37°.

Thin wire antennas (monopoles or dipoles) are generally narrowband antennas. A standing wave is formed along the antenna, and thus the antenna is a resonating device; it can be used in various modes such as quarter wavelength, three-quarter wavelengths, etc. At the feed point the voltage has its full amplitude; at the end points, the voltage is zero (as illustrated in Figure 16.1). Thus, an end-fed monopole will resonate when a quarter wave is set up along its length, the other quarter wave is provided by its image (see Figure 16.1). The next modes where the end point of the dipole has zero voltage occur at three-quarter wavelengths, five-quarter wavelengths, etc. A monopole or dipole generally has a doughnut-shaped radiation pattern with the radiated power concentrated broadside to the wire (see Figure 16.1); hence the preferred orientation of mounting the antenna is vertical. Examples are: 10.5-m (35-ft) or 5.1-m (17-ft) HF whips; a 5.1-m whip in quarter wavelength mode resonates at 20 m or 15 MHz. Whip antennas are shown in Figures 16.3 and

Figure 16.3   35 ft Fiberglass Whip AS-2537A (Courtesy U.S. Navy)

Figure 16.4   35 ft Aluminum Whip
AS-2807 (Courtesy U.S. Navy)

16.4. By detuning the antenna or loading it (with thick endplates), the antenna can be made more broadband. A typical approach of widening the bandwidth is using a conical dipole; Figures 16.5 and 16.6 show a discage (cage with counterweight on top) and a discone antenna (wire cone with counterweight on top). Another approach of widening the bandwidth is to use a thick wire or cylinder with a given length-to-diameter ratio where the voltage distribution is quite different, such antennas are primarily VHF and UHF antennas.

Figure 16.5   Discage Antenna AS-2802
(Courtesy U.S. Navy)

Figure 16.6   Discone Antenna AS-2865 (Courtesy U.S. Navy)

Figure 16.7   Helical Monopoles AS-2876 (Courtesy U.S. Navy)

Figure 16.8   Shipborne Log Periodic HF Antenna AS-2874 (Courtesy U.S. Navy)

Widening the bandwidth can also be accomplished by building a helical antenna such as the AS-2876/SRC, an antenna that is enclosed in a fiberglass cylinder and looks just like a thick monopole (see Figure 16.7). On the right side in the background of this figure, an AS-3018A UHF SATCOM transmitter antenna can be seen.

A broadband antenna with increased gain is formed by using several dipoles properly arranged in an array. Parasitic dipoles can be used as directors and reflectors to concentrate the transmitted energy into (or collect the received energy from) a cone- or club-shaped volume of space, hence the increased gain. By using multiple dipoles of different lengths the antenna gets a broadband frequency response. An optimum way of arranging the dipoles is in a logarithmic spacing, hence the antennas are called log-periodic arrays. They need to be rotatable in order to orient the aspect of highest gain into the required direction. The Navy's HF antenna of this type is the rotatable log-periodic array (RLPA), shown in Figure 16.8 on top of the mast. In addition this figure also shows a discage antenna (on the left side). Below it one can also see an AS-2815 loop antenna for the AN/SRR-1 UHF FLTSATCOM broadcast receiver.

Similar types of antennas (dipoles, thick stubs, helixes, biconical dipoles, discones, log-periodic arrays) are also used for VHF and UHF antennas. At these frequencies, the ground plane for a monopole needs to be supplied with the antenna. A typical example is the UHF AS-390/SRC antenna, shown in Figure 16.9.

As the frequency increases (high UHF, SHF, and EHF bands), communications antennas serve primarily satellite communications. High antenna gains are needed for satellite applications to cover the distances to the satellite (up to 42,000 km or about 26,000 miles when the satellite is 5 degrees above the horizon). SATCOM antennas also must be steerable to orient the antenna at the satellite and track the satellite through the ship's mo-

Figure 16.9 UHF Stub Antenna AS-390
(Courtesy U.S. Navy)

tion. For UHF, a dipole array over a ground plane is used, and for SHF and EHF, parabolic reflector antennas are used. These antennas use a paraboloid of rotation in the focus of which a feed is mounted. The paraboloid serves as a mirror that concentrates the radiation into a narrow beam, much like a flashlight concentrates light. The antenna surface must be precise to less than 1/16 of a wavelength, otherwise the mirroring property is not sufficient. The feed may be a microwave horn or a hyperbolic subreflector, the focus of which coincides with the focus of the paraboloid. The feed horn can then be at the apex of the paraboloid, thus having a superior mechanical mounting. The feed with a hyperbolic subreflector is known as a Cassegrainian feed. SATCOM reflector antennas are described in more detail in this book in the chapters on satellite communications.

REFERENCES

[JA–1] Jasik, H., *Antenna Engineering Handbook*, McGraw-Hill, New York, N.Y., 1961.

[LA–1] Law, Preston E., *Shipboard Antennas*, 2nd ed., ARTECH House, Dedham, Mass., 1986.

[SH–1] Shelkunoff, S. A. and H. T. Friis, *Antennas: Theory and Practice*, John Wiley & Sons, New York, N.Y., 1952.

17

Electromagnetic Compatibility (EMC)

The complexity and multitude of modern naval electronics, using the whole electromagnetic spectrum, has lead to a situation where the topside of a modern Navy ship now carries what is often referred to as an antenna farm. The antennas not only serve communications, they also support radar, electronic warfare (EW), weapon control, and navigation.

Figure 17.1 shows the topside of a naval vessel, illustrating its many antennas. On deck, there are helical monopoles; the yardarms on the structure seen in the figure carry various VHF and UHF antennas; on top of the mast is a steerable SATCOM antenna covered by a radome; lower, to its left, is a small UHF communications antenna; in the background are wires of additional HF antennas.

Antennas are generators and receptors of electromagnetic (EM) waves. As generators, antennas may interact with the ship's topside structure at certain wavelengths, and sometimes with the other antennas. As receptors, they often will receive the EM waves generated by antennas on the same ship: There will be unacceptable EM interference (EMI) unless very careful engineering is applied to the placement of antennas on the ship's topside. Additionally, in some cases, operational procedures are used to limit the EM interference; otherwise the communications, radars, weapon systems, and navigation may be rendered inoperable. Electromagnetic compatibility (EMC) between systems is, therefore, a vital aspect of shipboard communications.

Generally EMI is unavoidable on a naval vessel. Thus, EMC engineering must find

Figure 17.1   Antennas on the Topside of a
Modern Naval Vessel (Courtesy U.S. Navy)

an acceptable compromise such that the various systems will be able to coexist and operate to accomplish their missions in an acceptable manner [GR-1; OR-1; JU-1].

Often EMC engineering is performed as an afterthought or during ship overhaul, when new electronic systems are installed. The much-needed EMC engineering of new ships is often taken lightly. Correction is then required, which is a costly undertaking.

One of the author's real-life experiences may put this into perspective. For a small ship that had HF, SATCOM, navigation, VHF, and the required emergency communications, a topside design was performed without carefully assessing the HF antenna placement. A brief inspection by an EMC expert showed that the HF antenna would parasitically interact with parts of the topside structure. A model was built and tested, and a much superior placement for the antenna was found. This resulted in an engineering change (EC). As result of the EC, the ship hulls, as soon as they were at the pier of the shipyard where they were outfitted, had two holes cut into the deck—for the new antenna locations, and two holes welded shut—where the original antennas were to have been. This, of course, increased the cost. Careful EMC engineering at the time of the topside design would most likely have revealed the potential electromagnetic interactions and prevented the additional cost.

Often corrections of an EMI problem are not as simple as the example given here,

and major redesign at significant cost is required. An excellent in-depth coverage of shipboard EMI problems and EMC approaches is given by Law [LA-1].

17.1 DEFINITIONS

At this point, it is useful to present definitions of terms relating to EMC engineering.
These definitions are in accord with OPNAVINST 2410.4 [NA-1].

1. *Electromagnetic Compatibility (EMC).* Ability of the electronic/electrical equipments, subsystems, and systems to operate in their intended environments without suffering or causing unacceptable degradation because of electromagnetic
   radiation or response.

2. *Electromagnetic Interference (EMI).* Electromagnetic disturbance that inter-
   rupts, obstructs, or otherwise degrades or limits the effective performance of
   electronics/electrical equipment. It can be induced intentionally, as in some
   forms of electronic warfare, or unintentionally, as a result of spurious emissions
   and responses or intermodulation products. EMI may also be caused by atmospheric phenomena such as lightning and static and human-made electrical equipment such as motor vehicles and industrial machinery.

3. *Electromagnetic Environmental Effects (E^3).* The impact of the electromagnetic
   environment upon the operational capability of equipment, systems, and naval
   platforms. It encompasses all electromagnetic compatibility (EMC), electromagnetic interference (EMI) control, electromagnetic pulse (EMP), electronic
   counter-countermeasures (ECCM), and electromagnetic radiation hazards
   (RADHAZ) to personnel, ordnance, and volatile materials.

4. *Electromagnetic Pulse (EMP).* A large impulse type of electromagnetic field
   created by a nuclear explosion or by nonnuclear means.

5. *Electromagnetic Radiation.* Wanted or unwanted electromagnetic energy that
   propagates through space. Such an emission is called interference if it has undesired effects.

6. *Electronic Counter-Countermeasures (ECCM).* Actions taken to ensure friendly
   effective use of the electromagnetic spectrum despite the enemy's use of electronic warfare.

7. *Frequency Allocation.* Authorization for the operation of telecommunications
   equipment in a specific frequency band. Allocation does not authorize emission.

8. *Frequency Assignment.* Discrete frequency or frequencies where a telecommunications equipment or system is authorized to operate within its allocation and
   within the constraints delineated by the authorizing assignment.

9. *Electromagnetic Radiation Hazard (RADHAZ).* Potential for electromagnetic radiation to produce harmful biological effects in personnel (HERP), or to cause
   spark ignition in volatile combustibles such as fuels (HERF), or to adversely affect ordnance or electroexplosive devices (HERO).

17.2 DESCRIPTION OF EMI SOURCES

Shipboard EMI has natural and human-made sources. The natural sources are background noise, that is, atmospheric and galactic noise; the human-made noise comes from electrical machinery and electronic equipment. A typical plot of the various representative noise sources is given in Chapter 7 (Figure 7.1).

The main interference in the HF band is human-made noise, with atmospheric and cosmic noise as additional contributions. For the VHF, UHF, SHF, and EHF regions, where directional antennas are used in communications applications, the main background noise is galactic noise, solar noise, and atmospheric noise.[1]

The different types of noise have different characteristics. Atmospherics are typically generated by random impulses (lightning). Galactic and solar noise are best described as gaussian noise, that is, noise that is a stationary random process with the amplitude characterized by a gaussian probability distribution. Human-made noise is a mixture of impulse noise from machinery that in some cases also contains broadband components that may be described by a gaussian approximation.

In a shipboard environment, the human-made noise originates from various electrical machinery and apparatus and consists of motor and generator noise, engine-generated noise, and impulse noise from switches, circuit breakers and relays, and fluorescent lighting. Machinery will generate noise from arcing at brush contacts and slip rings. Gasoline engines that drive generators, pumps, and other equipment will produce ignition noise that is strong in the HF and VHF range; it is the same type of noise that will interfere with automobile radios if a car does not have proper EMI reduction filters in its ignition. Switches and circuit breakers will result in arcing that produces pulses that have a broadband spectrum; it is strong in the VHF to low UHF, around the 400-MHz region. Fluorescent lighting produces the same type of noise that is generated by gas discharge tubes. Because of widespread use of fluorescent lights, it will couple into and be radiated from the power distribution system onboard the ship.

In addition, one must understand that a modern naval vessel, especially a large one such as an aircraft carrier (CV/CVN), is almost like a miniature city with a barber shop, cafeterias, machine shops—all contributing electrical noise. Electrical components and appliances such as microwave ovens, data processing equipment, buzzers, and similar items will all contribute to the shipboard EMI noise environment.

The shipboard communications and radar equipment is, of course, one of the main sources of EMI on the topside structure. It covers the entire RF spectrum, from HF to EHF. The transmitters should, of course, be designed such that they radiate only in a defined and limited frequency band; however, even with the greatest care there will be effects that will result in interfering signals. The main reasons for interference signals are:

- Harmonic frequencies
- Sideband splatter

[1]For frequencies above 300 MHz the receiver front-end noise becomes increasingly important; this is shown in Chapter 12, where SATCOM link budgets are examined.

- Intermodulation (IM) and crossmodulation
- Parasitic oscillation
- Leakage.

Harmonic frequencies are the result of nonlinearities in transmitter output amplifiers; they are multiples of the *fundamental frequency* to which the transmitter is tuned. There are standards for suppression of the second and higher order harmonies (60 dB for the second harmonic, 80 dB for higher harmonics). Nevertheless, improper operation and insufficient maintenance often result in harmonics above these levels.

Sideband splatter is an interference similar to harmonic frequencies—it is composed of spurious components of the *modulation* outside of the radio bandwidth for which the amplifiers are designed. Overmodulation, poor tuning, and overdriving of amplifier stages are the main causes for sideband splatter.

Intermodulation is caused by the interaction of two or more signals in a nonlinear circuit. Often the nonlinearity stems from either an incorrectly grounded circuit, or a corroded grounding connection. Cross-modulation is the result of unwanted coupling from one transmitter to another. Often the closeness of antennas results in coupling, and the signal then feeds to the output stages where the cross-modulation is caused by the amplifier nonlinearities.

Parasitic oscillations can be caused by several effects. Antennas can excite elements of the topside structure, and amplifier stages may have feedback paths as a result of poor installation or moving of components during refurbishing or repair.

Leakage will occur at coaxial cable joints, or waveguide joints. Especially waveguides for high-power VHF and UHF transmitters and radars will contribute to leakage EMI.

In addition, there is the hull-generated intermodulation. The hull of a ship serves as the reference grounding point. However, considering the size of a ship's hull, the number of equipment items, the various power levels, and the location where the equipment is placed, it is obvious that one cannot find a single point of grounding. This will give rise to ground loops; furthermore, at the point where the various equipment items are grounded and the hull sections join, and because of the nature of the hull's steel itself, nonlinearities of junctions can develop as a result of corrosion. This is also known as the "rusty-bolt effect." This effect is exacerbated by the need to use various types of metals in a ship's construction; this will accelerate corrosion because of galvanic effects. Corrosion that will include nonlinear effects include corrosion at the attachment points of bolts, welding corrosion, crevice corrosion, and metal fatigue corrosion.

Nonlinearities will create intermodulation products when two signals of different frequencies are applied. Typically when signals of two frequencies f_1 and f_2 are applied to a nonlinearity, the resulting output will contain the harmonics ($2f_1$, $3f_1$, $4f_1$, ... $2f_2$, $3f_2$, $4f_2$, . . . etc.) and also modulation products of $af_1 \pm bf_2$, that is, $f_1 \pm f_2$, $2f_1 \pm f_2$, $2f_1 \pm 2f_2$, etc., where the sum $a + b$ is generally referred to as the order of the intermodulation product. If signals of more than two frequencies are applied, the products are $af_1 \pm bf_2 \pm cf_3 \pm df_4$ and so forth. As an example, if ten transmitters radiate simultaneously, there are theoretically 670 third-order intermodulation products, and close to a million tenth-order products. Very high order products (order 50 and above) have been measured in shipboard EMI tests. Therefore, it is very important to take great care of eliminating corrosion at grounding strap attachment points, in order to avoid IM interference resulting from nonlinearities at corroded grounding points.

Finally, reflected energy is a major topside EMI source. This reflected energy will create multipath effects, where EM waves will combine constructively and destructively, often setting up standing wave patterns. The topside arrangement of booms, masts, yardarms, davits, weapon systems, and many other types of structures will contribute to reflections. For HF, where the wavelengths are on the order of the lengths of shipboard structures, resonance effects will also exist, and antenna radiation patterns will be subject to the antenna arrangement. Again, very careful engineering on a system level will be required to achieve a workable compromise of EMI for shipboard communications.

17.3 EMC AND EMC ENGINEERING

EMC is engineering to meet the requirement that all shipboard electronic systems be compatible and not interfere with one another, that is, that there is minimum electromagnetic interference (EMI). EMC engineering is an intricate art to meet the requirement of not having or minimizing EMI.

EMC engineering requires a careful assessment of all electronic shipboard systems that radiate (or potentially radiate) electromagnetic energy, and all systems that receive energy. These systems include radar, navigation, weapons guidance, ECM, and communications systems.

The Navy has realized that EMI is a serious problem and has launched the Shipboard Electromagnetic Compatibility Improvement Program (SEMCIP) [MC-1]. SEMCIP covers three broad task areas:

- The design of ships with electronic systems to be electromagnetically compatible
- Identification of EMI problems of ships currently in the fleet
- Training of personnel involved in ship design to understand the EMC problems and to maintain EMC throughout a ship's life.

There are three elements of EMI: a source, a victim, and a coupling between source and victim. EMI results from RF energy generated by a source that interferes with the victim. The coupling can be via antennas of victims receiving the signal from the source (this is often referred to as "front-door" EMI) or by the victim receiving the interference internally such as feed-line cross-coupling or cross-coupling through a power supply (this is often referred to as "back-door" EMI). External coupling is usually occurring between different systems (intersystem interference). Internal cross-coupling may also exist within a system such as a radar system (intrasystem interference).

EMC engineering for shipboard communications considers primarily the interference with communications systems. The following aspects of ECM engineering are of primary importance:

- EMI assessment
- Topside integration
- Communications system design—EMC engineering of communications antennas and ECM engineering of below-deck signal routing
- EMC and EMI installation problems
- EMC specifications and standards.

17.3.1 EMI Assessment

The complexity and multitude of potential source-victim relationships onboard a naval ship is so large that it is not possible to attempt to control EMI by a project manager or a single engineer dedicated to the task of EMC engineering for a vessel. The U.S. Navy has established a policy of having an EMC Advisory Board (EMCAB) for large projects. The EMCAB is composed of EMI and EMC experts from naval laboratories and the project office; often specialists from universities and industry are also included. The EMCAB's task is to support the project manager by assisting with:

- Preparation of specifications
- System design analysis and EMC predictions
- Design reviews
- Preparation of test plans and evaluation of results
- System installation planning
- Ship construction planning.

An important task of the EMCAB is developing an EMI matrix that shows the potential sources and victims. The EMI matrix serves as a baseline for EMC problem identification analysis and resolution. Table 17.1 shows an example of an EMI matrix. The dots in the matrix identify potential interferers.

Based on the EMI assessment, the EMCAB will carry out its task. It will generate an EMI control plan that specifies practices to be followed during the design and installation for routing of cables and waveguides, shielding, grounding, bonding, quality control, inspections, and testing to demonstrate adequate suppression of EMI.

17.3.2 EMC Topside Integration

One of the more difficult aspects of shipboard EMC engineering is the arrangement of the antenna on the topside of the ship. For a large ship, one cannot expect to achieve freedom from interference but only to reach a workable compromise. The key problem is arranging a large number of various radiators and receiving antennas in the limited space available

TABLE 17.1 EXAMPLE OF AN EMI MATRIX

Source \ Victim	IFF	Air Radar (3D)	Surface Radar	UHF SATCOM	UHF LOS Communications	HF Communications	VHF LOS Communications	Bridge Communications	SATNAV	OMEGA Navigation	Ship-to-Air Communications
IFF	•			•							
Air Radar (3D)	•		•	•							
Surface Radar		•									
UHF SATCOM				•	•						
UHF LOS Communications				•	•				•		
HF Transmitter						•					
VHF Transmitter							•				
Bridge Communications											•
Ship-to-Air Communications								•			

topside. Not only need one consider the antenna arrangements on the available deck surface, but one also needs to consider the antennas in elevated positions on masts and yardarms.

The problem is complicated by the fact that, in an initial ship design, the available space is not fully defined, and the arrangement of antennas cannot be determined from the outset. In addition, there is a severe competition for topside space for weapons systems, radar, navigation, etc. Placing antennas optimally to accomplish the ship's overall mission is a very difficult task. As a result a special team, the Topside Integration Design Engineering Team (TIDET), has been established by NAVSEA. The TIDET is an engineering committee that addresses the topside engineering for new ship designs as well as major ship modernization. The TIDET must consider not only the antenna placement but also restrictions that are placed upon the certain deck locations such as helicopter landing pads, armaments, and missile launches. On the other hand, the TIDET must also address the problems of providing optimum coverage, acceptable weight distribution, and physical separation of the antennas. The TIDET and EMCAB cooperate in the topside antenna placement effort. The EMCAB provides the accumulated knowledge and the global outlook of EMC engineering, and the TIDET solves the engineering problems of the topside design.

Specific problems that need to be considered include:

- Blockage, that is, the shadowing of electromagnetic waves by elements of the superstructure.
- Coupling, that is, resonance effects of elements of the superstructure with antenna structures.
- RF emission, that is, unwanted EM radiation that may be generated, such as harmonics, intermodulation products, noise spikes, and broadband noise.
- High-level radiation, that is, radiation of a power level that poses biological hazards to personnel.

Communications antennas onboard a ship generally fall into the following categories:

- Omnidirectional receiver antennas that support reception independent of the ship's orientation; an example is a HF whip antenna.
- Directional transmitter and receiver antennas, for receiving spatially concentrated energy such as from satellites; an example is the AS-3399 SATCOM antenna for the AN/WSC-6.

HF antenna integration.   Special attention is required when an HF antenna is placed on a ship or retrofitted to an existing ship's topside during a ship alteration. The wavelengths of the HF frequencies used for communications (1 to 10 m) are in the range of the size of elements of a ship's superstructure. Also, the HF band antenna is basically a tuned element that operates most efficiently in a relatively narrow frequency band; covering the whole HF spectrum, from 2 to 30 MHz is not practical. Hence, the HF band is generally subdivided into overlapping segments, that is, 2–6 MHz, 4–12 MHz, and 10–30 MHz. By using broadband antennas and couplers in these regions, several receivers and transmitters may share an antenna. This reduces the number of antennas and tends to lessen interaction through multicouplers that are basically tuned devices, as discussed in more detail in a section to follow.

HF antenna placement is a complex art. Analytical methods can be used; they are discussed in a subsequent section. However, the most reliable method is modeling to determine the best alternative for HF antenna placement. The models are usually 1/48 scale, made of brass, and include all topside structures. The Navy maintains a test range at the Naval Command and Control and Ocean System Center (NCCOSC) RDT&E Division in San Diego, California, where extensive measurements are made on these models to determine an optimal HF antenna placement.

The greatest benefit from modeling is derived for transmitter antennas where a high efficiency is important for good performance. Shipboard receiving HF antennas are generally designed for low efficiency so that the atmospheric noise and the receiver noise generally match. The receiver antenna will thus not be a good victim for interference but will still be effective as an HF receptor.

VHF and UHF antenna integration.   The shipboard VHF and UHF antennas need to be placed where they are most effective. UHF antennas need to be located high on yardarms to provide a maximum line-of-sight (LOS) covering 360°. Several trade-off alternatives need to be considered, since the available space on yardarms is severely limited. In addition, isolation between antennas, located on the severely limited space, needs to be maximized. Clearly, only compromise solutions are possible that will minimize EMI as best as possible.

SATCOM antenna integration.   SATCOM antennas need to have hemispherical coverage. UHF SATCOM transmitter antennas are large and heavy—often several antennas are mounted at lower positions with electronic steering and hand-off from one antenna to the other to accomplish the hemispherical coverage. The SHF SATCOM terminal (AN/WSC-6) is built in two configurations: with a single antenna and with double antennas. If the antenna can be placed at a high elevation above the main deck of the ship, and there is no blockage, a single antenna configuration may be chosen. Otherwise, a dual antenna configuration is required, and the antennas need to be placed such that hemispherical coverage is accomplished. The AN/WSC-6 antenna control unit will switch between the dual antennas to provide full hemispherical coverage.

When all antenna placements are considered for a ship, candidate arrangements are developed. These candidate arrangements are then subjected to trade studies to find an arrangement that is optimal, that is, the compromises that will minimize the impact on operational effectiveness.

Antenna retrofitting.   During a major ship overhaul, a ship's topside design needs to be reexamined. Additions of structures and weapon systems, as well as modification of yardarm structures may severely impact the existing topside antenna arrangement. Again, an overall assessment of the topside arrangement must be made.

Generally, the effectiveness of the topside antenna arrangement will only become known in actual operations. Should the operational envelope not be satisfactory, changes are required—and such changes are costly. However, good initial planning, careful design and application of all available experience to accomplish a good EMC design should minimize occurrences of finding the operational effectiveness unsuitable.

17.4 SHIPBOARD INTERFERENCE ANALYSIS

The U.S. Navy has developed tools for analysis and design of shipboard exterior communications for RF EMC. The design process iteratively analyzes candidate system designs for their suitability [LI-1; LI-2]. Several computer codes have been developed for EMC analysis. These software design aids consist of several packages, one set for antenna modeling, such as the NEC[2] [LO-1; SI-1] (that includes codes using the method of moments, basic scattering, and analysis of reflector antennas), and a second set for communications

[2]NEC stands for Numerical Electromagnetic Code.

systems analysis (such as DECAL,[3] PECAL,[4] LINCAL,[5] and COSAM[6] II). These codes have been validated and applied successfully in designing and integrating a number of shipboard exterior communications and topside antenna systems.

17.4.1 Shipboard Exterior RF Communications System Design

The procedure for designing a shipboard exterior RF communications system for EMC uses iterations around two loops as shown in Figure 17.2. The first loop represents the RF system design, and the second loop represents antenna topside design.

The first loop uses the DECAL, and provides deterministic information about co-site interference relative to a desired threshold level. Given a preliminary design, a designer uses DECAL to determine, through iterative analysis, the best possible use of available communications-electronics equipment within limits of the equipment characteristics and other constraints. The output of this analysis provides the required antenna isolation. The antenna isolation is computed by a topside antenna analysis that determines the design parameters of the topside antenna element.

As a next step the analysis described in the second loop is carried out, in order to arrive at a link performance evaluation. The PECAL provides statistical information on the expected performance of the proposed intermediate design. The LINCAL calculates ground wave or sky wave propagation losses and received signal statistics for ranges of various input parameters (such as frequency, transmitter power, and distance).

17.4.2 EMI Control in Shipboard Multichannel HF Transmitter and Receiver System

Interference Sources.   Before we discuss EMI control methods for shipboard HF communications equipment, we describe interference sources and methods that quantify the interference level. Interference sources are:

- Internally generated IM distortion products between two or more colocated transmitters (indirect or backdoor IM).
- Externally generated IM products caused by a nonlinear mechanism, sometimes referred to as the "rusty bolt" problem.
- IM products generated at the victim receiver resulting from nonlinearities in the receiver.
- Broadband and sideband noise generated by the interfering transmitters that fall in the victim receiver's processing band.

[3]DECAL stands for Design Communications Algorithm.
[4]PECAL stands for Performance Evaluation Communications Algorithm.
[5]LINCAL stands for Link Communications Analysis Algorithm.
[6]COSAM stands for Co-Site Analysis Model.

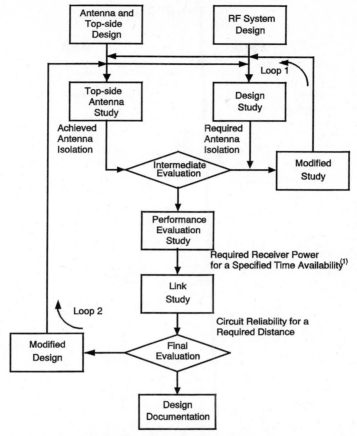

Figure 17.2   Design Procedure for a Shipboard Exterior RF Communications System (Courtesy of Naval Engineers Journal, S.T.Li, J.C. Logan, and J. W. Rockway, NCCOSC, San Diego, Calif.)

- Reciprocal mixing of the receiver local oscillator noise caused by an interfering signal in the receiver front-end passband.
- Front-end limiting (gain compression) caused by the presence of the strong interfering signal in the passband of the receiver front-end.

When two or more transmitters are active, a signal from one transmitter will modulate the output stage of another transmitter, either through the common antenna or by antenna-to-antenna coupling as shown in Figure 17.3. Direct coupling is interference to an

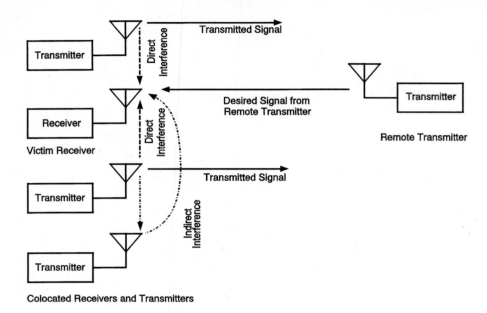

Figure 17.3   Colocation Interference

adjacent antenna; indirect coupling is interference from an antenna via a third antenna as shown in Figure 17.3. A large number of significant IM products may be generated in the power amplifiers, particularly if it is a wideband power amplifier. IM products will also be generated in the receiver, but a wideband power amplifier is likely to generate more significant IM products. An HF radio combining four exciters and driving a large Class-C amplifier generates a large number of IM products. It has been calculated that more than 12,000 significant IM products are produced and radiated to nearby receivers [KO-1].

Several studies have been conducted to determine shipboard interference sources. There are two available interference level measures available for the EMI designer. The first is the Quasi-Minimum Noise (QMN) [GU-1], and the other is the measurements of third-order IM products [SA-1]. The QMN represents typical interference relative to the thermal noise level in dB with respect to kT_0 (where T_0 is 300 K) during low-noise periods in low-noise environments. Table 17.2 lists the QMN spectral density as a function of frequency in the HF band. It was noted [DA-1] that actual shipboard interference levels may

TABLE 17.2   QUASI-MINIMUM NOISE,
SPECTRAL DENSITY RELATIVE TO kT_0

Frequency (MHz)	2	4	10	30
QMN (dB)	52	42	32	20

be below the QMN level defined in [GU-1]. However, the QMN is a useful approximation value for specifying the mitigation level for broadband HF transmitters and receivers.

The third-order IM products were measured on three U.S. Navy ships, USS Mount Whitney (LCC-20), USS Blue Ridge (LCC-19), and USS Iwo Jima (LPH-2) [SA-1]. Figure 17.4 shows the shipboard topside generated third-order IM products. These IM product measurements were taken by a receiver that was located near broadband transmitting antennas that were excited by a 1-kW transmitter. In order to isolate other interference induced by other shipboard emitters, the experiments were conducted on USS Mount Whitney (LCC-20), USS Blue Ridge (LCC-19), and USS Iwo Jima (LPH-2), which were designed with modern EMI control techniques. Such EMI control techniques make use of nonmetallic topside fixtures where possible; where metallic structures are required topside, bonding according to MIL-STD-1310 [DD-1] is employed.

Figure 17.4 shows the third-order IM products resulting from this measurement. The ratio of topside-generated third-order IM interference to QMN in the 3-kHz bandwidth at receiver antenna is plotted. Two sinusoidal tones are used as a source, each with 1-kW output power. These data are representative levels of IM products in shipboard

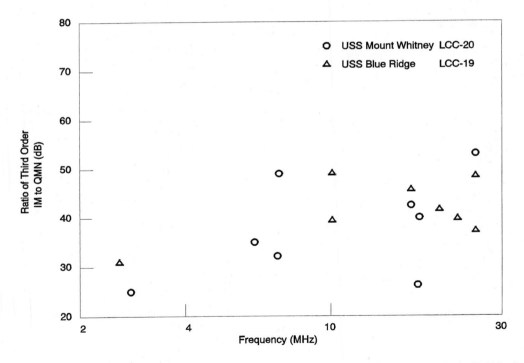

Figure 17.4 Ratio of Two-Tone Topside-Generated Third-Order IM Interference to QMN in the 3kHz Bandwidth at the Receiving Antenna: 1 kW per Tone (Reprinted with permission from Davis, Hobbis, and Royce, Copyright IEEE).

environments, which can be used for analyzing the EMI control methods of shipboard in-terference generated by HF radios.

EMI control methods.   Various EMI control methods have been used to con-trol EMI aboard ships. These methods are: (a) antenna separation, (b) using narrowband preselectors, (c) using self-steering filters, and (d) time discipline transmission.

Separating the transmitter and receiver antennas as much as possible is a difficult task aboard a ship because of space limitations; nevertheless it is one of the first steps taken in arriving at the topside antenna placement. Separating antennas will reduce the EMI by a few dB. Another step in determining the topside antenna placement is using the superstructure as a screen between transmitter and receiver antennas. The screening effect is more noticeable at frequencies above HF since signals at higher frequencies tend to be shaded by the ship's superstructure more sharply than signals at lower frequencies.

Another EMI control method is the use of narrowband receiver preselector filters, which are placed after the multicoupler as it is shown in Figure 17.5. Filtering the out-of-band signals for each receiver tends to improve the EMC.

More elaborate EMI control methods would use a self-steering filter which reduces

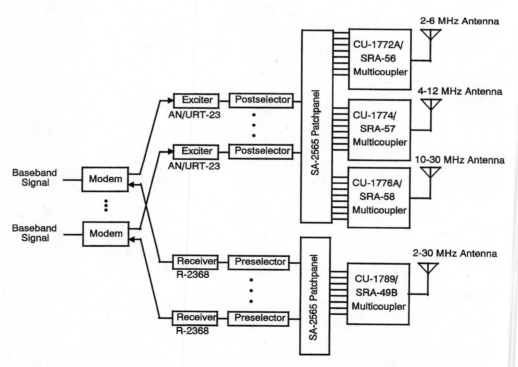

Figure 17.5   Multichannel Shipboard HF Communications System and Its EMI Control Method

direct (backdoor) IM [TR-1]. Figure 17.6 shows a block diagram of the self-steering broadband HF transmitter system. Several baseband signals are combined by a resistive summer. This signal is distributed to one of three filters (2–6 MHz, 4–12 MHz, and 10–30 MHz). These filters have a high Q and sharp skirts. A low-power preamplifier is placed between the resistive summer and three filters to compensate for the loss due to the filtering. The signal outputs from each filter are then excited by a highly linear power amplifier. The final step is to pass the signal through antenna couplers, which are also band-pass filters but with nominal cutoff characteristics.

Finally, one can control shipboard EMI by time-division multiplexed transmission. For example, consider a case where three signals are to be amplified by three exciters and transmitted by the same antenna, all in the 2–6 MHz band because of a MUF constraint. This tends to create IM products that may overload a particular victim receiver. One method to avoid such an interference is to transmit one signal at a time, thereby preventing IM products, but this method is very inefficient.

A more efficient method is to use a TDM method in which N time slots are generated at a data rate which is N time faster than that of each baseband data rate. This TDM scheme has the net effect of transmitting one signal at a time.

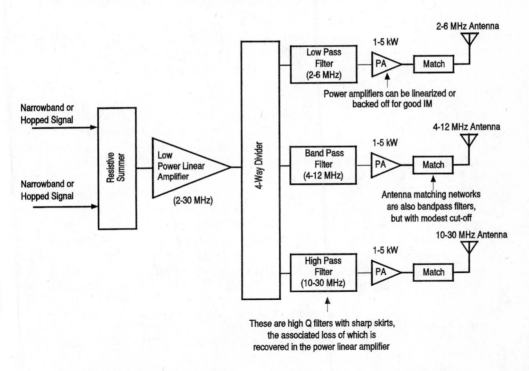

Figure 17.6   EMI Control Methods Applied to the HF Broadband AJ System (Reprinted with permission from Travis and Lenzing, Copyright 1988 IEEE)

Example 1: Conventional multichannel HF arrangement.   Figure 17.6 shows a typical shipboard multichannel HF transmitter (AN/URT-23) and receiver (R-2368) system, in which four or more baseband signals can be simultaneously transmitted and received. The modulated, amplified, and filtered signals are transmitted via one of three broadband antennas operating in separate bands (2–6 MHz, 4–12 MHz, and 10–30 MHz). HF signals received by a small broadband antenna are filtered, down-converted to IF frequency, and demodulated into baseband signals. Such a multichannel HF transmitter and receiver system is subject to interference between transmitters and receivers. In the transmitter system, narrowband filters (postselector) are used to reduce the transmitter-generated noise and harmonics. In the receiver system, filters (preselector) with narrow passbands and sharp cutoffs prevent overload or desensitization caused by strong transmitter signals.[7]

Multiple antennas are generally connected via transmitting multicouplers. The CU-1772A/SRA-56, CU-1774/SRA-57, CU-1776A/SRA-58 multicouplers allow up to eight exciters to feed a common antenna [WI-1; p. 682].[8] These three multicouplers are identical except for their operating frequencies: 2–6 MHz, 4–12 MHz, and 10–30 MHz respectively. A single receiver antenna is connected to four or more receivers via a CU-1789/SRA-49B multicoupler.[9]

Example 2: Broadband HFAJ radio.   In recent years, communications systems employing broadband HF have been the subject of considerable interest for shipboard communications due to ECCM (AJ) requirements. However, a radio like the multichannel HF transmitter and receiver system described in the previous section could not handle an ECCM modem because baseband signals utilize the entire HF band (2–30 MHz). Since the ECCM modem uses the frequency hopping technique, exciters cannot be keyed rapidly. A new broadband HF architecture was devised by allowing multiple exciter-amplifiers to feed to a common antenna with no tunable filters between the power amplifier and the antenna. One example of a system that employs a broadband HF is the AN/URC-109 radio, which can be used in two modes: ECCM (frequency hopping) mode or non-ECCM mode.[10]

Figure 17.7 depicts the key RF segment of the AN/URC-109 radio. In the transmitter subsystem, baseband signals are fed into the exciters in which the signals are converted to the selected HF frequencies. Baseband signals can either be in the non-ECCM or ECCM mode. The HF signals are then combined by exciter-combiners and amplified to an appropriate level by a low-power linear amplifier. The HF signals are automatically routed through the power bank of broadband amplifiers to the appropriate antennas covering the 2–9 MHz and 9–30 MHz range. The AN/URC-109 system eliminates high power

---

[7]Receiver overloading (or desensitization) is defined as a condition which is caused by a strong nearby transmitter reducing the receiver sensitivity.

[8]See section 18.2.2 for further discussion on multicouplers CU-1772A/SRC-56, CU-1774/SRC-57, CU-1776A/SRC-58.

[9]See section 18.2.2 for further discussion on the CU-1789/SRA-49B multicoupler.

[10]See section 10.4.6 for further discussion on the AN/URC-109. The AN/URC-109 is an integrated LF to HF radio covering the frequency region of 240 kHz to 30 MHz. In this section only characteristics related to the HF section are discussed.

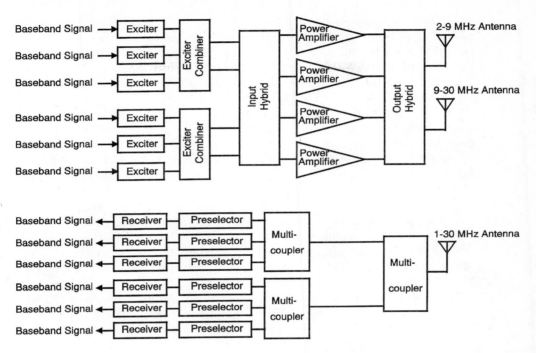

Figure 17.7   The AN/URC-109 HF Broadband Communications System

switching and tuning and achieves control of unwanted radiation by the use of highly linear power amplifiers and adaptive interference cancellation methods.

In the receiver subsystem, signals received by a small antenna are divided by two
layers of multicouplers, and filtered by narrowband preselectors, and then distributed to
multiple receivers.

This radio generates a significant amount of interference both in the non-ECCM or
ECCM modes [TR-1]. In the non-ECCM mode, this radio is required to operate in two HF
bands on two transmitter antennas simultaneously. Two transmitter antennas and a receiver antenna are colocated on a ship's topside in which the radiating elements of the
transmitter and receiver are a few wavelengths apart (the separation distance of transmitter and receiver is 50 m; the wavelength is 150 m at 2 MHz and 10 m at 30 MHz). This
situation creates a large amount of co-site interference. In the ECCM mode, the baseband
signal has the frequency hopping waveform. The frequency hopping waveform occupies a
much wider bandwidth, that is, four octaves from 2 to 32 MHz.

17.4.3 EMI Control in Shipboard VHF Radios

The SINCGARS shipboard VHF radio was discussed in section 11.4.1. Under normal
conditions shipboard SINCGARS may be subject to antenna co-site interference due to
simultaneous operation of up to four antennas. If three transmitter antennas and one

receiver antenna are colocated at the mast of a ship, the receiver may become the victim of the three transmitters as illustrated in Figure 17.3. The co-site interference problem of SINCGARS radios is not limited to those used in the shipboard environment. Army vehicles mounted with SINCGARS are plagued by similar problems [EC-1].

The shipboard SINCGARS radio is designed to operate either in single-channel mode or ECCM (AJ) mode. The ECCM mode uses the frequency hopping technique, and shipboard co-site interference becomes more prominent than in the single-channel mode. The shipboard VHF interference sources are somewhat similar to those of HF interference. Most interference sources are internally generated IM distortion products between two or more colocated transmitters (indirect or backdoor IM), IM products generated at the victim receiver due to nonlinearities in the receiver, or broadband and sideband noise generated by the interfering transmitters which fall in the victim receiver's processing band.

An EMI control method using a combination of frequency hopping filters and linear power amplifiers is shown in Figure 17.8 [IS-1]. Signals from four SINCGARS receiver-transmitter units are amplified by 50-W AM-7238/VRC power amplifiers. Amplified signals are filtered by frequency hopping bandpass filters. These signals are connected to the RF transmitter-receiver switch and fed into a single transmitter antenna via a hybrid combiner. Frequency data are distributed to the power amplifiers and the frequency hopping filters. The combination of frequency hopping filters and the hybrid combiner provides interference isolation of up to 20 dB between active transmitters. The isolation significantly reduces the interference caused by indirect IM products. However, the frequency hopping must operate with relatively high-power input signals (on the order of 50 W); the disadvantage is that high-power filters tend to be very expensive.

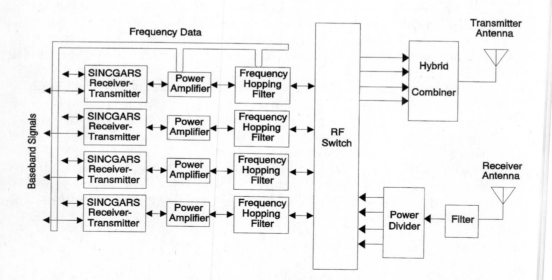

Figure 17.8 EMI Control Method for the Shipboard VHF SINCGARS Radio (Reprinted with permission from Isaacs, Robertson and Morrison, Copyright IEEE 1991)

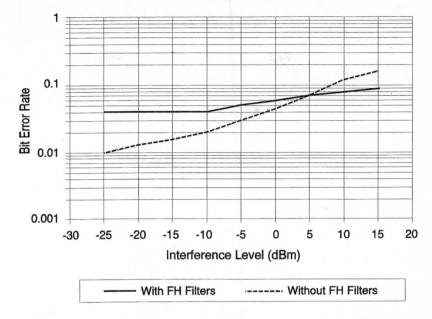

Figure 17.9   Bit-Error-Rate versus Interference Level of Shipboard SINCGARS Radios
with and without an EMI Control Method Using Frequency Hopping Filters (Reprinted
with permission from Isaacs, Robertson, and Morrison Copyright IEEE 1991)

The receiver signal is filtered by a highpass-lowpass filter and divided into four RF
signals. These signals are filtered by frequency hopping filters and demodulated into four
baseband signals.

Figure 17.9 shows the improvement in BER with and without frequency hopping
filters. The receive signal power level is assumed to be -100 dBm. The curves in Figure
17.9 are the BER versus interference level.

17.5 EMC APPLICATION TO BELOW-DECKS COMMUNICATIONS SYSTEM DESIGN

A significant part of shipboard communications system is below-decks. This includes transmitters, receivers, RF distribution networks, the demodulated information (often impulse-
like in nature and thus containing high-frequency components), the end instruments such as
digital data processing equipment, encryption equipment, and low-frequency equipment for
voice and teletypewriter applications. Thus, there are two key components of below-decks
EMI:

- The RF signals that have to be sent topside or brought down from the topside
- The signals, RF, impulse, and lower frequencies that represent the data and information to be used in operation.

The RF signals to and from the topside are of special concern. They have significantly different power levels: usually transmitter signals are strong, the received signals are weak and compete with the environmental noise (cosmic, atmospheric, and human-made, such as from arcing electrical machinery). In addition, these signals must penetrate the hull to connect the topside antennas and the below-decks receivers and transmitters. Two methods can be applied to reduce interference: filtering and shielding. Usually both must be applied in a suitable combination.

Shielding.   Shielding is a means of providing sufficient signal isolation between source and victim. It is accomplished by providing an enclosure (shield) for the signal-carrying conductor. This is not a simple and easy task; cable shields have to be properly terminated and electrically grounded to prevent leakage. Even with all precautions and the best engineering, applying EMI control practices according to MIL-STD-461 [DD-2], there may be EMI feeding from source to victim.

Shielding, in general, will reduce the level of the electric field. The shielding efficiency σ can be expressed as:

$$\sigma = 20\log_{10}(\frac{E_1}{E_2})\ [dB],$$

where E_2 is the field strength with shielding and E_1 is the field strength without shielding. The reduction of field strength is a result of the loss introduced by the thickness and the properties of the shielding material. A good summary can be found in Law [LA-1].

The best shields are completely enclosing metallic surfaces. Often, however, the material used for shielding may need to be perforated to allow for air flow for the purpose of cooling. The size of the openings in the perforation will make the shielding efficiency σ a function of frequency. For some frequencies (depending on the size of the openings), the openings will act as waveguides that inherently have filter characteristics. Usually perforated and honeycomb shielding is very effective at high (SHF) frequencies and reasonably effective at lower frequencies.

A shielding enclosure needs to be continuous, that is, without irregular openings or openings not included in the design. Boxes, for example, that contain electronics, need to have their covers closed with gaskets. Lead-in wires, themselves shielded, need to be properly connected with gaskets. One of the significant problems in naval EMC engineering is to provide such electronic gaskets and proper grounding of shields. The reason for this difficulty is corrosion resulting from salt spray in the sea air.

Grounding.   Grounding is particularly difficult. The bolts connecting an enclosure easily corrode, leading to the "rusty-bolt" problem. The rust not only makes poor contact, it also acts as a nonlinear element that produces harmonics of the frequency of the signal for which it needs to provide the ground.

The grounding reference is the ship's hull. All electronic equipment and installations must have a current-carrying capacity at least as large as the power supply lead bringing power to the equipment. Grounding connections should be welded or brazed; grounding connections must not exceed a resistance of 1 ohm.

Grounding requires particular attention in areas exposed to the elements, where the constant salt spray will corrode metallic connections, in particular connections of dissimilar metals. Consequently, ground straps tied to the hull on threaded studs or fasteners must be weather sealed using an approved sealing compound and painting.

Filtering.   Filters are used to protect potential victims by limiting the frequency band of the signals that may reach the victim. A typical application of low-pass filters is the use of the control of harmonic radiation at cable connectors. These harmonic radiations may come from digital equipment or harmonics generated by nonlinearities in transmitter equipment.

Another application of filtering is the blockage of EMI currents through the ship's power supply. Both the 400-Hz and the 60-Hz system can receive RF interference from one equipment item and carry it to another one. Filtering at the source and potential victim will reduce this type of EMI.

Multiple HF antennas are needed to cover the required RF band; their outputs are combined in multicouplers. At the multicouplers, filters are used to separate the signals. These multicoupler filters are most effective when tuned together with the transmitter and receiver tuners.

TEMPEST.   TEMPEST deals with the detection, evaluation, and control of encrypted signals in shipboard communications and data processing equipment. For security reasons, such signals must be entirely suppressed. The TEMPEST methods generally are the same as used for EMI reduction, that is, shielding, grounding, filtering, and signal isolation. Specific techniques are classified and therefore cannot be described here.

There are TEMPEST installation criteria and measures to minimize the emission of compromising electromagnetic signal emanation. Guidelines are given in DoD directive 5200.1 [DD-3].

Basically, there are Red zones (plain text) and Black zones (cipher text). Red zones are physically secured. Signal transfer from Red to Black zones is accomplished by cryptographic equipment. There is a Red-Black interface through the equipment. Electromagnetic signals from the Red zone to the Black zone must be eliminated.

Red and Black cabling must be segregated and virtually completely shielded. Optical signal coupling, using optocoupler devices, is used to separate Red and Black signal areas. In addition, using double shielding of cables and complete bonding of cable connectors at the backshell is practiced. Equipment racks for the Red and Black sides are strictly segregated.

These are just the general guidelines for TEMPEST engineering. TEMPEST engineering is a highly specialized discipline that is used in all installations where Red and Black signals exist.

17.6 EMC ENGINEERING DURING A SHIP'S LIFE CYCLE

A ship goes through several life cycles: design, construction, overhaul, and modernization. Independent of these life cycles, communications equipment is developed and must be introduced into the ship at the correct stages of the ship's life cycle.

During the design, generally a very dynamic process, the EMCAB is charged with the maintenance of sound EMC policies for the ship. The Navy has established a comprehensive EMC program [GA-1]. This is the most effective use of EMC engineering since at this time the design is still in flux and changes can be made before the design is final and ship construction begins. It is invariably more cost effective to have changes made during the design than during construction.

During construction, the EMCAB must operate as a watch committee to assure that the EMC design is maintained. Often changes after the design is finished and a design baseline has been established impact EMC heavily. It is one of the EMCAB's tasks to minimize such impact.

During ship modernization and refurbishment, new equipment is introduced to bring the ship up to the latest standard of technology. EMC conflicts will arise and must be resolved. The Navy's Waterfront Corrective Action Program (WCAP) is designed to accomplish the resolution of EMI problems during ship overhaul [WI-2; LA-2]. An EMC survey is a methodology to be used before modernization to assure EMC integrity since not only equipment changes but also ship structural changes are implemented during overhaul and modernization.

New equipment development generally is carried out on its own schedule. Fitting the equipment design schedule and matching it to a ship's life cycle is a special art, requiring meticulous planning. In a way, it is the task of getting two independently moving processes to unite at common points and interact positively with each other. New communications equipment is developed to a point where it can be fielded and can then be introduced successfully in a ship alteration (SHIPALT) program.

It is important that the equipment is developed with EMC goals established. Thus, the equipment will be less likely to generate EMI from the outset. Proper shielding, filtering, harmonic suppression, and spectrum control are then designed into the communications equipment from the start. The input for the proper design techniques should be taken from EMC studies that will highlight where the equipment induced EMI sources are and these studies should be used to establish the EMC design goals for new communications equipment.

Given the best equipment and a well-planned ship modernization plan for a particular vessel, an important step for a SHIPALT is performing an EMC survey. The survey should also note and use summaries from the SEMCIP records for that ship, should they exist. Then a shipboard EMC analysis for communications (SEMCAC) needs to be carried out to serve as a guideline for the communications refurbishment of the SHIPALT for the vessel. If required, the prediction and analysis techniques need to be applied, and the EMCAB needs to provide guidelines for the SHIPALT.

In general, after a SHIPALT, an EMC assessment should be made to ensure that newly introduced equipment does not create new EMI problems and that an acceptable EMI balance has been found.

In summary, EMC engineering for naval vessels will always be a compromise. The EMI cannot be eliminated but it should be reduced to levels where the equipment will operate within the mission parameters. Finding that compromise and maintaining an acceptable EMI balance onboard a ship is what EMC engineering needs to accomplish.

REFERENCES

[DA–1] Davis, J. R., C. E. Hobbis, and R. K. Royce, "A New Wide-Band System Architecture for Mobile High Frequency Communications Network," *IEEE Transactions on Communications*, September 1980, pp. 1580–1590.

[DD–1] "Shipboard Bonding, Grounding and other Techniques for Electromagnetic Compatibility and Safety," *Military Standard MIL-STD–1310F (Navy)*, U.S. Department of Defense, Naval Sea System Command, Washington, D.C. 20362, Dec. 30, 1990.

[DD–2] "Requirements for the Control of Electromagnetic Interference Emissions and Susceptibility," *Military Standard MIL-STD-461D*, U.S. Department of Defense, Space and Naval Warfare Command, Washington, D.C. 20363, Jan. 11, 1990.

[DD–3] *DoD Information Security Program Regulation*, DoD Directive 5200.1-R, Government Printing Office, Washington, D.C., June 1982.

[EC–1] Echevarria, R. and L. L. Taylor, "Co-Site Interference Tests of JTIDS, EPLARS, SINCGARS, and MSE (MSRT)," *Proceedings of the Tactical Communications Conference*, Fort Wayne, Ind., April 28–30, 1992, pp. 261–270.

[GA–1] Garrett, J. F., R. L. Hardie and P. A. Rogers, "Let's Design Out EMI!" *Naval Engineers Journal*, Feb. 1982, pp. 37–40.

[GR–1] Grich, R. J. and R. E. Bruninga, "Electromagnetic Environment Engineering—A Solution to the EMI Pandemic," *Naval Engineers Journal*, May 1987, pp. 202–209.

[GU–1] Gustafson, W. E. and W. M. Chase, "Shipboard HF Receiving Antenna System: Design Criteria," *Naval Electronics Laboratory Center Technical Report*, No. 1712, June 2, 1970.

[IS–1] Isaacs, J., W. Robertson, and R. Morrison, "A Co-Site Analysis Method for Frequency Hopping Radio Systems," *Conference Record of 1991 IEEE Military Communications Conference*, November 4–7, 1991, pp. 23.6.1–23.6.5.

[JU–1] Judson, H. M., G. R. Aschoff and J. W. Newcomb, "An Electromagnetic Environment System Engineering Process," *Naval Engineers Journal*, May 1987, pp. 218–226.

[KO–1] Koontz, F., "New Approach to Wideband Architecture for Shipboard Communications Systems," *Conference Record–1988 Military Communications Conference, Boston, MA*, pp. 4.7.1–4.7.5.

[LA–1] Law, Preston E., Jr., *Shipboard Electromagnetics*, Artech House, Norwood, Mass., 1987.

[LA–2] Layl, J. N. and G. D. Ellis, "A Fleet Oriented Electromagnetic Interference Control Program," *Naval Engineers Journal*, Feb. 1982, pp. 47–50.

[LI–1] Li, S. T., J. C. Logan and J. W. Rockway, "Automated Procedure for Shipboard Exterior Communication RF System Design," *IEEE Trans. on EMC*, November 1980.

[LI–2] Li, S. T., J. C. Logan, and J. W. Rockway, "Ship EM Design Technology," *Naval Engineers Journal*, May 1988, pp. 154–165.

[LO–1] Logan, J. C., et al., "Numerical and Physical Modelling at NOSC," *Proceedings of 1st Annual Review of Numerical Electromagnetic Code (NEC)*, Lawrence Livermore National Laboratory, March 19–21, 1985.

[MC–1] McEachen, J. C. P. and K. Mills, "The Shipboard Electromagnetic Compatibility Improvement Program (SEMCIP)—A Program for the Operating Fleet," *Naval Engineers Journal*, Oct. 1976, pp. 63–72.

[NA–1] "Electromagnetic Environmental Effects (E^3) Policy within the Naval Material Command, E^3 Definitions," *NAVELEX Instruction 2410.4*, Department of the Navy, Washington, D.C., 12. December 1984.

[OR–1] Orem, J. B., "The Impact of Electromagnetic Engineering on Warship Design," *Naval Engineers Journal*, May 1987, pp. 210–217.

[SA–1] Salisbury, G. C., "Top-side Intermodulation Interference aboard USS Mount Whitney (LCC 20), USS Blue Ridge (LCC 19), and USS Iwo Jima (LPH-2)," *Naval Electronics Laboratory Center Technical Report*, Document 206, December 22, 1972.

[SI–1] Sinnott, D. H., "Modelling By NEC a Dual-Fan Wide-Band HF Receiving Antenna," *Proceedings of 1st Annual Review of Numerical Electromagnetic Code (NEC)*, Lawrence Livermore National Laboratory, March 19–21, 1985.

[TR–1] Travis, G. W. and H. F. Lenzing, "Shipboard HF Interference: Problems and Mitigation," *Conference Record of 1989 IEEE Military Communications Conference*, October 15–18, 1989, Boston, Mass., pp. 3.7.1–3.7.5.

[WI–1] Williamson, J., ed., *Jane's Military Communications 1991–1992*, Jane's Information Group Inc., Alexandria, Va., 1991.

[WI–2] Winters, D. B. and R. V. Carstensen, "Waterfront Corrective Action Program (WCAP)/Pilot Shipyard Project: Resolving EMI Problems in Conjunction with the Ship Overhaul Process," *Naval Engineers Journal*, Feb. 1982, pp. 41–46.

18

Shipboard Technical Control and Interior Communications Systems

This chapter describes shipboard technical control and interior communications systems. Technical control functions include testing, patching, and monitoring of shipboard EXCOMM and interior communciations systems. Interior communications functions include: (a) the switching and distribution of RF signals between shipboard antennas, couplers, and radio equipment; and (b) the switching and distribution of baseband audio, data, and TTY from exterior communications systems and user terminal equipment.

Figure 18.1 shows a typical shipboard technical control facility and interior communications systems. The RF switching subsystem, shown on the left side of Figure 18.1, switches and distributes RF signals to and from antennas, multicouplers, receivers, and transmitters. On the baseband side, the interior communications system has a black baseband switch for encrypted signals and a red baseband switch for signals that need to be protected after decryption. Some signals are transmitted in plain-text format (i. e., not encrypted). The user terminal equipment may be audio devices or data processing sets; hence baseband switches are designed to accommodate both audio and data signals.

The data transfer for weapons systems such as radar and weapons control is not routed through the interior communications system. The interior communications system configuration shown in Figure 18.1 is typical for large combatant ships that have a large number of antennas, RF equipment, baseband terminals, and user terminal equipment. In medium-size ships, the configuration is similar, but there are fewer ports (i. e., signal connections terminals) for user subsystems. In small ships, one finds a simplified version

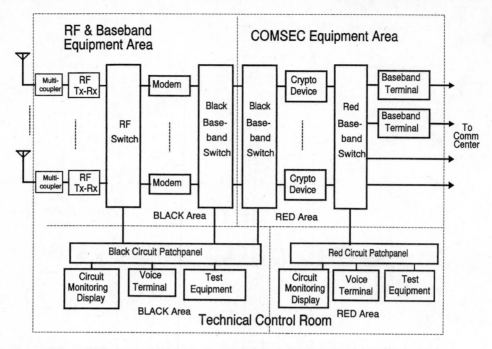

Figure 18.1 Block Diagram of a Typical Shipboard Technical Control Facility and Interior Communications Systems

with a manual RF patchpanel, and a baseband patchpanel for audio and data terminal equipment that can be controlled by a radio operator.

18.1 SHIPBOARD TECHNICAL CONTROL SYSTEMS

This section describes the shipboard technical control system, which provides the capability for:

- Monitoring circuit quality
- Patching and testing of communications security equipment
- Patching and testing of terminals

Radio operators perform technical control functions by using the equipment in the technical control room.

18.1.1 The Technical Control Facility (TCF)

The technical control (supervisory) room is the area where circuit and channel patchpanels are installed. Also provided are test equipment and monitoring facilities for a technical controller to perform the functions listed in Table 18.1.

TABLE 18.1 SHIPBOARD TECHNICAL CONTROL FUNCTIONS PERFORMED
IN THE TECHNICAL CONTROL ROOM

Function	Description
Technical Coordination	Performs technical administration among the connecting external technical control, the maintenance element, and the user
Technical Supervision	Exercises technical direction, coordination and supervision of: - transmission media - interface equipment in the patch and test facility - transmitter and receiver rooms that are not collocated in the technical control room
Restoration	Reassigns and restores disrupted communications circuits/links and circuit capability in accordance with a predetermined protocol
On-call Patching	Establishes on-call circuits
Testing	Performs quality control tests on all circuits passing through the technical control room
Activation and Deactivation	Establishes, activates or deactivates, rearranges, and discontinues circuits that enter and leave the platform
Report Generation	Reports circuit status to telecommunications supervisory organization for record purpose

18.1.2 The Communications Security Equipment Area

The COMSEC equipment area contains on-line cryptographic equipment with patchpanel, testing, and monitoring equipment so that personnel, under the direction of the technical control supervisor, can perform functions listed in Table 18.2.

TABLE 18.2 SHIPBOARD TECHNICAL CONTROL FUNCTIONS PERFORMED IN COMSEC
EQUIPMENT AREA

Function	Description
Trouble isolation	Isolation of equipment malfunctions occurring in the COMSEC equipment area
Patching	Substitution of equipment, channels, and circuits within the COMSEC equipment area
Testing	Performs back-to-back and loop-back tests of cryptographic equipment, channels, and circuits for maintenance purpose
Monitoring	Performs status checking and testing
Crypto timing	Performs crypto synchronization and resynchronization

TABLE 18.3   SHIPBOARD TECHNICAL CONTROL FUNCTIONS PERFORMED IN RF AND BASEBAND EQUIPMENT AREA

Function	Description
Circuit adjustment	Alignment and calibration of all circuits to meet transmission criteria
Restoration	Reestablishment of disrupted communications links
Patching	Replacement and/or bypassing of faulty RF equipment

18.1.3 The RF and Baseband Equipment Area

The RF and baseband equipment area houses modems, primary test equipment, and RF and baseband equipment, to permit operating personnel to perform the functions listed in Table 18.3.

18.1.4 The Surface Ship Exterior Communications Monitor and Control System (SSECMCS) AN/SQQ-33

The SSECMCS AN/SQQ-33 is an automated communications control and switching system for controlling and monitoring the operation of the EXCOMM system and routing communications signals to the appropriate systems [NS-3; sections 5–11, 5–12]. Figure 18.2 shows a typical shipboard exterior communications system with an AN/SQQ-33 Surface Ship Exterior Communications Monitor and Control System and an OJ-631 Technical Control Console. The SSECMCS provides the capabilities necessary to provide full supervisory control over the exterior communications system by two operators. The first operator performs normal technical control functions, the second operator performs backup functions and deals with surge traffic during busy hours. The SSECMCS is controlled from the control console and the data terminal set, both located in the communications technical control room. This system utilizes a computer and peripheral electronic data processing equipment to provide:

- Semi-automated control of all bus-controlled components of the radio communications system (RCS).
- All necessary data reduction and data processing needed to keep both operators informed of equipment status.
- A direct interface with the SSECMCS computer, and secure and nonsecure input from the consoles.
- A quality monitoring and test equipment suite, which enables the performance of real-time fault isolation, troubleshooting, and other circuit evaluation procedures.

The OJ-631 communications monitoring and control console (CMCC) is used to control, monitor, and supervise EXCOMM system operation. The OJ-631 performs

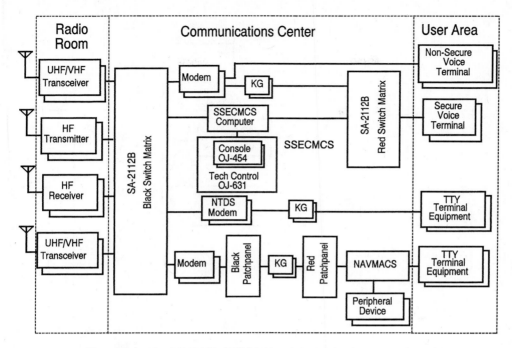

Figure 18.2 The SSECMCS AN/SQQ-33 and the SSECMCS Technical Control Console OJ-631 with a Typical Shipboard Interior Communications System

control of circuit configuration, frequency and mode selection, remote channel selection of UHF radios, circuit monitoring and fault detection and isolation. The OJ-631 provides mounting space for several types of equipment that assist the technical controller in manipulating the RCS efficiently. A variety of ancillary equipment is installed in the OJ-631, including the OJ-454 workstation and the Subscriber Interface Unit (SIU).

There are two OJ-454s workstations installed aboard a large ship. The first OJ-454 console is located in the CMCC and is operated by the chief technical controller. An assistant technical controller uses the second OJ-454 console that is located on a pedestal adjacent to the OJ-631 technical control console. The OJ-454 workstation consists of a CRT indicator panel, an electronic keyboard, and a slide-mounted logic board that includes an air inlet filter. The subassemblies work together to enable the technical controller and assistant technical controller to send and receive data from the AN/UYK-43 computer.

18.1.5 The Quality Monitoring Control System (QMCS)

The AN/SSQ-88 QMCS provides the equipment necessary for a radio operator to monitor the quality of communications circuits [NA-9]. Figure 18.3 is a block diagram showing the QMCS AN/SSQ-88 interfaces to shipboard EXCOMM and interior communications systems. Figure 18.4 shows the AN/SSQ-88 QMCS equipment.

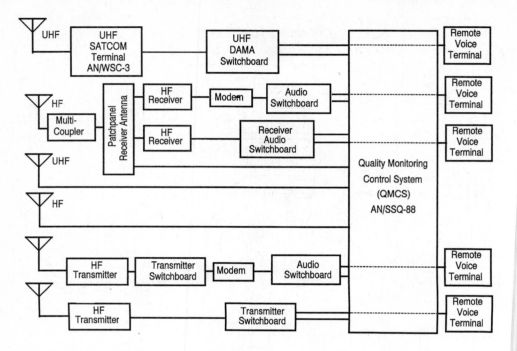

Figure 18.3 Block Diagram of Quality Monitoring Control System (QMCS) with Typical Shipboard EXCOMM and Interior Communications Systems

The QMCS is a system for on-line testing of communications circuits that does not interfere with normal circuit operation and quality. Off-line system, subsystem, and equipment tests can also be performed to determine the suitability of the system and equipment for use in communications circuits. This off-line capability allows the QMCS

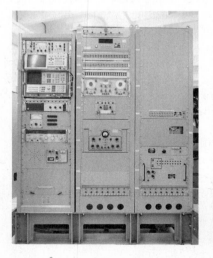

Figure 18.4 The AN/SSQ-88 Quality Monitoring and Control System (QMCS) (Courtesy of the U.S. Navy, Naval Electronic Systems Engineering Activities)

to carry out pre-deployment testing. Predeployment tests will certify the equipment's readiness, assist in calibration, and identify the need for repair. The AN/SSQ-88 is capable of, but not limited to, the following tests:

a. Data and teletypewriter signals
- Distortion measurement
- Speed error
- Signal shaping
- Signal level

b. Audio signals
- Level and linearity
- Frequency
- Harmonic products
- Aural monitoring

c. RF signals
- Signal and noise
- Frequency
- Carrier suppression
- Linearity of bandwidth
- Harmonic products
- Aural monitoring (off-the-air)

The AN/SSQ-88 QMCS is comprised of 11 test devices housed in a single CY-4516 electronic cabinet. Dedicated HF and UHF antennas are required for the QMCS. All necessary patch cords and connectors are provided as components of the QMCS.

The control and transfer group consists of switchboards, patchpanels, and the SA-2112 switching matrix. On a large ship (e.g., a CG-47 USS Ticonderoga class and an LHD-1 USS Wasp class), the equipment included in this group allows for a high degree of flexibility for connecting various functional groups to the required communication circuits. The types of signals which are switched by the control and transfer group are divided into three categories: Audio, DC, and data.

The SA-2112(V)6 Black Communications Switch (BCS) is the key element in the audio and transmitter control distribution portion of the control and transfer group. It is a programmable microprocessor-based switching matrix that can automatically connect a subscriber line to a transmitter, receiver, and/or transceiver. The BCS is functionally divided into four units: Audio switch, Controller, Built-In-Test (BIT), and Power. The audio switch routes audio and control signals between line and trunk section under the command of the controller.

The controller monitors and directs all BCS internal operations and external interfaces of the QMCS. This controller consists of two identical units of controllers; if one set fails, the other takes over control of the QMCS. The BIT unit determines which of the aforementioned two controller units is in command of the QMCS. It monitors the operation of the BCS and displays any indications of failure on the Control and Status Panel (CSP).

The power unit consists of two power supplies, two blower/line-filter assemblies, and a battery conditioner. The power supplies and blower are redundant, and their operation is monitored by the BIT section. The battery will supply power to BCS memory for 4 hours in case of a power failure. This BCS memory power supply ensures that critical circuit connection and frequency assignment information will be retained during a power failure.

A shipboard technical control facility needs an accurate and stable frequency standard to provide precise frequency signals to various transmitters, receivers, modems, and user terminals requiring an external frequency source. The Black Communications Switch (BCS) provides nonsecure switching of encrypted voice and data to data links, and modems to radio transceivers.[1] The BCS performs distribution of transmitter and receiver audio, and transmitter control signals to and from remote stations. The BCS provides the flexibility to change equipment configurations to accommodate changes in communications requirements. This is accomplished by connecting the remote stations to various transmitters in the ship. Although the interface switching takes place in the BCS, all circuits are routed through the Black Switch Matrix.

The BCS consists of a single cabinet. The front panel of this cabinet contains a control and status panel and a power supply circuit breaker panel. A battery and power conditioner unit is attached to the rear of the front panel door. The interior compartment houses circuit cards, redundant power supplies, and redundant blowers.

A nonblocking, solid-state Time Division Multiplex (TDM) switch matrix is used in the BCS to switch transmitter and receiver signals. Discrete control and status signals are multiplexed through the switch. The front Control and Status Panel (CSP) provides the operator interface to exercise complete control over all BCS functions. The CSP can be used to: (a) program independent receiver and transmitter interconnections, (b) display current interconnection status, (d) isolate system faults, and (e) monitor all circuits by using the panel-mounted handset.

18.1.6 Frequency Standards AN/URQ-23

The AN/URQ-23 is commonly used as shipboard frequency standard. Its functional block diagram is shown in Figure 18.5. It provides two frequency standard units: an on-line and an alternate unit, with provisions for switching when one unit fails. One of two on-line frequency standard units is selected by a manual frequency standard distribution panel. The selected on-line unit then provides three choices of 5.0 MHz, 1.0 MHz, and 0.1 MHz to the RF distribution amplifier. The RF distribution amplifier is connected to three Frequency Standard Signal Distribution Units which supply the standard frequency signal to shipboard communications systems [NA-8]. Figure 18.6 shows the AN/URQ-23 Frequency Standard.

---

[1]See Figure 18.19 for the relationship of the BCS to secure and nonsecure voice terminals.

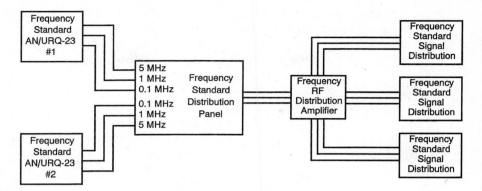

Figure 18.5   Block Diagram of the AN/URQ-23 Frequency Standard

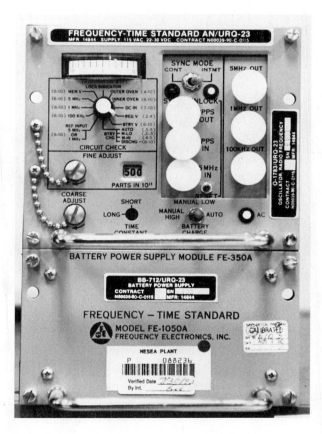

Figure 18.6   AN/URQ-23 Frequency Standard (Courtesy of the U.S. Navy, Naval Electronic Systems Engineering Activities)

18.1.7 Tactical Frequency Management Systems AN/TRQ-35 and AN/TRQ-42

The AN/TRQ-35 and AN/TRQ-42 Tactical Frequency Management Systems (TFMSs) provide effective real-time frequency management of HF radio circuits by enabling frequency selection on a quantitative basis [BR-1; WI-3]. The AN/TRQ-35 and AN/TRQ-42 consist of three equipment items: a Chirpsounder® transmitter and receiver to measure propagation conditions, and a spectrum monitor to measure interference and noise conditions. The system provides the radio operator with instrumentation which measures and displays an optimum frequency for HF operations. In section 10.2.4, we have discussed the AN/TRQ-35 and AN/TRQ-42 Chirpsounders® as HF sounding in ionospheric propagation. Figures 18.7 and 18.8 show the AN/TRQ-42 Chirpsounder® Transmitter and Receiver.

18.2 RF SWITCHING SUBSYSTEMS

Shipboard RF switching subsystems provide the interface between the shipboard exterior communications system and the shipboard technical control and interior communications system. Patchpanels, such as the ones shown in Figure 18.9, allow for flexibility in the RF distribution system. The following units are terminated at the RF patchpanel:

Figure 18.7   AN/TRQ-42 Chirpsounder® Transmitter (Reprinted courtesy or BR Communications Inc.)

Figure 18.8   AN/TRQ-42 Chirpsounder® Receiver (Reprinted courtesy of BR Communications Inc.)

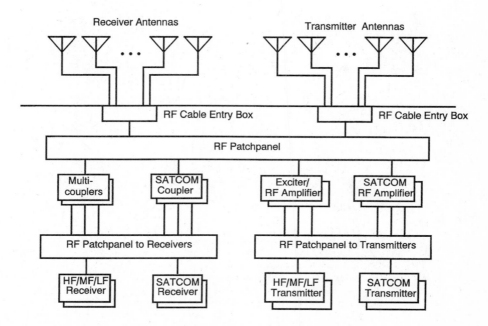

Figure 18.9   Block Diagram of a Typical RF Distribution Subsystem

- Each transmitter antenna input and each receiver antenna output
- Each multicoupler input and output
- Each receiver RF input and transmitter RF output

Radio operators use the patchpanel for performing technical control of the exterior communications system. The radio operator can control the system by substituting and bypassing equipment as required to insure a smooth operation.

18.2.1 Transmitter and Receiver Antenna RF Distribution

Transmitter antenna inputs and receiver antenna outputs are normally terminated at the RF patchpanel as indicated in Figure 18.9. RF shielding is provided so that the high-powered transmitter signals do not interfere with the receiver section of the patchpanel. Patching equipment and RF multicouplers are normally colocated in the radio room. All antenna transmission lines enter the radio room through the RF cable entry and connection box, where cable size or cable type are often changed because RF distribution cables in the radio room are thinner than cables connecting the antenna to the below deck cable entry box. Table 18.4 lists commonly used shipboard RF patchpanel.

Antennas, transmitters, receivers, and intermediate equipment currently in use are designed for 50Ω RF impedance matching. However, on many ships the existing radio rooms have older 70Ω and 75Ω coaxial cables in their RF distribution systems. When new 50Ω equipment is installed on platforms using 70Ω and 75Ω cabling systems, the older cables are replaced by 50Ω cables in order to avoid impedance mismatch and further deterioration due to aging.

18.2.2 Antenna Multicouplers

Receiver multicouplers facilitate the use of a single antenna as signal source for several receivers. Occasionally, passive multicouplers are used, but most multicouplers are amplifying devices that raise the output of each channel to the level required by each individual receiver.

It is an accepted practice to install multicouplers in a cascaded, "normal-through" arrangement when operations require more than eight outputs from a single antenna.

TABLE 18.4   SHIPBOARD RF PATCHPANELS

Nomenclature	Name
SB-3648 [NA-11]	Shipboard RF Test Instrument Patchpanel
SB-4249 [NA-12]	Shipboard RF Patchpanel
SA-2561 [NA-13]	Shipboard Frequency Standard Patchpanel
CA-1100 [NS-2]	Shipboard RF and Control Circuit Patchpanel

Up to five multicouplers can be cascaded without degrading the signal below acceptable limits. For each installation, at least one spare multicoupler must be provided as a stand-by unit. This spare multicoupler also permits manual cascading. If more than five multicouplers are needed, one spare should be installed for every ten units. All multi-couplers should follow the distribution as shown in Figure 18.10. When multicouplers are installed in cascade, the input to the cascaded multicouplers does not appear on the RF patchpanel.

The CU-1772A/SRA-56, CU-1774/SRA-57, and CU-1776A/SRA-58 ship-board HF transmitter multicouplers. The CU-1772A/SRA-56, CU-1774/SRA-57, and CU-1776A/SRA-58 HF transmitter multicouplers provide the capability of simul-taneously connecting up to four 1kW transmitters into a single broadband antenna, thus reducing the number of HF antennas required for effective communication [WI-1].

The CU-1772A/SRA-56 can connect simultaneously up to four broadband transmit-ters to a single antenna in the 2 MHz to 6 MHz band. The CU-1774/SRA-57 can also con-nect up to four broadband transmitters to an antenna; it tunes to the 4 MHz to 12 MHz band. The CU-1776A/SRA-58 can couple as many as four broadband transmitters to an antenna in the 10 MHz to 30 MHz band. For each of these couplers the input and output impedances are 50Ω unbalanced for coaxial cable interconnection to the transmitters and the antenna. The multicouplers can protect against overload by removing the RF power from the transmitter in the event that the network reflected power exceeds 250 W (3:1 VSWR) at a 1 kW input level, or if the network resonator voltages become too great.

Figure 18.11 shows the CU-1774/SRA-57 HF Transmitter Multicoupler. The other two HF multicouplers, the CU-1772A/SRA-56 and the CU-1776A/SRA-58, look the same as the CU-1774/SRA-57.

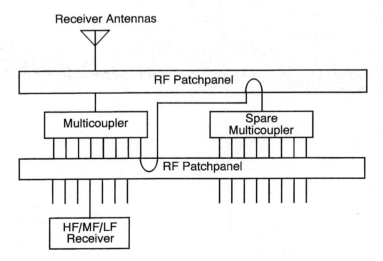

Figure 18.10   Shipboard Receiver Spare Multicoupler Connections

Figure 18.11   CU-1774/SRA-57 HF Transmitter Multicoupler (Courtesy of the U.S. Navy, Naval Electronic Systems Engineering Activities)

The CU-1789/SRA-49B shipboard HF receiver multicouplers.   The CU-1789/SRA-49B HF Receiver Multicoupler provides the capability of simultaneously connecting HF receivers such as the R-1051 and the R-2368 into a single broadband antenna. Figure 18.12 shows the CU-1789/SRA-49B HF Receiver Multicoupler.

18.2.3 RF Patchpanels

RF patchpanels are a device that distributes antenna outputs, multicoupler inputs and outputs, and receiver RF inputs, thereby providing for flexibility in the RF distribution system. These RF patchpanels are used by shipboard radio operators to provide access to signals, patch spare equipment items on-line quickly, and perform troubleshooting and testing.

The RF patchpanel SB-4249.   Figure 18.13 shows the connection for the SB-4249 RF patchpanel [NA-12]. For each connection, the patchpanel has two connectors in column. The upper connector is wired to the source (also marked "trunk" or "comm") and the lower to the load (also marked "equipment" or "modem"). Normally, an equipment is connected to a source; this is the normal pass-through connection. In Figure 18.13, load A is connected to source A, and so forth. When a patch cord is inserted (plugged) the normal pass-through connection is opened, so that source and load are disconnected. Thus, when a patch cord is inserted into the upper connection of one source (say B) and the lower connector of another load (for example, load C), source B and load C are connected as shown in Figure 18.13.

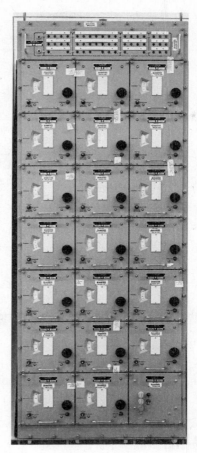

Figure 18.12   CU-1789/SRA-49B HF Receiver Multicouplers (Courtesy of the U.S. Navy, Naval Electronic Systems Engineering Activities)

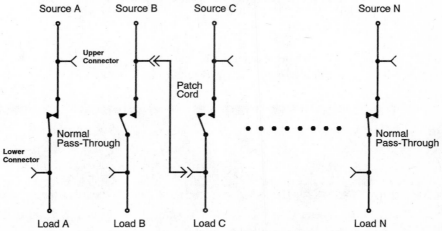

Figure 18.13   Wiring Diagram of the SB-4249 RF Patchpanel

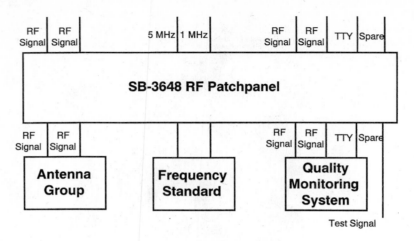

Figure 18.14 SB-3648 Patchpanel Block Diagram

The SB-3648 RF patchpanel. The SB-3648 is a special type of shipboard RF patchpanel that distributes test signal sources to the equipment to be tested [NA-11]. The three types of test signal sources include a frequency standard, test signals from a quality monitor system, and RF signals from an antenna test group. These RF test signal sources are connected to the rear of the RF patchpanel as shown in Figure 18.14. These test signals are connected to the equipment through the patch on the panel by manual patch cords. The antenna test group provides two RF signals; the frequency standard produces 1 MHz and 5 MHz frequency tones; the quality monitor system sends out two RF signals, a TTY test signal and a spare. Figure 18.15 shows the front of the SB-3648 RF patchpanel.

The CA-1100 RF patchpanel. The CA-1100 is a system of RF and control-circuit patchpanels which provides a simple means of connecting radio transmitters to their associated antennas, couplers, or dummy loads. Figure 18.16 shows the CA-1100 RF patchpanel.

The system includes transmitter panels and load panels, which can be mounted on a standard rack or in a cabinet. Each transmitter panel can accommodate from one to eight

Figure 18.15 Front Panel View of the SB-3648/UR RF Patchpanel (Courtesy of the U.S. Navy, Naval Electronic Systems Engineering Activities)

Figure 18.16   CA-1100 RF Patchpanel (Courtesy of D&M/Chu Technology, Inc)

transmitters, and each load panel accommodates up to five loads. The capacity of the system can be expanded by increasing the number of panels to meet the total interconnection requirements. The system is designed to prevent crossconnection between sources and loads. Each plug is arranged so that the RF connection is completed before the control contact is made and so that contact is broken before the RF line is opened. Control lines can be used to prevent transmitter operation when not connected to a load.

The SA-2561 RF frequency standard patchpanel.   The SA-2561 is a shipboard RF patchpanel that distributes three frequency test standards (100 kHz, 1 MHz, and 5 MHz) from two AN/URQ-23 Frequency and Time Standard units to RF equipment. Figure 18.17 is a block diagram of the SA-2561 RF Frequency Standard Patchpanel [NA-13]. Figure 18.18 shows the SA-2561 RF Panel.

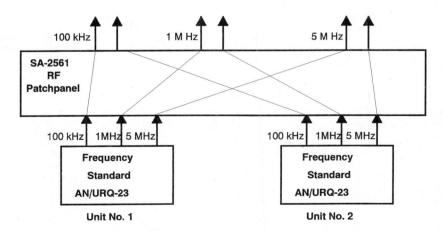

Figure 18.17   Block Diagram of the SA-2561 RF Frequency Standard Patchpanel

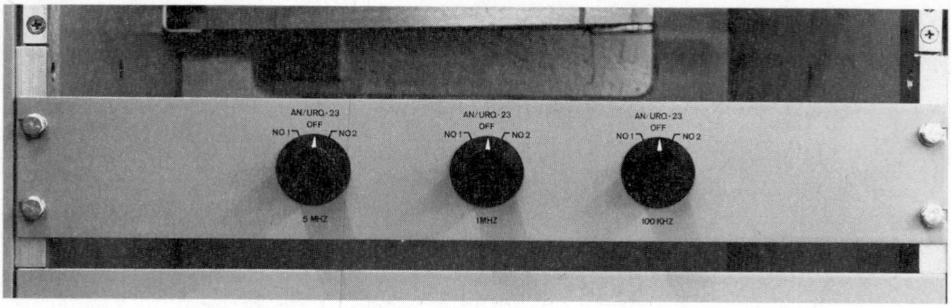

Figure 18.18 SA-2561 Frequency Standard Patchpanel (Courtesy of the U.S. Navy, Naval Electronic Systems Engineering Activities)

18.3 COMMUNICATIONS SWITCHING SYSTEMS

The control and transfer of audio and data communications between various shipboard users (e.g., the bridge, the radio room, the combat information center, and the exterior communications system) are provided by a baseband communications switching system. A functional block diagram of the typical shipboard interior switching system using the SA-2112A(V) 6 Black Communications Switch and SA-2112(V) 8 Red Communications Switch is shown in Figure 18.19. This shipboard interior communications switch system

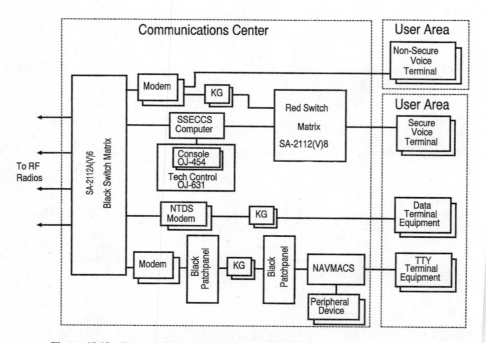

Figure 18.19 Typical Shipboard Interior Switching System Using a SA-2112A(V)6 Black Communications Switch and SA-2112(V)8 Red Communications Switch

TABLE 18.5   SHIPBOARD BASEBAND SWITCHING SUBSYSTEMS

Nomenclature	Name
SA-2112A(V)6 [NA-3]	Shipboard Black Communications Switch
SA-2112(V)8 [NA-3]	Shipboard Red Communications Switch
SB-4268 [NA-14]	Shipboard Audio Frequency (AF) Patchpanel (28 Jack Set)
SB-4269	Shipboard AF Patchpanel (20 Jack Set)

connects lines to their shipboard users on demand by EXCOMM radio circuits. Functionally, it is similar to a commercial telephone switching system. But unlike a telephone system, where there are thousand of connected lines, a shipboard interior switching system connects a few hundred lines at most.

A typical shipboard baseband communications switching system consists of switchboards, patchpanels, and the switching matrix. Table 18.5 shows a list of shipboard baseband switching subsystems. The signals that are switched by the control and transfer group are divided into three main categories: audio and transmitter control, DC, and data. The control and switch group provides nonsecure switching of audio and data to radio transceivers from cryptographic equipment, data links, and modems.

Some examples of three representative communications switching systems are described below. They are the SA-2112A(V)6/STQ Black Communications Switch, the SA-2112(V)8/STQ Red Communications Switch, and the MCS-2000 Interior Communications System. The SA-2112A(V)6/STQ Black Communications Switch manages the control and switching of non-secure audio and data terminals only. The SA-2112(V)8/STQ Red Communications Switch manages the control and switching of both nonsecure and secure audio and data terminals, and is often referred to as the Single Audio System (SAS). The MCS-2000 is designed to manage control and switching of interior audio and data terminals and the exterior communications system.

18.3.1 The Black Communications Switch SA-2112A(V)6/STQ

The SA-2112A(V)6/STQ Black Communications Switch is an automated analog radio-telephone switching system that controls and switches nonsecure audio and data terminals [NA-3]. The SA-2112A(V)6/STQ is generally located in the communications center of a ship as shown in Figure 18.19. The SA-2112A(V)6/STQ Black Communications Switch and SA-2112(V)8/STQ Red Communications Switch enable large ships to talk with other afloat units and shore stations. The SA-2112A(V)6/STQ Black Switch equipment is shown in Figure 18.20.

The SA-2112A(V)6/STQ can operate in local mode or remote mode. In remote mode, the SA-2112A(V)6/STQ can be switched from the technical control console.[2] The

---

[2]An example of the technical control system is the Surface Ship Exterior Communications Monitoring and Control System (SSECMCS) AN/SSQ-33 (see section 18.1.4), which includes a shipboard computer (AN/UYK-43).

Figure 18.20   The SA-2112A(V)6 Black Communications Switch (Courtesy of the U.S. Navy, Naval Electronic Systems Engineering Activities)

local mode allows the operator to program the SA-2112A(V)6/STQ manually using front panel controls on the control status panel.

The SA-2112A(V)6 Remote Programming Unit (RPU) enables control of the SA-2112A(V)6/STQ. The RPU facilitates the initial loading of the SA-2112A(V)6/STQ line and trunk assignments, modification of these assignments, storage of assignments on magnetic tape for subsequent use, and basic status monitoring of the SA-2112(V)6/STQ. Figure 18.21 shows the SA-2112(V)RPU, which includes a printer, CRT display, keyboard, and cassette tape drive.

18.3.2 The SA-2112(V)8/STQ Red Communications Switch

The SA-2112(V)8/STQ Red Communications Switch, also referred to as the Single Audio System (SAS), is a digital switch that manages the control and switching of plain and cipher audio and data terminals and external radio circuits. The SA-2112(V)8/STQ, as shown in Figure 18.22, works in conjunction with Black Communications Switch SA-2112A(V)6/STQ. The SAS features allow remote operating positions to select either

Figure 18.21   SA-2112A(V)6 Remote Programming Unit (Courtesy of the U.S. Navy, Naval Electronics Systems Engineering Activities)

secure or nonsecure handsets without reconfiguring the existing system. Also, the SAS features allow the automatic switching between remote handsets and shipboard external communications units, or the cryptographic equipment. In addition, it controls and monitors the system using an AN/UYK-43 computer and the Surface Ship Exterior Communi-

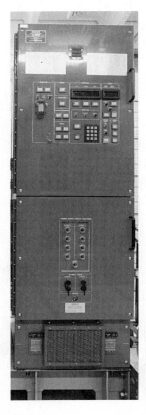

Figure 18.22   SA-2112(V)8 Red Digital Switch (Courtesy of the U.S. Navy, Naval Electronic Systems Engineering Activities)

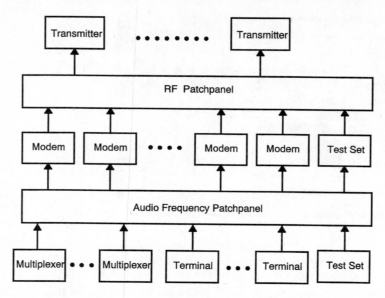

Figure 18.23   Typical Applications of a Shipboard Audio Frequency Patchpanel

cations Monitoring and Control System (SSECMCS) technical control console or by manual control from the front panel (SA-2112A(V)6/STQ).

The SA-2112(V)8/STQ system capabilities are as follows:

- It cross-connects the audio and discrete keying signals from various user terminals to various security equipment as shown in Figure 18.19.
- It can serve 117 user terminals and 54 security devices.
- It can connect 114 lines and 78 trunks.
- Patch cords can be used for casualty interconnection.
- It provides line interfaces for:
  (1) 95 telephones TA-970[3] [NA-6],
  (2) 83 control indicators C-10276, and
  (3) 22 access lines to the ON-201.

18.3.3 The SB-4268 Audio Frequency Patchpanel

An Audio Frequency (AF) patchpanel provides means to interconnect, test, and monitor the same or similar types of equipment in audio frequency bands. Figure 18.23 shows a typical application of a shipboard AF patchpanel. The AF band normally covers from 30 Hz to 15,000 Hz. Audio frequency lines are interconnected at an AF patchpanel by a tech-

---

[3]See section 18.5.4 (TA-970 Telephone Set).

nical control operator. The most commonly used method of connecting various lines served by an AF patchpanel is by plugs (connectors), jacks, and patch cords. In the cord-type patchpanel, the lines to be switched are terminated by jacks, and connection between them is made by patchpanel plugs and cords. Figure 18.24 shows the frequently used shipboard Audio Frequency patchpanels SB-4268 (with 28 sets of jack holes) [NA-14] and SB-4269 (with 20 sets of jack holes). Also shown in Figure 18.24 are three test instruments which are attached to the SB-4268 and SB-4269 AF patchpanels. The test instruments are two AF oscillators, which provide AF signals, and an AF monitor.

18.3.4 The MCS-2000 Interior Communications System

The MCS-2000 Interior Communications System provides control and switching of the interior audio and data terminals and the exterior communications system in a manner similar to the SA-2112A(V)6/STQ and SA-2112(V)8/STQ Communications Switches [WI-2]. It automatically routes communication inputs, both internal and external, to the designated stations throughout the ship as shown in Figure 18.25. The MCS-2000 enables shipboard users to talk with each other and also to other ships through an automated, centralized switching system. Shipboard users can be interconnected with single audio terminal access to sound-powered audio and radio communications. Each terminal can be configured for specific user requirements using cassette tapes that can modify up to 12 individual functions. Conferencing, one of the functions that can be customized for each

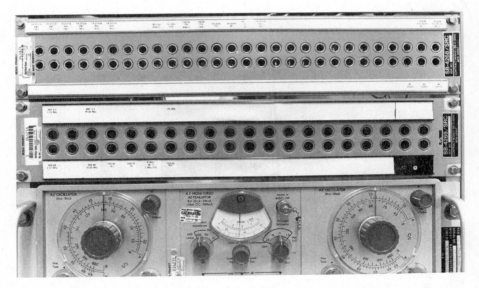

Figure 18.24   SB-4268 and SB-4269 Audio Frequency Patchpanels with AF Oscillators and an AF Monitor Set (Courtesy of the U.S. Navy, Naval Electronic Systems Engineering Activities)

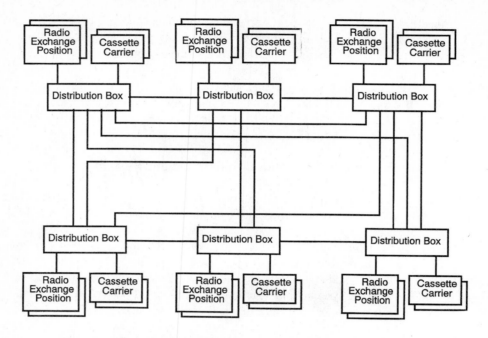

Figure 18.25   Block Diagram of the MCS-2000 Interior Communications System

user, is provided for a maximum of 10 participants. The MCS-2000 can accommodate up to 45 full capability user stations (i. e., user stations with a full complement of sound-powered voice, direct radio access, 12-function cassette tapes, and conferencing capabilities) or a larger number of less capable terminals.

18.4 DATA SWITCHING SYSTEMS

There are several shipboard switching systems that provide control and transfer of data communications: (a) the SA-4176 Data Transfer Switch, (b) the NAVMACS-V AN/SYQ-7 system, and (c) the ON-143 Interconnecting Group. The SA-4176 Data Transfer Switch performs switchboard functions for shipboard tactical data links and exterior communications radios. The NAVMACS-V AN/SYQ-7 is an automated message processing and distribution system for the fleet broadcast and ship/shore, point-to-point messages. The messages are transferred from radio subsystems, processed, stored, and distributed to remote terminals and then relayed to other units. The ON-143(V)6 Interconnecting Group is a versatile interior communications and control system that interfaces various FLTSATCOM exterior radio system and user terminals [NA-5].

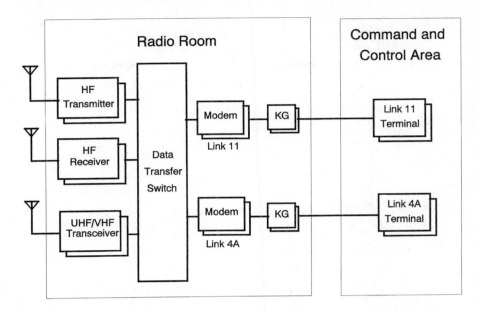

Figure 18.26   Block Diagram of the SB-4176 Data Transfer Switch

18.4.1 The SA-4176 Data Transfer Switch

The Data Transfer Switch SB-4176 is a data switch which interconnects shipboard tactical data links (Link 11[4] and Link 4A[5]) to appropriate EXCOMM radios such as the AN/URT-23 HF transmitter, the R-1051(G) HF receiver, and the AN/URC-93 VHF/UHF transceivers, as shown in Figure 18.26. The SB-4176 has the capability to select up to seven radios and up to five tactical data terminals [NA-10].

18.4.2 The Navy Modular Automated Communications System (NAVMACS-V) AN/SYQ-7

The NAVMACS-V [NT-1] is an automated message processing and distribution system in which the fleet broadcast and ship/shore point-to-point messages are transferred to and received from radio subsystems, then are processed, stored, and distributed to remote terminals and relayed to other afloat units.[6] The NAVMACS-V is also used to assist the naval message drafters, releasers, and readers by (1) aiding in composition and editing, (2) electronically delivering messages from operational spaces to the message processing

---

[4]Link 11 is described in section 10.3.

[5]Link 4A is described in section 11.4.8.

[6]The FLTSATCOM CUDIXS, which uses the NAVMACS, is described in section 13.2.2.

center, and (3) electronically formatting and selectively distributing priority messages to readers in operational spaces. The NAVMACS-V is also used by radio operators for automatically screening incoming messages, detecting duplicate messages, determining distribution, filing for retrieval, and sending messages. The NAVMACS-V is controlled from terminals located in the Joint Message Center. Messages are entered and received at remote located terminals.

Figure 18.27 shows a NAVMACS-V block diagram. The NAVMACS-V controls the following four types of channels:

- On-line Fleet Multichannel Broadcast channels
- On-line full-period terminal channels
- CUDIXS Net primary and special subscriber channel
- Off-line (via torn tape) channels that can be received via ship-to-ship or ship-to-shore netted, full-period terminal, or broadcast radios

Information is exchanged through the CUDIXS by interfacing through the ON-143(V)6 Interconnecting Group which contains the encryption device and the AN/WSC-3 UHF SATCOM radio. The broadcast is received either via conventional HF equipment (AN/UCC-1) or via a satellite receiver (AN/SRR-1). Messages from full-period circuits and other ship-to-ship and ship-to-shore-to-ship circuits are transferred by conventional HF transmitters and receivers. Figure 18.28 shows the NAVMACS-V system.

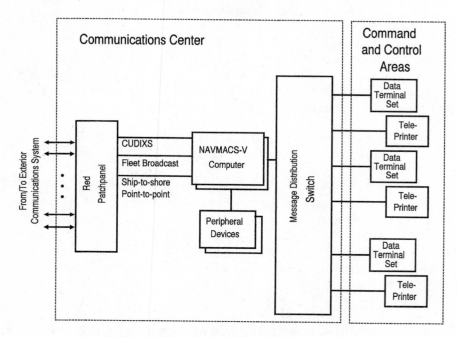

Figure 18.27   Block Diagram of the NAVMACS-V

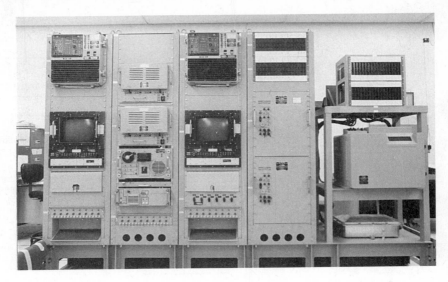

Figure 18.28 AN/SYQ-7 Navy Modular Automated Communications System (NAVMACS-V) (Courtesy of the U.S. Navy, Naval Electronic Systems Engineering Activities)

18.4.3 The ON-143(V)6 Interconnecting Group

The ON-143(V)6 Interconnecting Group is an electronic device used in interfacing various equipment in a Fleet Satellite Communications System installation [NA-5].[7] It is used in SSIXS, OTCIXS, and TADIXS installations, and it provides support for TTY message traffic. It has the capability to operate in OTCIXS and SSIXS circuits, and to support SECVOX circuits. The ON-143(V)6 is a multipurpose interface and data distribution and switching system functioning between the AN/UYK-20 or AN/UYK-43 data processing set and satellite communications equipment. Figure 18.29 is a simplified block diagram showing the ON-143(V)6 and its connections to the data communications equipment (i. e., radio equipment) and the data terminal equipment (i. e., user data processing equipment). The functions of the ON-143(V)6 are:

- Cryptographic auxiliary unit for red-black isolation
- Timing generation and transmit delays
- Receiver synchronization timing
- Cryptographic test signal generation
- System equipment interface

[7]For a discussion of the FLTSATCOM subsystems, see Chapter 13.

Red/Black Isolation

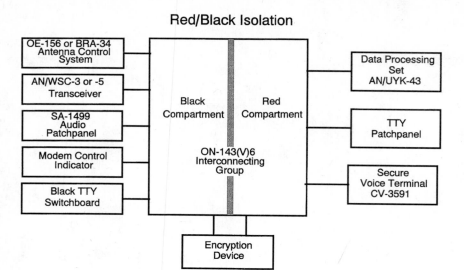

Figure 18.29   ON-143(V)6 Interconnecting Group Block Diagram

Figure 18.30   ON-143(V)6 Interconnecting Group (Courtesy of Allied Signal Inc., Ocean Systems)

- Signal distribution
- Level conversion to the MIL-STD-188C standard

The flexibility of the ON-143(V)6 is attributable to the use of dynamically adaptive receiver/transmitter (DART) circuit boards. These boards are processor controlled interface boards. The ON-143(V)6 has DART boards for VOCODER, crypto, control panel, TTY, data links, and black data. When powered-up (i.e., on initial application of power) information contained in a programmable read-only memory (PROM) chip is downloaded to the DART boards to prepare the ON-143(V)6 for its operation in the particular circuit it supports (i.e., OTCIXS, SSIXS, or SECVOX).

At base, the ON-143(V)6 consists of two compartments; the first is a compartmented, black card cage, and the second is a red card cage. The two compartments are RF shielded from each other through the use of sophisticated technology (i. e., fiber optics, electrostatic relays, and heavy shielding). Signals within the unit can pass through the red-black partition, yet the signals are kept strictly separated, thus providing an certified red-black signal separation. Figure 18.30 shows the ON-143(V)6 Interconnecting Group.

18.5 OTHER INTERIOR COMMUNICATIONS SYSTEMS

There are other frequently used special-purpose, shipboard interior communications systems. The following four systems are discussed in this section:

- Flight Deck Communications System (FDCS)
- Sound-powered telephone system
- Announcing System
- Telephone set.

18.5.1 The Flight Deck Communications System (FDCS)

An important special-purpose, shipboard interior communications system is a flight deck communications system, called the FDCS [WI-4]. The FDCS is a multichannel UHF radio that is capable of providing voice transmission in both secure and nonsecure modes. It provides for the transmission of orders and information to various locations, simultaneously or selectively, by means of portable radios and a base station. Figure 18.31 shows the AN/SRC-47 FDCS.

Figure 18.32 shows how components of the AN/SRC-47 FDCS System are interconnected. The base station equipment is located below deck. The RF repeater RT-1372 serves as a low-power relay between the base station and mobile units. The signal data converter CV-3979 interfaces all base station equipment. The control indicator C-10907 establishes communications between interior and exterior mobile units. Several AT-150 deck-mounted antennas are used to coordinate communications between the flight deck,

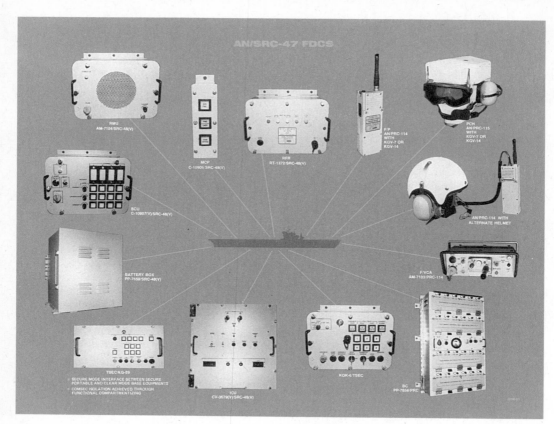

Figure 18.31 AN/SRC-47 FDCS Equipment Suite (Courtesy of GTE Government Systems Corporation)

hangar, and announcing systems. The mobile units are the AN/PRC-114 Man-on-the-Move radio and the AN/PRC-115 helmet-mounted radio.

18.5.2 Sound-powered Telephone Systems

Shipboard sound-powered telephone systems (e.g., AN/WTC-2) consist of up to 100 individual sound-powered telephone circuits. A sound-powered telephone circuit operates without any external power. It provides a reliable and battle damage-resistant means of communications between various shipboard stations.

Sound-powered telephones are an electromagnetic device. The telephone transmitter in a handset contains a diaphragm, an armature, a permanent magnet, and a coil. Sound pressure moves in the diaphragm, to which the armature is connected. As the armature moves, the coil wrapped around the permanent magnet produces electric current. Up to 10 mA current is produced, which allows sound-powered telephones to be used over dis-

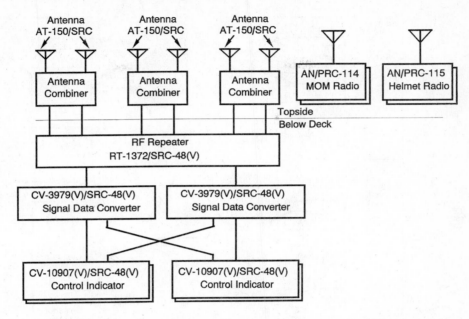

Figure 18.32   AN/SRC-47 Flight Deck Communications System Block Diagram

tances up to 1.5 km. Ringing current required for signaling in shipboard telephone networks are provided by a hand generator.

There are three levels of the sound-powered telephone circuits:

- Primary circuits, which provide communications for essential functions such as ship control, weapon control, aircraft control, engineering, and damage control.

- Secondary circuits, which provide an alternate means of communications for selected primary circuits.

- Supplementary circuits, which provide means of communications for subordinate control, operating and service functions.

Sound-powered circuits are configured to link a switchboard or switchbox by direct point-to-point wire connections. A switchboard consists of a large number of line cutout switches mounted in an action cutout (ACO) switchboard, connecting incoming lines to a common circuit bus and allowing circuits to be paralleled, tied together, or disconnected in case of casualty. A switchbox circuit consists only of several line cutout switches mounted in a switchbox that functions as a small ACO switchboard.

Figure 18.33 shows a diagram illustrating how sound-powered telephone circuits are connected. Each station is equipped with a sound-powered telephone H-202 headset, H-200 chestset, or a H-203 handset. These telephone sets are connected to several, redundant ACO switchboxes. Each switchboard is connected by a parallel bus to all other

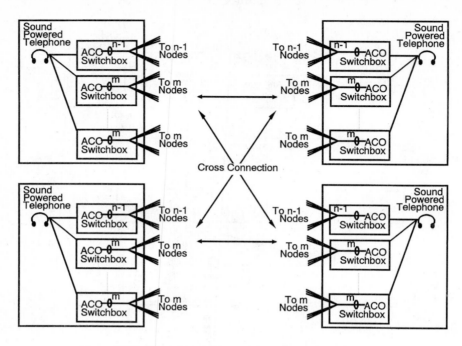

Figure 18.33   Block Diagram of a Sound-Powered Telephone System

switchboards. There is a large number of wires in the switchboard. Similarly, one headset, chestset, or handset is connected to several ACO switchboxes. Each switchbox is connected by a parallel bus to all other switchboxes. There are fewer wires in the switchbox than in the switchboard. A sound-powered telephone selector switch and handset is in-

Figure 18.34   Sound-Powered Telephone Set (Side Crank) AN/WTC-2A(V) (Courtesy of Dynalec Corporation)

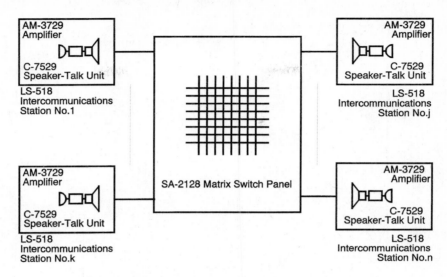

Figure 18.35   Block Diagram of a Two-Way Announcing System LS-474/U

stalled for the officer-in-charge or station supervisor at control and operating stations served by more than one sound-powered telephone circuit. Figure 18.34 is the Sound-Powered Telephone Set AN/WTC-2A(V) with a side crank, located at various locations throughout the ship. The physical dimension of AN/WTC-2A(V) is 20 cm (7.75 in.) high, 16 cm (6.32 in.) wide, and 12 cm (4.75 in.) deep.

18.5.3 The Announcing System LS-474/U

A shipboard announcing system is an electrically powered voice system providing one-way or two-way communications, either throughout the ship or in selected locations. In addition, nonvoice emergency alarms can be broadcast over the announcing system.

The LS-474/U two-way announcing system provides shipboard intercommunicating capabilities for selected watch stations and command and control spaces by a talk-listen loudspeaker unit [NA-1]. Each loudspeaker unit is connected to all other units on the circuit, with calling access available by pushbutton selection of the called station. Figure 18.35 shows a diagram of the two-way announcing system LS-474/U.

The one-way announcing system provides the capabilities for shipboard broadcast information of general interests and audio emergency announcement throughout the ship, or to selected zones from fixed microphone stations. Figure 18.36 illustrates how a one-way announcing system is connected.

18.5.4 The TA-970 Telephone Set

The TA-970/U telephone set is part of the automated Single Audio System (SAS), and is able to send and receive single-channel ciphered and plain audio signals in a shipboard interior communications system [NA-6]. A typical shipboard SAS is the

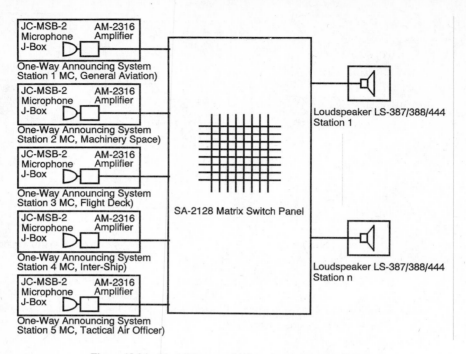

Figure 18.36 Block Diagram of a One-way Announcing System

Figure 18.37 The TA/970 Telephone Set
(Courtesy of the U.S. Navy, Naval
Electronic Systems Engineering Activities)

SA-2112(V)8/STQ Red Communciations Switch, which is described in section 18.3.2. The TA-970/U telephone set consists of a handset with associated circuits and controls. Figure 18.37 shows the TA-970/U. The TA-970/U control panel has several control switches and the plain and cipher indicator lamps. For example, an amber indicator lamp, labeled DETECT, is mounted on the control panel to show that the cryptographic device is processing incoming secure voice signal.

REFERENCES

[BR–1]   *BR Communications, Inc. Data Sheet on AN/TRQ-42 Chirpsounder*, Sunnyvale, Calif. 94089.

[NA–1]   *Loudspeaker LS-474/U*, Technical Manual, SPAWAR-0967-306-1010, Space and Naval Warfare Systems Command, Washington, D.C., 20363-5100.

[NA–2]   *Interconnecting Group ON-143(V)4/USQ*, Technical Manual, SPAWAR-0967-LP-614-7010, Space and Naval Warfare Systems Command, Washington, D.C., 20363-5100, September 1977.

[NA–3]   *Communications Switch Matrix SA-2112(V)6*, Technical Manual, SPAWAR-EE682-AD-OMI-110/W152, Space and Naval Warfare Systems Command, Washington, D.C., 20363-5100, August 1988.

[NA–4]   *Switch Matrix SA-2112(V)/STQ and Control-Indicator C-10276/SSC*, Technical Manual, SPAWAR-EE107-AA-OMI-010/E110-SA2112, Space and Naval Warfare Systems Command, Washington, D.C., 20363-5100.

[NA–5]   *Interconnecting Group ON-143(V)6*, Technical Manual, SPAWAR-EE130-AC-OMI-010/ON-143(V) 6, Space and Naval Warfare Systems Command, Washington, D.C., 20363-5100, October 31, 1992.

[NA–6]   *Telephone Set TA-970/U*, Technical Manual, SPAWAR-EE165-GB-OMI-010/W100-TA970,80,90, Space and Naval Warfare Systems Command, Washington, D.C., 20363-5100, March 3, 1983.

[NA–7]   *Radio Frequency Patchpanel SB-4267A/SRC*, Technical Manual, SPAWAR-EE167-ST-OMI-010/W152-SB4267A/0913-LP-279-3400, Space and Naval Warfare Systems Command, Washington, D.C., 20363-5100, May 24, 1987.

[NA–8]   *Frequency Time Standard AN/URQ-23*, Technical Manual, SPAWAR-ET710-AA-OPI-010/5102, Space and Naval Warfare Systems Command, Washington, D.C., 20363-5100.

[NA–9]   *Quality Monitoring Set AN/SSQ-88*, Technical Manual, SPAWAR-EE822-AA-OMP-010/P630 SSQ-88, Space and Naval Warfare Systems Command, Washington, D.C., 20363-5100, October 20, 1985.

[NA–10] *Transfer Switchboard SB-4176/U*, Technical Manual, SPAWAR-SE-676-AW-MMO-010/SB4176, Space and Naval Warfare Systems Command, Washington, D.C., 20363-5100.

[NA–11] *SB-3648 RF Patchpanel*, Operations and Maintenance Manual, EE167-SE-OMI/SB-3648, Space and Naval Warfare Systems Command, Washington, D.C., 20363-5100, February 15, 1991.

[NA–12] *SB-4249 RF Patchpanel*, Operations and Maintenance Manual, EE679-HA-OMI-

010/W110/SB-3648, Space and Naval Warfare Systems Command, Washington, D.C., 20363-5100, February 15, 1989.

[NA–13] *SA-2561 RF Patchpanel*, Operations and Maintenance Manual, EE119-ST-OMP-010/W152/SA-2561, Space and Naval Warfare Systems Command, Washington, D.C., 20363-5100, May 15, 1986.

[NA–14] *SB-4268 AF Patchpanel*, Operations and Maintenance Manual, EE679-LC-OMP-010/W152-SB-4268, Space and Naval Warfare Systems Command, Washington, D.C., 20363-5100, June 30, 1986.

[NS–1]   *Antenna Coupler Groups AN/SRA-56, AN/SRA-57, AN/SRA-58*, NAVSEA-0967-284-6010, Technical Manual, Naval Ship Systems Command (now Naval Sea Systems Command), Washington, D.C., 20362, March 27, 1969.

[NS–2]   *Radio Frequency and Control Circuit Patch System, Chu Model CA-1100*, Technical Manual, NAVSEA-0967-LP-619-6010, Naval Sea Systems Command, Washington, D.C. 20362.

[NS–3]   *LHD-1 Class Combat System Elements Description Handbook*, Naval Sea Systems Command (PMS-377), Washington, D.C. 20362, May 30, 1987.

[NT–1]   *Naval Modular Automated Computer System AN/SYQ-7(V)3*, System Operator's Manual, NTSIC 307.10B, Naval Telecommunications Systems Integration Center, Cheltenham, Md.

[WI–1]   Williamson, J., ed., "AN/SRA-56/57/58 HF Transmitting Multicoupler," *Jane's Military Communications 1989*, Jane's Information Group, Alexandria, Va. 22314-1651, 1989, p. 692.

[WI–2]   Williamson, J., ed., "MCS 2000 Communication System," *Jane's Military Communications 1991–1992*, Jane's Information Group, Alexandria, Va. 22314-1651, 1991, p. 377.

[WI–3]   Williamson, J., ed., "AN/TRQ-35(V)2 Tactical Frequency Management System," *Jane's Military Communications 1991–1992*, Jane's Information Group, Alexandria, Va. 22314-1651, 1991, pp. 112–114.

[WI–4]   Williamson, J., ed., "AN/SRC-47(V) Man-On-The-Move Communications System," *Jane's Military Communications 1991–1992*, Jane's Information Group, Alexandria, Va. 22314-1651, 1991, p. 266.

19

Shipboard Communications Protocols

Shipboard mission subsystems such as combat control, intelligence, weapons systems, and command systems, communicate with each other and the exterior world. Information is transferred to shipboard combat systems by communications subsystems such as the Link 11 [NA-1; DD-2], FLTSATCOM terminals [NC-1], Link 11, and OTCIXS. In order to insure successful transfer of information from outside the ship to the shipboard combat systems, all subsystems must adhere to information transfer conventions. Such conventions are called protocols.

Several international and military organizations are responsible for the standardization of communications protocols. These governing bodies include the International Standard Organization (ISO), the Consultative Committee of International Telephone and Telegraph (CCITT), the Electronic Industries Association (EIA), the Institute of Electrical and Electronic Engineers (IEEE), and the Defense Information Systems Agency (DISA), formerly Defence Communications Agency (DCA).

The key elements of a protocol are syntax, semantics, and timing. Syntax includes data format and signal levels; semantics include control information for coordination, error detection, and correction; timing includes data rate and synchronization information. In this chapter, protocols used in shipboard communications are discussed.

19.1 SHIPBOARD COMBAT SYSTEMS

A layout of typical shipboard combat systems that shows the sources and destination of information will be used to give a synopsis of shipboard communications protocols. Signal types and their characteristics and relationships to each other will also be discussed here.

Early shipboard combat systems were an aggregate of data processing and communications subsystems that were connected by cables, usually twisted pair and coaxial type. A modern shipboard combat system is much more complex; it consists of data processing subsystems, sensor subsystems, weapon subsystems, electronic support measure (ESM) subsystems, and communications subsystems. The method of interconnecting these subsystems has evolved from using cables to using local area networks (LANs). In order to interact effectively, many Navy processing systems use standardized data transfer methods such as the Navy Digital System Interconnect Standard (MIL-STD-1397A) [NS-1], the Navy Aircraft System Data Bus Standard (MIL-STD-1553B) [DD-1; DD-4; IL-1], and the NATO Aircraft Data Bus Standard (STANAG 3838) [GR-1; NS-1; NT-1].

19.1.1 Description of Typical Shipboard Combat Systems

Figure 19.1 shows an aggregate of typical shipboard combat systems for surface ships such as guided missile cruisers, destroyers, and frigates [GR-1; FL-1]. The major elements of shipboard combat systems include:

- Radar systems
- Identification systems
- EW systems
- Navigation systems
- Underwater systems
- Command and decision systems
- Weapon control systems
- Weapon systems
- Telemetry systems
- Exterior communications systems
- Interior communications systems
- Meteorological systems
- Training systems

These elements perform the basic functions of detection, control, and engagement in anti-air, anti-submarine, anti-surface warfare, and amphibious warfare as well as the combat support functions of navigation, communications, meteorological support, and training.

The left-most column in Figure 19.1 shows shipboard sensor systems which include radar, IFF, EW, sonar, sensor data link, and optical systems. The second column shows the command and decision system and the weapons control system. The third column shows combat support systems such as the navigation, interior communications, combat training

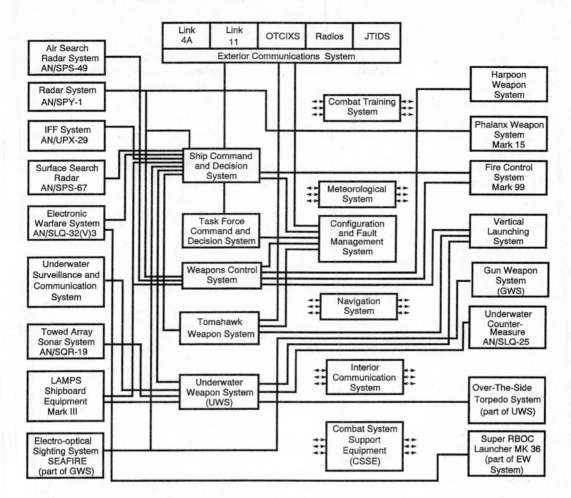

Figure 19.1   Typical Shipboard Combat System (Reprinted with permission from D. Green and D. Marlow, "Application of LAN Standards to the Navy's Combat System," *Naval Engineers Journal,* May 1990, pp. 146–153.)

systems. The top of the figure shows typical shipboard exterior communications systems that provide communications from the ship to other ships and to shore nodes. In Appendix B, each element of the combat system shown in Figure 19.1 is described in detail.

19.1.2 Shipboard Signals and Interfaces

The information transferred in a typical shipboard combat system is summarized in Table 19.1. The table shows that both analog and digital signals are used; signal formats are different, and they are matched to the type of information to be transferred.

In order to accommodate efficient transfer of signals in shipboard combat systems, standardized signals and interfaces are used. One of the most widely used digital data interfaces is the NTDS interface, as specified in the Navy Digital System Interconnect Standard MIL-STD-1397A [NS-1]. MIL-STD-1397A defines the interface characteristics (physical, functional, and electrical) and the standard interfaces between digital equipment, that is, (a) computer to peripheral, (b) computer to computer, and (c) peripheral to peripheral. There are three types of NTDS formats to accommodate different data transfer needs: (1) NTDS slow, (2) NTDS fast, and (3) NTDS serial.

Another widely used digital interface standard is the Aircraft System Data Bus Standard, defined in MIL-STD-1553B [DD-1; DD-4; IL-1]. The MIL-STD-1553B bus is a tri-service data bus interface; it is used by the Navy for avionics subsystems, and it is also applied in shipboard subsystems.

19.1.3 Requirements for Shipboard System Interconnection

The shipboard combat subsystems shown in Figure 19.1 need to interact with each other. Shipboard combat subsystems are diverse in nature; the salient points are summarized in Table 19.2.

The interconnection of the subsystems is subject to several conflicting requirements that are often difficult to reconcile, and thus subsystem connection via onboard communications is a complex design task. A shipboard LAN interconnecting the various subsystems must satisfy a wide range of the resulting requirements since it is an integral component of a combat system with diverse applications.

Interconnection of the diverse elements of a combat system must overcome conflicts arising from:

- The independent development of subsystems
- The installation of those subsystems on a variety of different platforms
- The combination of different generations of combat subsystems arising from differences in platforms and equipment life cycles

In order to accommodate a wide range of user requirements, shipboard LAN systems are structured for use of more than one protocol suite [DD-3]. Multiple protocol options allow for balancing of the conflicting needs of interoperability and performance characteristics in the design of shipboard combat systems. A LAN must have a broad spectrum of functions and operating modes and must be capable of adapting to the specific needs of individual applications [GR-1].

In combat systems, the data types to be transferred include a variety of discrete and analog signals as summarized in Table 19.1. Onboard computers and associated systems are rapidly evolving. The number of interconnected computer nodes in a combat system can vary from ten to several hundred. Growth to more than a thousand nodes [GR-1] in a large combatant ship is a future possibility. Each computer node will typically operate multiple software packages. These software packages must be able to transfer data with each other and various specialized devices, and they must provide operator-to-computer-interfaces (OCIs) to the personnel who operate the subsystems.

TABLE 19.1   SHIPBOARD SIGNAL TYPES FOR SURFACE
COMBATANTS

Information	Type of Signal
Intercomputer connection	Digital data
Display	Digital data; Digital video; Analog video
Radar and infrared imaging	Analog video; Digital video
TV	Analog video; Digital video
Voice	Digital data; Analog data
Navigation	Analog (event) data; Digital data
Propulsion and machine control	Analog (event) data; Digital data
Damage control	Analog (event) data; Digital data

In order for all subsystems to transfer data, a resilient high data rate LAN is needed, and the intercommunications protocols must be carefully chosen. The SAFENET, to be discussed later, is a LAN with these characteristics.

19.1.4.   External Shipboard Communications Connectivity

Figure 19.1 also identifies shipboard EXCOMM systems, for example, Link 4A, Link 11, OTCIXS, JTIDS, and other radios.[1] These EXCOMM systems connect the ship to shore stations and other ships. Each of these communications systems uses specific signal

TABLE 19.2   CHARACTERISTICS OF SHIPBOARD COMBAT SYSTEM ELEMENTS

Features	Characteristics
Independently developed system	• Each system developed by different requirements and missions • Each system developed by different program sponsors
Variety of different platforms	• Wide range of functions (mission planning to weapons control) • Variety of sizes of platforms (tens of meters for small craft to thousands of meters for aircraft carriers) • Common design for multiple platforms (one to a few hundred)
Life cycle of combat systems and platforms	• Life of the combat system requirement is on the order of years to tens of years • Life of the platform may range from twenty to forty years with upgrades

[1]Although this chapter is focused on shipboard communications protocols, shipboard EXCOMM systems are shown in Figure 19.1 to show the relationships between shipboard combat systems and shipboard EXCOMM systems.

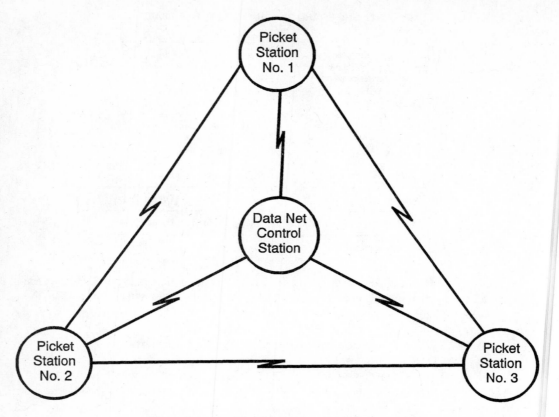

Figure 19.2   Simplified Connectivity of Link 11 Net Using TADIL-A

formats and protocols. Certain systems such as the OTCIXS and JTIDS have been discussed in sections 13.2.3 and 11.4.7. A more detailed description of the TADIL-A, the protocol for Link 11, is given in this chapter.

19.2 OVERVIEW OF LINK 11 MESSAGE FORMATS (TADIL-A)

The Link 11 tactical digital communications net is used to communicate among aircraft, ships, submarines, and shore facilities. A simplified Link 11 net is shown in Figure 19.2 [DD-2]. The Link 11 uses a standard message format for the exchange of digital information among the data subsystems. The digital information is generated by radar and sonar sensors and transmitted to all participants in the net primarily via HF but also by UHF links. The Link 11 utilizes a time division multiplexing approach to share a common

a. Roll Call (DNCS Interrogation Message)

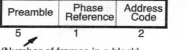

Preamble	Phase Reference	Address Code
5	1	2

(Number of frames in a block)

b. Roll Call (Picket Reply Message)

Preamble	Phase Reference	Start Code	Data	Picket Stop Code
5	1	2	Variable	2

c. Roll Call (DNCS Interrogation with Message)

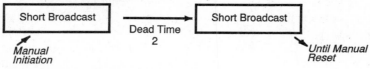

Preamble	Phase Reference	Start Code	Data	Control Stop Code	Address Code
5	1	2	Variable	2	2

d. Short Broadcast (by DNCS and Picket)

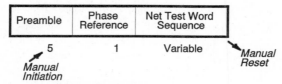

Preamble	Phase Reference	Start Code	Data	Picket Stop Code
5	1	2	Variable	2

Manual Initiation

e. Long Broadcast (by DNCS and Picket)

Short Broadcast	→	Short Broadcast

Dead Time
2

Manual Initiation   *Until Manual Reset*

f. Net Test (by DNCS only)

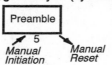

Preamble	Phase Reference	Net Test Word Sequence
5	1	Variable

Manual Reset

Manual Initiation

g. Net Sync (by DNCS only)

Preamble
5

Manual Initiation   *Manual Reset*

h. Picket Silence (Receive only)

Figure 19.3   TADIL-A Message Formats

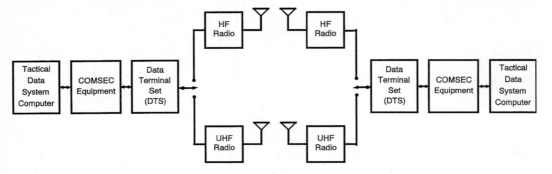

Figure 19.4   Typical Link 11 Supporting NTDS

frequency spectrum. The net operations enable all participants to receive current tactical situation information.

The Link 11 employs a well-defined protocol (TADIL-A) in order to exchange tactical information efficiently among the net participants. This permits only one station to transmit a specially formatted message at a given time. Any nontransmitting station monitors the frequency for messages from other stations. One station is designated as the data net control station (DNCS), and all other net units are participating picket stations.

The most frequently used mode of net operation is designated as the roll-call mode. Each station is identified by a unique address code. The DNCS establishes a roll-call sequence for all of the address code. Each station has a time slot allocated during the roll-call sequence. In normal net operation, the DNCS sends an interrogation message to a station, inquiring whether it has any data to transmit to the net. The interrogation message includes the address code of the next station to be called. If the called station has data to send, it responds to the DNCS by sending data in the specified format. If the station has more data to send than its allocated time allows, it stops and waits for the next roll-call. If a station has no data to send, it notifies the DNCS accordingly. If the DNCS has a tactical message to transmit to a participating picket station, then the DNCS will combine its own tactical message with the roll-call message.

Figure 19.3 shows data formats of the TADIL-A used for Link-11. In addition to the roll-call protocol, there are other operating modes for a Link 11 net such as the test and broadcast modes. When tactical data are received, a station transfers the tactical data (if any) to its tactical computer and updates the situation information in the computer.

Figure 19.4 shows a typical Link 11 supporting NTDS, which consists of a tactical data processor (TDP) computer, COMSEC equipment, data terminal set (DTS), and HF or UHF radio. The TDP computer processes C2 functions of the Naval Tactical Data System (NTDS). In the HF band, the radio equipment is tuned to any multiple of 100 Hz in the 2 MHz to 30 MHz frequency range; in the UHF band, the radio equipment is tuned to any multiple of 25 kHz in the 225 MHz to 400 MHz frequency range.

TABLE 19.3   TRANSMISSION MEDIA FOR RING LANs

Transmission media	Data rate (Mbps)	Range (km)	Number of taps
Twisted pair	4	0.1	72
Baseband coaxial cable	16	1.0	250
Optical fiber	100	2.0	240

19.3 LOCAL AREA NETWORKS (LANS)

In a modern combatant ship the interconnection of the subsystems, shown in Figure 19.1, is accomplished by LANs. Two types of LANs are common: Ethernet (IEEE standard 802.3)[2] [IE-2] which uses a random access method (nominal capacity of 10 Mbps, but in practice the throughput is much less) and high-capacity token rings (capacity up to 100 Mbps). The military versions of high-capacity LANs are the SAFENET, which follows the IEEE standard 802.5 [IE-1], and the Fiber Distributed Data Interface (FDDI) [AN-1; RO-1]. The FDDI is a high-speed fiber optic cable token ring using a protocol compatible with the IEEE standard 802.5. SAFENET implementations may use either coaxial cable or fiber optics.

19.3.1 Shipboard LANs

A LAN generally connects a wide variety of data processing subsystems such as computers, terminals, peripherals, and others. For shipboard applications, the LAN providing local communications among the subsystems shown in Figure 19.1 carries the signals listed in Table 19.1 (data, voice, video, and graphics). The connectivity structure provided by a LAN can be a bus, tree, or ring. Early combat systems used bus or tree networks, but current shipboard LANs predominantly utilize rings wired by coaxial or fiber optic cables.

The standard governing the ring LAN is the Token Ring Protocol (IEEE 802.5) which also applies to the Fiber Distributed Data Interface (FDDI). Table 19.3 shows how data rates compare for different implementations of ring LANs. Optical fiber cables can accommodate data rates of 100 Mbps, and these rates are compatible with the data rate requirements in a typical modern shipboard combat system. Transmission range is another factor in choosing the cable type for a shipboard environment: 22-gauge twisted pair wire has a range of 0.1 km; baseband coaxial cable, 1 km; optical fiber, 2 km.

Figure 19.5 shows the attenuation of typical guided transmission media, specifically of 22-gauge twisted pair wire, 3/8-inch coaxial cable, and optical fiber, as a function of operating frequencies. An optical fiber cable has the lowest attenuation, 0.2 dB/km to 0.8 dB/km.

---

[2]The IEEE Standard 802.3 is for the Carrier Sensor Multiple Access with Collision Detection (CSMA/CD) protocol.

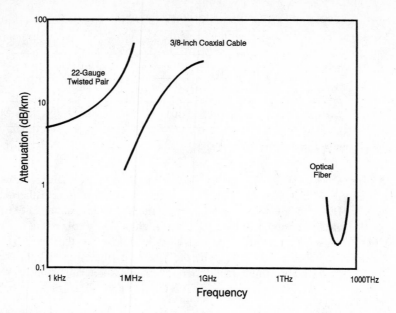

Figure 19.5 Attenuation of Typical Guided Transmission Media (Reprinted with permission of Macmillan Publishing Co.; W. Stallings, *Data and Computer Communications,* Figure 2–15, p. 49.]

19.3.2 Token Rings

A token ring connects elements of a network in a ring, as shown in Figure 19.6. The token ring protocol is defined by IEEE Standard 802.5.

In accessing a LAN, problems of contention among the users arise. In a token ring, this is resolved by using a token (a data frame with a special bit pattern), and the token

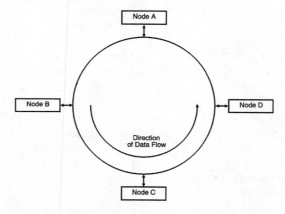

Figure 19.6 Token Ring LAN

moves along the ring in a round-robin fashion. When the token arrives at a node, the node can seize the token and send data. The LAN protocol limits the amount of time the node can transmit data frames by means of a Token Ring Timer (TRT).

When a node has data to transmit and receives the token, it seizes it and sends data frames until either all data are transmitted or the TRT indicates that the time has expired. The data circle the ring and are received at the destination. The data continue to circle until they arrive at the originating node. While data are circling the ring, no other node can access the ring. The originating node compares the returned data to the transmitted data. If no error is detected, the data frame is removed, and the token is put back on the ring. The ring then becomes free for other nodes to use.

This method keeps the ring occupied when a single node sends data, hence, the LAN is not very efficient. To improve efficiency another method was devised for some token rings. The first method (described above) is called Regular Token Passing (RTP); the second method is called Early Token Release (ETR). In the RTP method, the sending node holds a token until the data have circled the ring and have been checked at the sending node, which then releases the token at the end of this process. In the ETR method, the sending node releases the token immediately after it completes transmitting its packets. This allows multiple nodes to send data and access the ring simultaneously.

When a node is idle, it reads the addresses of frames circling the ring; thus it monitors the destination address of each frame. When the node detects its own address, it copies the frame to a local buffer for further use such as transmission quality monitoring.

There is a token ring characteristic called latency. Latency is a delay caused by nodes on the ring checking the data of passing frames and repeating the data. This delay may vary from one to several bit times.

For most applications, the throughput of a token ring LAN and its inherent delay, caused by the limitation of copper wires, is not a problem. Where greater speed is required, the use of fiber optic cable using the Token Ring protocol can solve the problem.

19.3.3 Fiber Distributed Data Interface (FDDI)

The Fiber Distributed Data Interface (FDDI) is a high speed token ring using a protocol compatible with the IEEE standard 802.5. The FDDI is designed for high-speed networks and uses 100 Mbps optical fiber cables. The FDDI protocol is defined by ANSI standard X3.139-1987 [AN-1].

In the idle state of the FDDI, a token frame circles the ring. When a node has data to send it seizes the token from the ring as it passes by and starts to send data. Upon completion of data transmission, the node places a new token on the ring. The FDDI protocol also limits the amount of time in which the node can transmit data frames by a Token Holding Timer (THT).

Other nodes on the ring read and repeat passing data. Thus each node introduces a small delay as it repeats frames. The data circle until they arrive at the originating node. If a node detects its own address it copies the data, compares it to the data sent, and if there is no error, removes the token frame from the ring.

In the FDDI, the data to be transferred may be of extensive length or may come in

TABLE 19.4   ADVANTAGES OF USING FIBER OPTICS IN SURFACE SHIPS

Technical Features	Benefits
Electromagnetic effects	• Not susceptible to EMI since it is a passive component • No crosstalk • No ground loops
Survivability	• Battle and fire damage reduced because of redundant paths • Not susceptible to EMP • Isolation of data from power source
Functional requirements	• Occupies less space than other cables • Low weight • High data rate
Flexibility	• Long cable runs possible and few restrictions on cable routing • Supports high-speed LANs • Spare fiber will support system growth
Cost	• Reduced cost for installation and maintenance
Components	• Simplified technology reduces required components

bursts. In order to accommodate both long data streams and data bursts, the FDDI protocol uses a complex capacity allocation method. The capacity allocation method has two FDDI transmission modes, synchronous or asynchronous. In the synchronous mode, the LAN is made available to designated nodes each time they receive a token for sending data; transmission time duration allocation may vary among nodes. A node that does not have a synchronous allocation may send data frames only asynchronously, that is, when it has data ready to send; the LAN is not automatically available to this node when a token is received. When the node has data and receives a token, then asynchronous data frames can be transmitted according to a timed priority scheme.

19.3.4 Use of Fiber Optics in Shipboard Combat Systems

With the rapidly increasing use of fiber optics in commercial telecommunications, there has also been an increase in use of fiber optics aboard naval ships. Using fiber optics aboard naval ships offers several advantages. Table 19.4 lists these advantages.

19.4 OPEN SYSTEMS INTERCONNECTION (OSI) STANDARDS

The communications protocol standards which have direct bearing on the design of shipboard communications networks are contained in the International Standard Organization's (ISO) Open Systems Interconnection (OSI) reference model. The government's standard for Open System Interconnection is the Government Open System Interconnection Profile (GOSIP).

All government and defense systems are conforming to GOSIP. The standards used before GOSIP was introduced were the DoD Transport Control Protocol (TCP) and the DoD Internet Protocol (IP). Since older systems still in use adhere to the DoD TCP/IP standards for communications, these will be also described.

19.4.1 Open Systems Interconnection (OSI) Model

The International Standard Organization adopted the OSI reference model ISO 84 and ISO 7498, and generated standards for data interconnection communications systems [KN-1]. Within the OSI model, the communications functions are partitioned into a set of layers. Each layer performs a related subset of functions required to communicate with another system. The OSI model provides the basis for connecting computers and peripherals with minimum restrictions, as long as the systems meet the structural requirements of the model. Because each layer requires processing, the OSI model, while providing flexibility and interoperability, also introduces overhead.

Each of the seven layers must adhere to a standardized protocol. Through the protocol a block of data is encapsulated by control information (or header) of each layer. Data are accepted or generated by a user terminal and encapsulated into a message, containing data plus control information. The control information (or header) may include the Applications Header (AH), Presentation Header (PH), Session Header (SH), Transport Header (TH), Network Header (NH), and Data Link Header (DLH). As an example, the transport layer may need 48 bytes of TH, which include source address, destination address, sequence number, acknowledge number, data offset, window, checksum, urgent pointer, options, and others. The control information represents an overhead; the overhead can be substantial (up to 15%).

Figures 19.7 and 19.8 show the communications systems services and the functions provided by each layer:

- The physical layer is the bottom layer of the protocol architecture. It controls the physical media for interconnection with different control procedures. Examples of the standards at this layer are RS-232, RS-449, RS-422, RS-423, V.24, V.25, portions of X.21, and LAN standards 802.3, 802.4, and 802.5.
- The data link layer is the next higher protocol layer. It includes the error detection and correction techniques to insure that physical communications media (e.g., twisted wire pairs or fiber optics) transmit data reliably between subsystems. Examples of the standards at this layer are HDLC, LAP-B, LLC, and LAP-D.
- The network layer is the third layer. It provides a connection path between a pair of transport entities and also a path between two intermediate nodes. There are also connection-oriented services where a connection is established and the arrival of the data is verified. Another service is connectionless where datagrams are sent.
- The transport layer is the fourth layer. It provides the control of data transport from source end-systems to destination end-systems. This transport service relieves higher layer entities from any concern with the transportation of data between them. Services provided in this layer are either connection-oriented or connectionless.

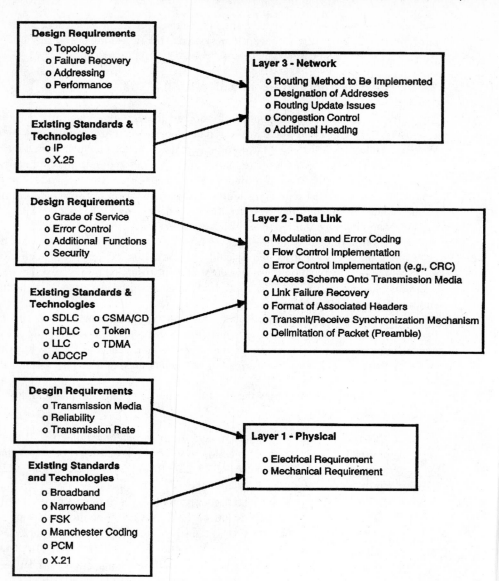

Figure 19.7   OSI Model, Layers 1 to 3

- The session layer is the fifth layer. This layer is needed to organize and synchronize dialogue and to manage the exchange of data.
- The presentation layer is the sixth layer. This layer is needed for presentation and manipulation of structured data for the benefit of application programs. Examples of presentation protocols are the encryption and virtual terminal protocols.

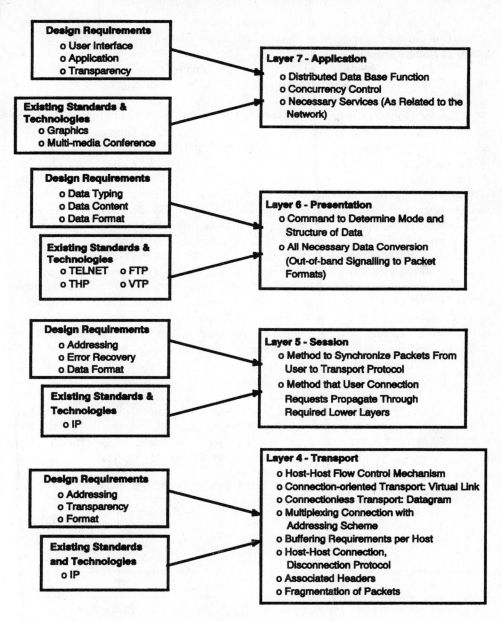

Figure 19.8   OSI Model, Layers 4 to 7

- The application layer is the highest layer. This layer provides interface for user applications. Examples of protocols at this level are the file transfer and electronic mail protocols.

The purpose of adopting this ISO reference model is to provide a basis for the coordination of standard development of communications system interconnection, while allowing existing standards.

19.4.2 Government Open System Interconnection Profile

GOSIP is the Open System profile adopted by the federal government to introduce OSI standards into data and communications systems used by the US Government. GOSIP is administered by the National Institute for Standards and Technology (NIST), which has issued GOSIP as Federal Information Processing Standard FIPS-146 [FI-1]. The DoD also adheres to GOSIP for its systems; consequently new systems to be deployed will adhere to the OSI principle through GOSIP. The DoD, through the Defense Information System Agency (DISA) (formerly DCA), is working to identify OSI applications for naval communications systems through a GOSIP application effort.

Figure 19.9 is an overview of the GOSIP protocols. The figure also identifies the current standards applicable to various layers of GOSIP. OSI and GOSIP are still being developed in some areas of the architecture, notably in layers 3 to 6.

The primary benefit of adopting GOSIP for naval applications is that commercial off-the-shelf (COTS) products can be used in shipboard communications design. While many COTS products can be used for military applications, some special naval requirements can only be met by tailored products. An example is the SAFENET that will provide shipboard interconnection for data processing and communications subsystems.

19.4.3 DoD Military Protocol Standards

Within the DoD, DISA has developed military standard protocols to meet all DoD requirements. Table 19.5 lists DoD military protocol standards. The Internet Protocol (IP), defined by MIL-STD-1777, provides for connectionless network service. The Transport Control Protocol (TCP), defined by MIL-STD-1778, provides for connection-oriented

TABLE 19.5 COMPARISON BETWEEN MILITARY AND ISO PROTOCOL STANDARDS

Military Standards	International Standards
MIL-STD-1777, Internet Protocol (IP)	DIS 8473, Connectionless mode network service protocol
MIL-STD-1778, Transport Control Protocol (TCP)	ISO 8073, Connection-oriented transport protocol
MIL-STD-1780, File Transfer Protocol (FTP)	DIS 8671, File Transfer, Access, and Management (FTAM) protocol
MID-STD-1981, Simple Mail Transfer Protocol (SMTP)	X.400, Message handling system protocol
MIL-STD-1782, TELENET Protocol	DIS, 9041, Virtual Terminal Protocol (VTP)

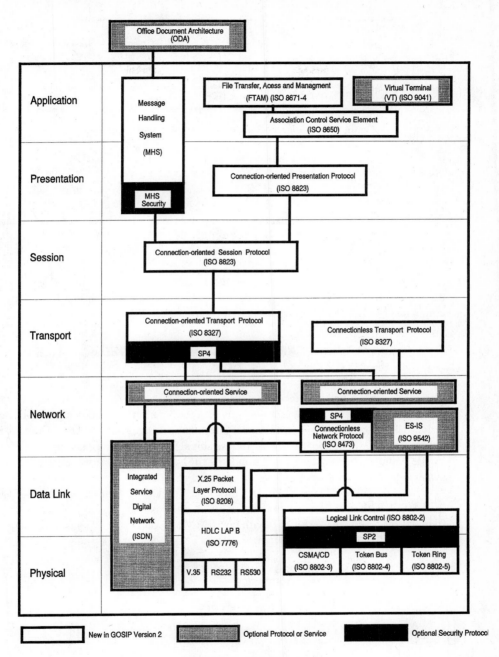

Figure 19.9   GOSIP Protocols

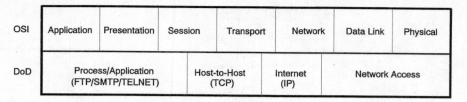

OSI	Application	Presentation	Session	Transport	Network	Data Link	Physical
DoD	Process/Application (FTP/SMTP/TELNET)			Host-to-Host (TCP)	Internet (IP)	Network Access	

Figure 19.10　Comparison of the OSI and DoD Communications Architecture

transport services. The File Transfer Protocol (FTP), defined by MIL-STD-1780, is an application layer protocol for file transfer service in a network. The Simple Mail Transfer Protocol (SMTP), defined by MIL-STD-1981, is the protocol for simple electronic message handling systems; it is an application layer protocol. The TELENET Protocol, defined by MIL-STD-1782, is also an application layer protocol, and sets specification for virtual terminals (VT) in TELENET operations. The FTP, FTAM, and VT are compatible with the OSI protocols and are contained in GOSIP. TCP and IP will be replaced in developing systems layers 3 and 4 of the GOSIP.

The TCP/IP protocols have been used extensively in DoD military systems. Both the OSI and TCP/IP protocols will coexist until the TCP/IP is phased out. Figure 19.10 compares the DoD communications architecture to the OSI reference model.

19.5 THE SURVIVABLE ADAPTABLE FIBER OPTIC EMBEDDED NETWORK (SAFENET)

The Survivable Adaptable Fiber Optic Embedded Network (SAFENET) effort was initiated as a follow-up to a program that was developing a serial, electrical point-to-point interface (based on the NATO Low Level Interface specification) for Aegis Combat System computer equipment aboard DDG-51 class ships. The SAFENET standards were adopted by the Navy as the standard LAN standards for combat and platform control related subsystems on Navy combatant ships.

The SAFENET uses token ring protocol. SAFENET I has adopted the IEEE standard 802.5 with a 16 Mbps data rate; SAFENET II has adopted the ANSI FDDI standard (compatible with IEEE 802.5) with a 100 Mbps data rate. Table 19.6 summarizes SAFENET I and SAFENET II characteristics.

TABLE 19.6　SAFENET I AND SAFENET II
CHARACTERISTICS

Features	SAFENET I	SAFENET II
LAN Standards	IEEE 802.5	ANSI X3.139
Data Rate	16 Mbps	100 Mbps
Topology	Token ring	Token ring

The SAFENET II standard is published in Navy Handbook MIL-HDBK 0036 [DD-3]. The key requirements for SAFENET design are:

- High reliability and survivability for shipboard internal communications,
- Rapid and frequent automated fault detection, isolation, and recovery,
- Rapid reconfiguration, in fractions of a second, to meet changing mission requirements or to recover from component failure.

To meet the reliability requirements of SAFENET, individual SAFENET components can withstand the severe physical environments found in combat systems. SAFENET optical fiber cables are more reliable than commercially available optical fiber cables. In addition, SAFENET uses a dual path configuration, as shown in Figure 19.11.
The survivability of the SAFENET is insured by the following design criteria:

(a) Use of separate and redundant cable paths,
(b) Use of distributed processing, distributed systems, and local control,
(c) Incorporation of backup, local control capability in all major subsystems,
(d) Location of the communications center in a protected area away from high-probability hit zones,
(e) Provision for a separate emergency radio room with data and hard copy traffic capabilities.

The SAFENET profile contains two basic protocol suites: the OSI suite and the lightweight suite. The SAFENET OSI protocol contains all services for the seven layers of the ISO standards as interpreted by GOSIP. The SAFENET lightweight suite is used where the ISO standards are not suitable for Navy requirements. Figure 19.12 shows the definition of the SAFENET protocol profile.

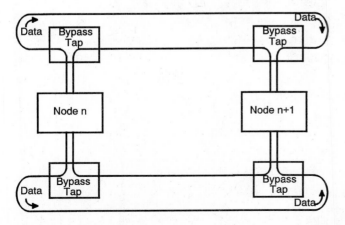

Figure 19.11   Dual Path Configuration for SAFENET Cable

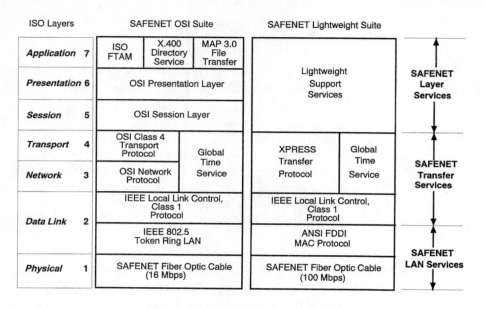

Figure 19.12 SAFENET Protocol Profile

19.5.1 The SAFENET OSI Suite

The SAFENET OSI suite provides interoperable protocols for OSI users and is defined in terms of the following seven layers of OSI protocol standards:

Layer 1 - Fiber optic cable is used as the physical layer, and SAFENET standards are applied with a data rate of 16 Mbps.

Layer 2 - For the data link layer, two sublayers are defined:

- The IEEE 802.5 token ring protocol used as a LAN protocol
- The IEEE logical link, Class 1 protocol

Layer 3 - For the network layer, two concurrent layers are defined:

- The OSI network protocol
- A SAFENET-specific layer called global time service; this will provide a time code from an external time reference such as a ship's GPS. This protocol also includes segments of layer 4

Layer 4 - The transport layer uses the OSI Class 4 transport protocol.

Layer 5 - The session layer uses the OSI session protocol.

Layer 6 - The presentation layer uses the OSI presentation protocol.

Layer 7 - For the application layer, the following three concurrent protocols are defined:

- For file transfer applications, the ISO File Transfer and Access Management (FTAM) protocol is used; the ISO FTAM performs three application layer functions, that is, virtual file store, file service, and file protocol.
- For electronic mail directory service and message transfer, the ISO defined X.400 electronic mail and directory services protocol are used.
- For manufacturing applications, a multinational interoperability standard called the Manufacturing Automation Protocol (MAP) 3.0 is used; MAP 3.0 is a standard set by an industry group led by General Motors and Boeing.

As can be seen in the application layer protocol, the SAFENET OSI suite is designed to maximize the interoperability of diverse users in shipboard computer systems.

19.5.2 The SAFENET Lightweight Suite

One of the important requirements of shipboard combat systems is the limited response time to perform sensor processing and weapons system control. The SAFENET OSI suite tends to add significant time delay by processing through all seven layers of the SAFENET OSI protocol suite.

One way to decrease the delay is to combine several layers of the OSI model. However, this layer combination is only possible if all participant processing subsystems are a part of a distributed architecture with the same operating system, using the same compiler; then the processing of the combined layers is performed as a front-end processor function.

The SAFENET lightweight protocol suite uses special services for a new class of user nodes. It adheres to the OSI protocols where feasible. The functions and services are combined into four layers.

Layer 1 - The physical layer is fiber optic cable, and the SAFENET standards are used with a data rate of 100 Mbps.

Layer 2 - The data link layer has two defined sublayers:

- The ANSI FDDI token ring protocol
- The IEEE logical link control, Class 1

Layer 3 - For the network and transport layers, a SAFENET-specific protocol called the XPRESS Transfer Protocol (XTP) is defined; this layer combines layers 3 and 4 of the OSI protocol. This will significantly reduce the time required to transfer data over the LAN. In addition, the global time service protocol using GPS timing data is adopted in the SAFENET lightweight suite.

Layer 4 - For the session, presentation, and application layers, another SAFENET-specific protocol called the Lightweight Support Services protocol is defined. This combines layers 5, 6, and 7 of the OSI protocols for reducing the time required to process the data over the SAFENET; this layer is tailored to meet the needs of the

SAFENET. It is in this layer where the operating systems of applications need to be the same.

In the two superlayers (lightweight support service and XTP protocol), the SAFENET lightweight suite is designed to minimize delays of the LAN while sacrificing interoperability since only computers with the same operating system can be connected and communicated with on this LAN.

The following systems use SAFENET: the Navy Command and Control System (NCCS) Ashore, the Flag Data Display System (FDDS), the Afloat Correlation System (ACS), the Navy Intelligence Processing System (NIPS), the Naval Tactical Command System Afloat (NTCS-A), and the Tactical Environmental Support System (TESS-3).

19.6 THE MIL-STD-1553B BUS

The MIL-STD-1553B bus was initially designed to be an extremely robust means of communication between computers onboard an aircraft avionics system. From avionics, it is extensively applied to shipboard and submarine weapon systems, electronic pods carried on aircraft, and missile systems.

The MIL-STD-1553B data stream is time-division multiplexed and transferred serially over two twisted-shielded copper wires. Information is coded into 2-bit words (16 bits of data). The transmitted waveform is Manchester II (biphase) operating at a bit rate of 1 Mbps. Acting as a local area network, the MIL-STD-1553B bus is very efficient in transferring short messages from one device to another in a transparent and secure manner.

The bus has four main elements as shown in Figure 19.13: a bus controller that manages the information flow, remote terminals that interface one or more simple subsystems to the data bus and respond to commands from the bus controller (embedded remote

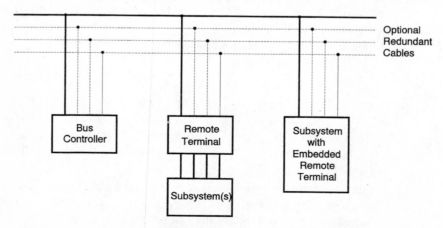

Figure 19.13   Block Diagram of a MIL-STD-1553B Bus

terminals may be used for more complex subsystems), the bus monitor, which is used for data bus test, and the data bus shielded twisted pair wire, isolation resistors and transformers required to provide a single data path between the terminal.

The comparable NATO digital interface standard is the STANAG 3838 [NT-1], which defines the standard interface between direct memory access and shipboard computers via a data bus.

REFERENCES

[AN–1] *Fiber Distributed Data Interface (FDDI) Token Ring Media Access Control*, American National Standard Institute (ANSI) Standard X3.139-1987, November 5, 1986.

[DD–1] *Digital Time Division Command/Response Multiplex*, MIL-STD-1553B, Department of Defense, Washington, D. C. 20350-2000, September 21, 1978.

[DD–2] *Interoperability and Performance Standards for Tactical Digital Information Link (TADIL) A*, MIL-STD-188-203-1A, Department of Defense, Washington, D. C. 20350-2000, March 1, 1987.

[DD–3] *Survivable Adaptable Fiber Optic Embedded Network II (SAFENET II)* , MIL-HDBK-0036 (Draft), Department of Defense, Washington, D. C. 20350-2000, March 1, 1990.

[DD–4] *MIL-STD-1553B Bus Handbook*, MIL-HDBK-1553, Department of Defense, Washington, D. C. 20350-2000, Sept. 24, 1986.

[FI–1] *Government Open Systems Interconnect Profile (GOSIP)* , Federal Information Processing Standard FIPS-146, National Institute of Standards and Technology, Gaithersburg, Md.

[FL–1] Flanagan, J. D. and G. W. Luke, "Aegis: Newest Line of Navy Defense," *Johns Hopkins University APL Technical Digest*, October–December, 1981, Laurel, Md. 20707, pp. 237–242.

[GR–1] Green, D. and D. Marlow, "Application of LAN Standards to the Navy's Combat System," *Naval Engineers Journal*, May 1990, pp. 146–153.

[IE–1] *Token Ring Access Network Method and Physical Layer Specifications*, IEEE Standard 803.5, January 25, 1990.

[IE–2] *Carrier Sensor Multiple Access with Collision Detection (CSMA/CD) and Physical Interface Specification*, IEEE Standard 803.3.

[IL–1] *MIL-STD-1553 Designer's Guide*, Second ed., ILC Data Device Corporation, Bohemia, N.Y., 1988.

[KN–1] Knightson, K. G., T. Knowledge, and J. Larmouth, Standards for Open Systems Interconnection, McGraw-Hill, New York, N.Y. 1985.

[NA–1] *US Navy TADIL A (Link 11) Standard Operating Procedures*, OPNAVINST C2308.1, Department of the Navy, Washington, D. C. 20350-2000.

[NC–1] *Navy UHF Satellite Communications System Description*, NSHFC 301, NOSC (now NCCOSC RDT&E Div.), San Diego, CA 92152-5185. December 31, 1991.

[NS–1] *Input/Output Interfaces, Standard Digital Data, Navy System*, MIL-STD-1397A (Navy), Naval Sea System Command, Washington, D. C. 20362, January 7, 1993.

[NT–1] *Aircraft Internal Time Division Command/Response Multiplex Data Bus*, NATO

STANAG 3838, National Communications Systems Office (NCS-TS), Arlington, Va. 22204-2198.

[RO–1] Ross, F., "An Overview of FDDI: The Fiber Distributed Data Interface," *IEEE Journal on Selected Areas in Communications*, September 1989.

20

Trends in Shipboard Communications

With the rapid progress made in communications and computer technology in recent years, Navy shipboard communications programs have been taking advantage of and adapting to these new technologies in the areas of network management, equipment automation, multimedia transmission, and automatic encryption key management. Prior to the emergence of these technology applications, Navy shipboard communications programs have had a long history of their own technological innovations that have now enabled the Navy to embrace these new advances.

20.1 CHRONOLOGICAL REVIEW OF NAVAL TELECOMMUNICATIONS AUTOMATION PROGRAMS

In this section, we review the evolution of various U.S. Navy shipboard communications automation programs. Figure 20.1 shows a progression of the major programs undertaken since 1960. Shipboard communications automation began with the automation of the radio room and has developed into full-scale resource management of shipboard communications, culminating in using shipboard local area networks that utilize coaxial and fiber optic cables, and fully automated communication support systems. In the following sections, the communications automation programs, shown in Figure 20.1, will be discussed.

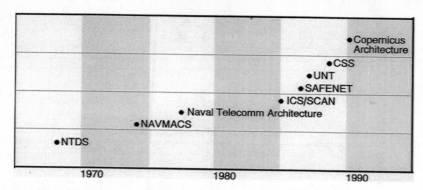

Figure 20.1   Evolution of Navy Shipboard Communications Automation

20.1.1 Naval Tactical Data System (NTDS)

The first automated shipboard and airborne communications system, the Naval Tactical Data System (NTDS), was developed in 1961 to solve the anti-air warfare problem by computerizing the shipboard Combat Information Center (CIC) [WA-1]. The NTDS is presently onboard more than 200 active ships in the fleet, including aircraft carriers, cruisers, destroyers, frigates, and amphibious ships. With the NTDS came the introduction of the structured protocol that uses Tactical Digital Link A (TADIL-A), a polled net control. NTDS is supported by the following communications links:

- Link 11, a two-way, real-time, encrypted data link operating at HF or UHF used by naval surface ships and combat aircraft for computer-to-computer data interface

- Link 4A, a two-way UHF data link for operations with combat aircraft, enabling the fighter to transfer or download track and own ship status data to the air controller

- Link 14, a one-way HF or UHF link which enables NTDS equipped ships to communicate with non-NTDS equipped ships.

A detailed description of the NTDS is presented in section 10.3.

20.1.2 Naval Modular Automated Communication System (NAVMACS)

The next major advance in shipboard communications automation was the Naval Modular Automated Communications System (NAVMACS), a message processing system, which handles the large record traffic volume aboard major combatant ships and ashore nodes. NAVMACS provides the capability for automatic reception and remote printing of message traffic at a number of remote printers. Several of these remote sites also include message entry capabilities, because increasingly the burden of message preparation is being pushed further back from the radio room toward the original drafter of the message [NC-2].

In shore nodes, a comparable system called the Naval Communications Processing and Routing System (NAVCOMPARS) has been developed to automatically process naval messages exchanged between ship-to-shore nodes [WA-1]. The Naval Computer and Telecommunications Area Master Station (NCTAMS) generates the broadcast for its area of communications coverage using NAVCOMPARS equipment that automates the message dissemination process. The NAVCOMPARS also serves as the broadcast injection point for FLTSATCOM transmission.

The NAVCOMPARS and the NAVMACS were developed to meet the need of the DoD-wide automatic digital network AUTODIN system. The Navy's portion of the AUTODIN system is called the Local Digital Message Exchange System (LDMX), which is a base level message concentrator and automated message handling system. The ship termination of a circuit may be connected to NAVCOMPARS and operated as though the ship was a remote terminal. The availability of NAVCOMPARS significantly increased the capability to cope with large message volume.

20.1.3 Naval Telecommunications System (NTS) Architecture

In 1976, the Naval Electronic Systems Command (NAVELEX), later renamed Space and Naval Warfare Systems Command (SPAWAR), developed a Naval Telecommunications System (NTS) architecture, in which the problems associated with intra Task Force and long-haul shipboard communications were addressed. One may consider the NTS architecture as the forerunner of the Communication Support System (CSS) which is discussed in section 20.2. The NTS architecture defines a basic platform subsystem arrangement with RF subsystems, user subsystems, and quality monitoring and control subsystems.

Although the NTS architecture was not fully implemented, it has significantly influenced the formulation of later programs. The concept of an afloat node with UHF SATCOM, SHF SATCOM, HF, and UHF LOS as controllable links was introduced in this architecture. To facilitate these link resources, the architecture included switches and controls, quality monitoring control systems (QMCS) and control processor subsystems; section 18.1.5 describes the QMCS.

20.1.4 ICS/SCAN

In 1981, the Naval Sea Systems Command (NAVSEA) developed a system called the Integrated Communications System/Shipboard Communications Area Net (ICS/SCAN), which was another major step to the modernization of shipboard communications suites. The ICS/SCAN was developed for the DDG-51 Arleigh Burke class destroyers [BR-1].

The ICS/SCAN architecture uses a distributed local area network (LAN) designed for reliability and survivability. It is a dual redundant broadband cable network. Using coaxial cable for 450 MHz bandwidth, the Frequency Division Multiplex (FDM) shipboard LAN supports hundreds of channels while simultaneously providing the remote control and status signals necessary for fully automated shipboard communications. Furthermore, the ICS/SCAN serves the end users of tactical communications circuits. The

ICS/SCAN includes a remote console and status displays situated in the CIC as well as in the communications center for selected control and real-time display of communications status as configured by the ship's communications plan (COMMPLAN). The ICS/SCAN uses a Cable Adaptor Unit (CAU), which is functionally similar to the bus interface unit of fiber optic cable LANs.

20.1.5 Survivable Adaptable Fiber Embedded Network (SAFENET)

Another important step in automating shipboard communications was the introduction of a fiber optic Fiber Digital Data Interface (FDDI) LAN. The SAFENET was initiated as a follow-up effort to the ICS/SCAN.[1] Two primary technologies investigated by NAVSEA were a fiber optic based point-to-point interface and a local area network. As a result, the LAN standard using FDDI for data transfer among computer equipment was adopted in 1984 [BR-1; GR-1].

The SAFENET program also conformed to the International Standard Organization (ISO) Open System Interconnection (OSI) basic reference model. Current SAFENET standards specify two suites of interfaces: the ISO suite and the lightweight suite. These two suites provide flexibility in interconnecting most shipboard resources; they are described in detail in section 19.5.

20.1.6 Unified Networking Technology (UNT)

In the mid-1980s, the U. S. Navy saw the need for multimedia communications for packet switched radio transmission in order to utilize available shipboard radios more efficiently. The same approach is also applicable to shipboard LANs such as the SAFENET. The concept of multimedia radios that share several transmission media (e.g., HF, UHF SAT-COM) was explored. The result of this effort was the Unified Networking Technology (UNT), which was jointly developed by the Naval Research Laboratory (NRL), Naval Ocean Systems Center (NOSC), and Space and Naval Warfare Systems Command (SPAWAR) in 1986 [NC-1; CA-1]. Packet switched radio transmission was explored by a Defense Advanced Research Project Agency (DARPA) project known as Aloha. The UNT program combines the concept of a packet switching system with efficient information transfer through various subsystems. The UNT represents a step to serve several user systems aboard a ship by a LAN.

The primary objective of the UNT program was to demonstrate, in a scenario and in a fleet exercise, the operational benefits of new networking technology and the impact of this technology on naval communications services. The initial focus of UNT was the development of survivable HF and UHF communications networks, that could be integrated into a multi-network architecture. The multimedia network architecture offers a wide variety of multicast and point-to-point services over nets that are essentially transparent to the user.

---

[1]The SAFENET was initially a Naval Sea Systems Command (NAVSEA) program to develop a serial, electrical point-to-point interface of computer equipment for shipboard combat systems aboard DDG-51 Arleigh Burke Class ships.

Another focus of the UNT was the use of a packet switched network that grew out of a need to allow computer users to have access to resources beyond those available in a single communications network, because the resources of a single network were often inadequate to meet users' needs. Hence, several networks were interconnected to enhance the capability of data communications. This interconnected network was termed an Internet. The Internet is controlled by a handshake method called Internet Protocol (IP). Communications circuits combined by the UNT are shipboard HF and UHF LOS circuits such as the Link 11.

20.1.7 Communication Support System (CSS)

The concept of the CSS was evolved from UNT discussed above. The goal of the CSS program is the automation of shipboard communications and resource management [SN-1]. Two devices, called the link controller and the subscriber interface unit, were introduced to facilitate multimedia networking and efficient management of shipboard communications assets (e.g., modems, cryptographic equipment, receivers, and transmitters). The CSS will be discussed in more detail in section 20.2.

The CSS provides virtual network connectivity between two platforms through different available communications services. These communications services are similar to ship-to-ship or ship-to-shore circuits. CSS services are defined by a set of parameters such as the operating frequency, the operating time period, the data type, the security label, and the quality of information transfer.

20.2 DESCRIPTION OF THE COMMUNICATION SUPPORT SYSTEM (CSS)

The goal of the CSS was to automate all shipboard communications and to provide reliable, secure, and survivable message, voice, and data communications among afloat nodes or among afloat and ashore nodes. Conventional shipboard exterior communications (EXCOMM) systems are limited in their operations because they consist of stand-alone and independent communications links, and do not optimize communications resource management. As a result, these shipboard EXCOMM systems, while individually optimized, provide disjointed communications services and limit the efficiency of exploiting the circuit available aboard afloat nodes.

20.2.1 CSS Concepts

The CSS concept is intended to overcome the deficiencies summarized in Table 20.1 and to support the following shipboard EXCOMM systems requirements:

- Receipt and transfer of higher volumes of data among afloat nodes
- Increased and more efficient processing to reduce reaction times
- More efficient networking capability for shipboard combat systems.

TABLE 20.1 PROBLEMS OF CONVENTIONAL SHIPBOARD
EXCOMM SYSTEMS

- The shipboard EXCOMM system is fragmented; when stressed by jamming, a lost EXCOMM circuit is difficult to restore by other circuits in the node.
- The shipboard EXCOMM system is dedicated to a specific user.
- The shipboard EXCOMM system is difficult to adapt to changing traffic loads.
- The shipboard EXCOMM system is not sufficiently flexible to accommodate to changing operational situations.
- The shipboard EXCOMM system is not flexible enough to accommodate new users without costly hardware and software changes.

Figure 20.2 illustrates an example of conventional EXCOMM systems that supports:

- Tactical information systems such as NTDS and Joint Operational Tactical System (JOTS)
- Command display systems such as Flag Data Display System (FDDS)
- Intelligence system such as Afloat Correlation Systems (ACS) and Prototype Ocean Surveillance System (POST)
- Electronic warfare systems such as Electronic Warfare Control System (EWCS).

Note that the communications for NTDS and FDDS could only be provided by Link 11; the communications for POST and EWCS could only be supported by UHF SATCOM.

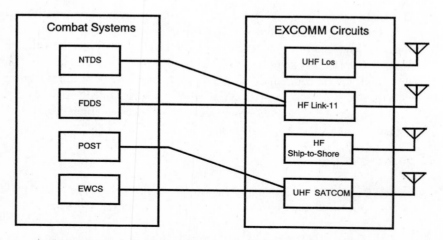

Figure 20.2 Example of Allocating Multiple Combat Systems to a Communications Resource

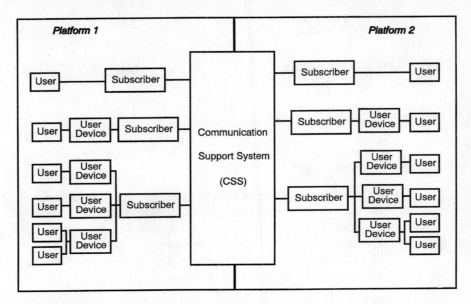

Figure 20.3   Block Diagram of the CSS Architecture Showing Subscribers and Users in Two Platforms

20.2.2 Concept of the CSS Communications Services

The concept of the CSS is illustrated by a block diagram of a shipboard communications model shown in Figure 20.3. The block diagram shows the relationship among the following entities:

- User—sender and receiver of information
- User Device—device to convert user information from digital phones and computers
- Subscriber—device to access CSS
- CSS—communications automation device to transfer information and process communications to support information transfer.[2]

Through the CSS, entities called the services are introduced. A service is defined for each type of information transfer by a set of parameters shown in Table 20.2.

The services delineate how the CSS provides communications to subscribers and users using predetermined procedures. The manner in which a communications service is provided in CSS is similar to the manner in which an EXCOMM circuit is established

[2]In a conventional shipboard EXCOMM system, information is transferred by a circuit such as Link 11 or the Officer-in-Tactical Command Information Exchange Subsystem (OTCIXS).

TABLE 20.2   PARAMETERS OF CSS SERVICE

- Operating frequency
- Operating time period
- Data type
- Security label
- Quality of information transfer
- Participants

in the current shipboard system. However, the CSS services are different from the EXCOMM circuit in the following ways:

- The CSS service does not specify operating frequencies and may utilize several frequencies as the optimum frequency changes with time.
- The CSS service does not specify specific modulation and access protocols. These communication techniques are transparent to the user device.
- The CSS service must define delivery time requirements (for priority decisions) and features (for reliability and redundancy).

Figure 20.4 shows how a subscriber views the communications services in CSS. A CSS service can be thought of as a virtual circuit that can accept outgoing information and deliver incoming information. The subscriber knows that the virtual circuit has certain characteristics so that:

- The subscriber is attached to a known set of users.
- The subscriber can adequately handle a certain volume of information for users under normal conditions.
- The subscriber has a known data format, reliability, and security characteristics.

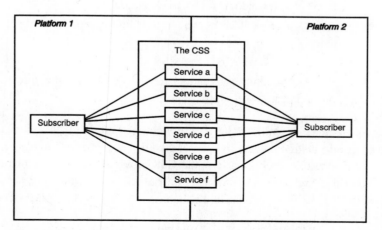

Figure 20.4   CSS Services for Subscribers on Two Platforms

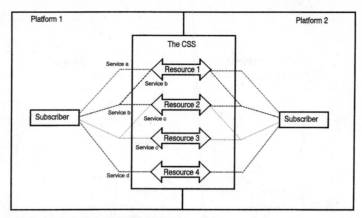

Each subscriber in Platform 1 has 4 services: a, b, c, and d.
In this example, the primary connection for Service c is via Resource 3;
the backup connection is via Resource 2.

Figure 20.5   Example of a CSS Resource Implementation of Four Services for Subscribers on Two Platforms

20.2.3 Implementing CSS Services with EXCOMM Resources

The shipboard resources available for implementing communications services are radios, modems, encryption equipment, and a CSS controller. The CSS service is dictated by COMMPLAN, that is, the communications resource plan in support of a mission. Figure 20.5 shows an example of selection of a set of shipboard resources by two platforms to accomodate a CSS service. The CSS architecture is designed to implement services and resources by using:

- Modular building blocks such as modems, radios, and cryptos
- Interoperable interfacing component hardware and software systems that conform to ISO standards

For example, for the subscriber on platform 1, the primary connection for service c is via the resource 3; the backup connection for service c is via the resource 2. Service c in this example could be the NTDS. The primary for the NTDS connection is via the UHF Link 11 circuit; the backup connection is via the UHF Link 11 circuit. Another example would be service b where the primary connection for service b is via resource 1; the backup connection for service b is via resource 2.

Figure 20.6 illustrates how CSS can be implemented by using current Navy shipboard EXCOMM systems in transferring information among three platforms. On each platform the users are connected via a LAN; the same LAN also connects the users to the EXCOMM circuit. Thus, the user system D (e.g., NTDS) on platform 2 and the user system D (e.g., NTDS) are connected through a virtual path linking LANs on platforms 2 and 3 via EXCOMM circuits.

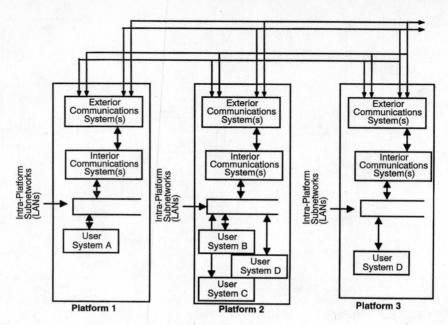

Figure 20.6   Example of Current Navy Shipboard EXCOMM Systems on Three Afloat Platforms for Use in Implementing the CSS Architecture

20.3 COPERNICUS ARCHITECTURE

The CSS became a key component in yet another new Navy initiative called the Copernicus Architecture. The U.S. Navy developed the Copernicus Architecture for communications systems ashore and afloat and for shore-based and tactical afloat command centers [NA-1; SN-1]. This naval C2 and communications architecture consists of four entities:

- Global Information Exchange Systems (GLOBIXS),
- CINC Command Complex (CCC),
- Tactical Data Information Exchange System (TADIXS), and
- Tactical Command Center (TCC).

These four entities are intended to restructure the existing command and control process serving the Navy tactical commanders afloat, the numbered fleet commanders, and other command support nodes ashore.

The challenge that lies ahead for future Navy tactical communications is to effectively utilize the existing resources, merge them with rapidly progressing technology, and support the emerging naval communications architecture in an economical manner.

REFERENCES

[BR–1] Bruninga, R. E., "Toward a Survivable Radio Room," *Naval Engineers Journal*, September 1990, pp. 27–31.

[CA–1] Casey, K. R. et al., "Networking Technology for Integrated Naval Tactical Communications," *1987 IEEE MILCOM Conference Proceedings*, October 19–22, 1987, Washington, D. C., pp. 241–248.

[GR–1] Green, D. T. and D. T. Marlow, "Application of LAN Standards to the Navy's Combat Systems," *Naval Engineers Journal*, May 1990, pp. 146–153.

[NA–1] *Copernicus Architecture-Phase I Requirements Definition*, U.S. Navy (OP-094), Washington, D. C. 20350-2000, August 1991.

[NC–1] *Communications Support System (CSS) Overview*, Technical Report, CSS-10001-00, NOSC (now NCCOSC RDT&E Div.), San Diego, Calif. 92152-5185, July 26, 1990.

[NC–2] *Navy UHF Satellite Communications System Description*, NSHFC 301, NOSC (now NCCOSC RDT&E Div.), San Diego, Calif. 92152-5185, December 31, 1991.

[NC–3] *UHF LOS Network Architecture Specification-For the Unified Networking Technology (UNT) Program*. Specification, 250-U-0002, NOSC (now NCCOSC RDT&E Div.), San Diego, Calif. 92152-5185, June 1, 1988.

[SN–1] *Copernicus Architecture-Initial Implementation Plan for Phase II*, Space and Naval Warfare Systems Command, Washington, D. C. 20363-5100, December 1991.

[WA–1] Walsh, E. J., "Navy's Tactical Data Link Advances Fleet Technology," *Signal*, February 1990.

Appendix A

Abbreviations and Acronyms

AAW	anti-air warfare
A/C	aircraft
ACK	acknowledgment
ACO	action cutout
ACOC	area coordination operations center
ACP	allied communications policy
ACS	afloat correlation system
ACTS	Aegis combat training system
A/D	analog-to-digital
ADCCP	advanced data communications control procedure
ADMIN	administration
AEW	airborne early warning
AF	audio frequency
AFCEA	Armed Forces Communications and Electronics Association
AFNET	Air Force network
AFSATCOM	Air Force satellite communications
AFSCN	Air Force satellite control network
AH	application header
AI	articulation index
AJ	anti-jamming
ALE	automatic link establishment
AM	amplitude modulation

AMCE	available MILSTAR control element
AMW	amphibious warfare
ANDVT	advanced narrowband digital voice terminal
ANSI	American National Standards Institute
AOR	Atlantic ocean region
ARQ	automatic repeat request
ARRL	American Radio Relay League
ASCII	American standard computer information interchange
ASK	amplitude-shift-keying
ASuW	anti-surface warfare
ASW	anti-submarine warfare
ASWCCS	anti-submarine warfare command and control system
ASWM	anti-submarine warfare module
ASWOC	anti-submarine warfare operation center
AT&T	American Telephone and Telegraph Company
ATDS	air tactical data system
ATF	amphibious task force
AUTODIN	automatic digital network
AUTOSEVOCOM	automatic secure voice communications
AUTOVON	automatic voice network
AWACS	airborne warning and control system
AZ	azimuth
BCA	Broadcast Control Authority
BCS	black communications switch
BG	battle group
BER	bit error rate
BILP	beginning of intermediate line pair
BIT	built-in-test
BITS	base information transfer system

BLACKER	an encryption system
BLOS	below line-of-sight
BOLP	beginning of line pair
bps	bits per second
BPSK	binary phase-shift-keying
BTU	basic terminal unit
CAFE	communications automation follow-on effort
CAINS	carrier aircraft inertial navigation system
Cal	calendar time
CAP	combat air patrol
CAS	close air support
C&R	contact and reporting
CATF	Commander, amphibious task force
CCC	communications control channel
CCC	CINC command complex
CCITT	Consultative Committee for International Telephone and Telegraph
CCOW	channel control order wire
CCTV	closed circuit television
CCS	communication control system
CCS	Constellation Control Station
CCSC	combat cryptological support console
CDMA	code division multiple access
CEG	communications equipment group
CELP	code-excited linear predictive
CES	coastal earth station
CG	cruiser, guided missile
CGN	cruiser, guided missile, nuclear
CHBDL-ST	common high bandwidth data link-ship terminal

CIA	Central Intelligence Agency
CIC	combat information center
CINC	Commander-in-Chief
CIU	cable interface unit
CLF	Commander, landing force
CMCC	communication monitoring and control console
CMD	command
CNO	Chief of Naval Operations
CO	commanding officer
COMMPLAN	communication plan
COMNAVSPACECOM	Commander, Naval Space Command
COMNAVTELCOM	Commander, Naval Telecommunications Command
COMSAT	communications satellite
COMSEC	communications security
COMSECONDFLT	Commander, Second Fleet
COMSPAWARSYSCOM	Commander, Space and Naval Warfare Systems Command
CONUS	Continental United States
COSAM	co-site analysis model
COTS	commercial off-the-shelf
CRC	cyclic redundancy code
CRPL	Central Radio Propagation Laboratory
CSTS	combat simulation test system
CSMA/CD	carrier sense multiple access with collision detection
CSOC	consolidated space operations center
CSP	control status panel
ctrl	control
CUDIXS	common user digital information exchange subsystem

CV	aircraft carrier
CVIC	aircraft carrier intelligence center
CVLF	compact very low frequency
CVN	aircraft carrier, nuclear
CVSD	continuously variable slope delta
CW	continuous wave
CWC	composite warfare commander
C2	command and control
C3	command, control, and communications
C3I	command, control, communications, and intelligence
DAMA	demand assigned multiple access
DARPA	Defense Advanced Research Project Agency
DART	dynamically adaptive receiver/transmitter
dB	decibel
dBm	decibel, referenced to 1 mW
DBPSK	differentially encoded binary phase-shift-keying
dBW	decibel, referenced to 1 W
DCA	Defense Communications Agency
DCS	defense communications system
DCT	digital communications terminal
DCTN	defense commercial telecommunications network
DD	destroyer
DDG	destroyer, guided missile
DDN	defense digital network
DDSP	distributed digital signal processor
DEA	Drug Enforcement Agency
DECAL	design communications algorithm
DECCO	defense commercial communciations office
DEFSECDEF	Deputy Secretary of Defense

DF	direction finder
DISA	Defense Information Systems Agency
DISN	defense integrated services network
DLCS	data link control system
DLED	digital link encryption device
DLH	data link header
DMS	defense meteorological satellite
DMS1, DMS2	defense message system phase 1, phase 2
DNCS	data net control station
DoD	Department of Defense
DPSK	differential phase-shift-keying
DS	direct sequence
DSB	double-sideband
DSBAM	double-sideband amplitude modulation
DSN	defense switched network
DSNET 1, 2, and 3	defense secure network 1, 2, and 3
DSCS II, and III	defense satellite communciations system phase II, and phase III
DSCSOC	defense satellite communications system operation center
DSP	digital signal processing
DTG	date time group
DTS	diplomatic telecommunications system
DTS	data terminal set
DSVT	digital secure voice terminal
EAM	emergency action message
EASTPAC	Eastern Pacific
EC	Earth coverage
EC	engineering change
E^3 (E-cubed)	electromagnetic environmental effect

ECCM	electronic counter-countermeasures
ECM	electronic countermeasures
EHF	extremely high frequency
EIA	Electronics Industry Association
EIRP	effective isotropic radiated power
EL	elevation
ELANT	Eastern Atlantic
ELF	extremely low frequency
ELINT	electronic intelligence
ELOS	extended line-of-sight
EM	electromagnetic
E-Mail	electronic mail
EMC	electromagnetic compatibility
EMCAB	electromagnetic compatibility advisory board
EMCON	emission control
EMI	electromagnetic interference
EMP	electromagnetic pulse
EOL	end of line
EOM	end of message
EPAC	Eastern Pacific
ES/IS	end system/intermediate system
ESM	electronic support measure
ETR	early token release
EXCOMM	exterior communications
EW	electronic warfare (also: early warning)
EWC	electronic warfare commander
EWCS	electronic warfare control system
FCC	Federal Communications Commission
FCC	fleet command center

FDCS	flight deck communications system
FDDI	fiber distributed data interface
FDDS	fleet data display system
FDM	frequency division multiplex
FDMA	frequency division multiple access
FEP	fleet Satellite EHF Package
FF	frigate
FFG	frigate, guided missile
FH	frequency hopping
FLETAC	fleet tactical circuit
FLIR	forward looking infra-red
FLT	fleet
FLTBROADCAST	fleet broadcast
FLTCINC	Fleet Commander-in-Chief
FLTSAT	fleet satellite
FLTSATCOM	fleet satellite communications
FM	frequency modulation
FOSIC	fleet ocean surveillance information center
FOSIF	fleet ocean surveillance information facility
FOT	fréquence optimum de travail (optimum operating frequency)
FSK	frequency-shift-keying
FTAM	file transfer access and management
FTP	file transfer protocol
FTS2000	Federal Telephone System 2000
GENSER	general service
GLOBIXS	global information exchange system
GMF	ground mobile force
GMT	Greenwich mean time
GOSIP	government open system interconnection profile

GPS	global positioning system
HAVE QUICK	VHF radio
HDLC	high-level data link control
HELO	helicopter
HERF	hazard to electromagnetic radiation to fuel
HERO	hazard to electromagnetic radiation to ordnance
HERP	hazard to electromagnetic radiation to personnel
HESSA	high efficiency solid state amplifier
HF	high frequency
HFAJ	high frequency antijamming
HFIP	high frequency improvement program
HFMR	HF modem replacement
HHR	high hop rate
HICOM	high command communications
HPA	high power amplifier
HSFB	high speed fleet broadcast
HW	hardware
Hz	Hertz
IABS	Integrated Apogee Booster System
ICS-3	integrated communications system-3
ICS/SCAN	integrated communications system/shipboard communciations area net
ICW	interrupted continuous wave
ID	identification
IEEE	Institute of Electrical and Electronic Engineers
IF	intermediate frequency
IFF	identification of friend or foe
IM	intermodulation
INFO	information
INMARSAT	International Maritime Satellite

INS	inertial navigation system
INSICOM	integrated SI communications
INTELCAST	intelligence broadcast net
INTEL	intelligence
INTELNET	intelligence net
INTELSAT	International Telecommunications Satellite
IO	intelligence officer
I/O	input/output
IOR	Indian ocean region
IP	internet protocol
IR	infra-red
IRE	Institute of Radio Engineers
ISB	independent sideband
ISO	International Standards Organization
ISDN	integrated services digital network
ITU	International Telecommunications Union
IUSS	integrated undersea surveillance system
IXS	information exchange subsystem
JANAP	Joint Army, Navy and Air Force publication
JCS	Joint Chiefs of Staff
JINTACCS	joint interoperability of command and control system
JLE	jammer locating equipment
JRSC	jam-resistant secure communications
JTIDS	joint tactical information distribution system
JOTS	joint operational tactical system
LAMPS	light airborne multi-purpose systems
LAN	local area network
LANT	Atlantic
LAP-B	link access protocol-balanced (related to HDLC)

LCC	amphibious command ship
LCAC	landing craft, air cushion
LDMX	local digital message exchange
LDR	low data rate
LEASAT	leased Satellite
LEI	link eleven improvement
LF	low frequency
LF	landing force
LFSR	linear feedback shift register
LHA	amphibious assault ship
LHD	amphibious assault ship
LINCAL	link communications analysis algorithm
LLC	logical link control
LLTV	low light television
LNA	low noise amplifier
LORAN	long range navigation
LOS	line-of-sight
LPC	linear predictive coding
LPD	low probability of detection
LPH	amphibious assault ship
LPI	low probability of intercept
LRU	line replace unit
LST	landing ship, tank
LWCA	light weight communications antenna
MAC	media access control
MAP	manufacturing automation protocol
MARECS	French INMARSAT
MBA	multibeam antenna
MCC	mission control complex

MCE	MILSTAR control element
MCM	mine countermeasure
MCS	master control system
MDF	main distribution frame
MDR	medium data rate
MED	Mediterranean
MEDVAC	medical evacuation
MEECN	minimum essential emergency communications network
MF	medium frequency
MFSK	multiple frequency-shift-keying
MGR	military grid reference
MHS	message handling system
MILDEP	military department
MILNET	military network
MILSTAR	military satellite relay system
MIL-STD	military standard
MIDL	modular interoperable data link
MIDS	multifunctional information distribution system
MIST	modular interoperable surface terminal
MLDT	mean logistic delay time
MLP	multi-link processor
MOC	MILSTAR operations center
MOF	maximum observable frequency
MPA	maritime patrol aircraft
MPDS	message processing and distribution system
MSG	message
MSK	minimum-shift-keying
MTBF	mean time between failure
MTSC	modified transmit signal transient and preamble characteristics
MTTR	mean time to repair

MUF	maximum usable frequency
MUX	multiplex
NAS	Naval air station
NATO	North Atlantic Treaty Organization
NAV	Naval or Navy
NAVCOMMSTA	Naval communications station
NAVCAMS	Naval communications area master station
NAVCOMPARS	Naval communications processing and routing system
NAVELEX	Naval Electronic Systems Command
NAVEUR	Navy European Command
NAVMACS	Naval modular automated computer system
NAVNET	Navy network
NAVSAT	navigation satellite (See TRANSIT)
NAVSEA	Naval Sea Systems Command
NAVSTAR	navigation satellite timing and ranging (See GPS)
NCA	National Command Authority
NCCOSC	Naval Command and Control and Ocean Systems Center
NCCS	Naval command and control system
NCS	network coordination station
NCTAMS	Naval computer and telecommunications area master station
NDI	nondevelopment item
NEC	numerical electromagnetic code
NEEACT PAC	Naval electronics engineering activity-Pacific
NEOM	not end of message
NESP	Navy EHF satellite program
NGFS	naval gunfire support
NH	network header
NIPS	Naval intelligence processing system
NKDS	Navy key distribution system

NOSC	Naval Ocean Systems Center
NPG	network participant group
NRADWARM	Naval Research and Development, Warminster
NRL	Naval Research Laboratory
NRZ	nonreturn to zero
NS	net synchronization
NSA	National Security Agency
NSCS	Naval satellite control station
NST	Navy standard terminal
NT	net test
NTCC	Naval telecommunications center
NTCOC	Naval telecommunication command operations center
NTDS	Naval tactical data system
NTS	Naval telecommunications system
NTSC	National Television Standard Committee
NVIS	near vertical incidence skywave
OBT	onboard training system
OCI	operator-to-computer interface
ODA	office document architecture
OMEGA	LF navigation system
OOB	order of battle
OOK	on-off-keying
ops	operations
OQPSK	offset-quadrature-phase-shift-keying
OR	operational requirements
ORESTES	TTY circuit
OSI	open system interconnect
OSIS	ocean surveillance information system
OTC	officer-in-tactical command

OTCIXS	officer-in-tactical command information exchange subsystem
OTH	over-the-horizon
OTH-T	over-the-horizon targeting
OWF	optimum working frequency
PAC	Pacific
PB	phonetically balanced
PCM	pulse code modulation
PECAL	performance evaluation communications algorithm
PH	presentation header
PLA	plain language address
PN	pseudo noise
POR	Pacific ocean region
POST	prototype ocean surveillance terminal
PPLI	precision participant location and identification
PRITAC	primary tactical circuit
PROM	programmable read-only memory
PSK	phase-shift-keying
PSTN	public switched telephone network
QMCS	quality monitor and control system
QNM	quasi-maximum noise
QMS	quality monitoring system
QPSK	quadrature phase-shift-keying
QUICKSAT	Navy SHF satellite system
R	rate
RADHAZ	radiation hazard
RATT	radio TTY
RC	roll call
RCCOW	return channel control order wire
RCS	radio communications system

RDF	radio direction finder
RDT&E	research, development, test and evaluation
RECCE	reconnaissance
RF	radio frequency
RFI	radio frequency interference
RI	routing indicator
RIXT	remote information exchange terminal
RLPA	rotatable log periodic array
RPU	remote programming unit
RPV	remote pilotless vehicle
RT	receiver-transmitter
RTC	return of control
RTT	round trip timing
SAFENET	survivable adaptable fiber embedded network
SAFS	SSN AFSATCOM interface
SAM	surface-to-air missile
SANGUINE	ELF communications system
SAR	search and rescue
SAR	synthetic aperture radar
SAS	single audio system
SATCOM	satellite communications
SBC	short broadcast
SC	single channel
SCI	special compartmented intelligence
SCT	single channel transponder
SDLC	synchronous data link control
SDP	service deliver point
SDS	surveillance direction system
SECVOX	secure voice subsystem

SEMCAC	shipboard electromagnetic compatibility analysis for communications
SEMCIP	ship's electromagnetic compatibility improvement program
SES	ship earth station
SEW	space and electronic warfare
SGLS	satellite ground link system
SH	session header
SHF	super high frequency
SHIPALT	ship alteration
SI	special intelligence
SIGINT	signal intelligence
SINCGARS	single channel ground to airborne radio system
SIOP	single integrated operational plan
SITREP	situation report
SIU	subscriber interface unit
SLCM	ship launched cruise missile
SM-2	standard missile-2
SMTP	simple mail transfer protocol
SNR	signal-to-noise ratio
SOM	start of message
sr	steradian
SRBOC	super rapid bloom offbound chaff
SSB	single sideband
SSBAM	single sideband amplitude modulation
SSBN	ballistic missile submarine, nuclear
SSECMS	surface ship exterior communications monitor and control system
SSES	special signal exploitation system
SSIXS	submarine satellite information exchange subsystem
SSN	attack submarine, nuclear

SSW	strategic submarine warfare
STANAG	standardization agreement (allied military)
STDL	submarine tactical data link
STT	shore targeting terminal
STU-III	secure telephone unit III
STW	strike warfare
SUBJ	subject
SUBOPAUTH	submarine operational authority
SUPPLOT	supporting plot
SURTASS	surveillance towed array sensor system
SVS	secure voice system
SW	software
TACAMO	take charge and move out
TAC	tactical aircraft
TACAN	tactical air navigation
TACC	tactical air control center
TACINTEL	tactical intelligence information exchange subsystem
TACNET	tactical network
TACSAT	tactical satellite
TADIL-A	tactical digital information link A
TADIL-B	tactical digital information link B
TADIL-C	tactical digital information link C
TADIL-J	tactical digital information link J
TADIXS	tactical data information exchange subsystem
TAO	tactical action officer
TCC	tactical command center
TCF	technical control facility
TCP	transport control protocol
TDM	time division multiplexing

TDMA	time division multiple access
TDP	target data processor
TE$_{m,n}$	transversal electric mode m, n
TELNET	terminal-remote host protocol developed by DARPA
TEMPEST	detection, evaluation, and control of classified signals
TERCOM	terrain contour matching
TESS	tactical environmental support system
TFMS	tactical frequency management system
TF	tactical force
TFCC	tactical force command center
TG	tactical group
TGO	tactical group ORESTES
THP	transport header protocol
THT	token holding timer
TIDET	topside integration and design engineering team
TLM	telemetry link monitor
TM$_{m,n}$	transversal magnetic mode m, n
TRANSPT	transport
TRANSEC	transmission security
TRANSIT	navigation satellite system (See NAVSAT)
TRE	tactical receive equipment
TRI-TAC	tri-service tactical
TRT	token ring timer
TT&C, TTC	telemetry, tracking and command
TTY	teletypewriter
TWT	traveling wave tube
UAV	unmanned aerial vehicle
UFO	UHF follow-on
UHF	ultra high frequency

UNT	unified networking technology
UNCLASS	unclassified
USMTF	United States message text format
USS	United States ship
Verdin	VLF circuit
VF	voice frequency
VHF	very high frequency
VLF	very low frequency
VLS	vertical launch system
VOCODER	voice coder
VP	shore based patrol aircraft
VS	carrier based search and attack aircraft
VSWR	voltage standing wave ratio
VT	virtual terminal
VTP	virtual terminal protocol
WAWS	Washington area wideband service
WCAP	waterfront corrective action program
WESTPAC	Western Pacific
WLANT	Western Atlantic
WPAC	Western Pacific
WS	workstation
WWMCCS	worldwide military command and control system
XTP	Xpress transport protocol
Zulu time	Greenwich mean time (GMT)

Appendix B

A Description of Typical Shipboard Combat Systems

In section 19.1.1, typical shipboard combat systems are presented with a block diagram (Figure B.1). The following thirteen elements of typical shipboard combat systems are described below:

- Radar systems
- Identification systems
- EW systems
- Navigation systems
- Underwater systems
- Command and decision systems
- Weapon control systems
- Weapon systems
- Telemetry systems
- Exterior communications systems
- Interior communications systems
- Meteorological systems
- Training systems

B.1 RADAR SYSTEM

The shipboard radar system is used primarily to detect and track targets, and to compare and correlate with other information obtained either from other sensors on the ship or from outside sources. Shipboard surveillance radars shown in this figure include the AN/SPS-49 air search radar, the AN/SPY-1 radar [PH-1], AN/SPS-67 surface search radar. Other shipboard radar systems are long range 2-dimensional air surveillance radars (AN/SPS-48 and AN/SPS-40), 3-dimensional air surveillance radar (AN/SPS-72), aircraft control radars (AN/SPN-43 and AN/SPN-41), and weapons firing radars (AN/SPW-2 and AN/SPQ-55).

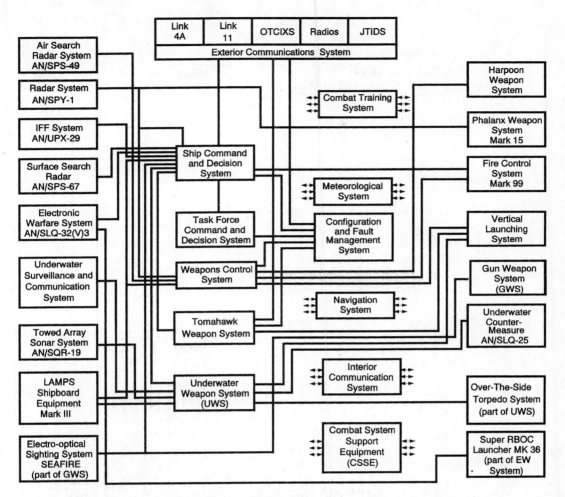

Figure B.1 Typical Shipboard Combat System (Reprinted with permission from D. Green and D. Marlow, "Application of LAN Standards to the Navy's Combat System," *Naval Engineers Journal,* May 1990, pp. 146–153.)

B.2 IDENTIFICATION SYSTEM

The identification, friend or foe (IFF) system is a means of obtaining identity provided by a type of coded secondary radar. The IFF system transmits signals which cause IFF transponders to transmit a coded signal in response. Such transponders are fitted in friendly units. The IFF system shown in Figure 19.1 is the AN/UPX-29. Other shipboard IFF systems are AN/UPX-24 and AN/UPX-25 [BL-3].

B.3 EW SYSTEM

Electronic warfare (EW) is the exploitation of an enemy's use of electronics, and counter-measures of this use of such techniques. Essentially electronic warfare consists of three elements: ESM, ECM, and ECCM [KI-1]. The EW system shown in this figure is the AN/SLQ-32 EW system [LA-1] [BL-9]. Other EW systems are AN/WLR-1, AN/WLR-11A, and MK-36 Super Rapid Bloom Offbound Chaff (SRBOC) launcher (a deck-mounted, mortar-type countermeasure system).

B.4 NAVIGATION SYSTEM

The second column from the right shows additional essential support subsystems such as the navigation system. Shipboard navigation systems include the Navigation Satellite Timing and Ranging (NAVSTAR), Global Postioning System (GPS), the AN/SPS-64(V) radar, harbor navigation radars (AN/SPS-64 and AN/SPS-66) and the AN/WSN-5 inertial navigation system.

The NAVSTAR GPS is a space-based radio navigation system which provides highly accurate and jam-resistant position, velocity, and time information to users any-where if equipped with GPS receivers, on a continuous basis.

B.5 UNDERWATER SYSTEM

The underwater system includes sonars, underwater weapons, countermeasures, and un-derwater communications. Two underwater systems, that is, the towed array, passive, long-range sonar system AN/SQR-19 and the underwater surveillance and communica-tions system, are shown in Figure 19.1. The shipboard sonar system detects, classifies, and tracks enemy submarines by active or passive sensors. Other shipboard sonar systems are the AN/SQS-26 echo-ranging sonar for surface ships, the AN/SQS-53B/C long range, active/passive sonar [BL-1], and the AN/SQR-35 towed sonar system for surface ships [BL-2] [UR-1]. The AN/SQQ-89 is a sonar suite which consists of the AN/SQS-53B/C active sonar, the AN/SQR-19 towed array sonar, and the AN/SQQ-28 sonobuoy processor linked to the LAMPS III ASW system [SN-1; NL-2].

The AN/SLQ-25 underwater torpedo countermeasure (also known as Nixie) is an active transducer which is used as noise augmentation to confuse enemy homing torpe-does for the purpose of ship's defense [BL-5].

The underwater communications system is an acoustically propagated system using transducers at two frequency bands: the lower frequency band of 1.45 kHz–3.1 kHz and the higher frequency band of 8.3 kHz–11.1 kHz. The system carries both voice and CW signals. Currently the AN/WQC-2A sonar communication set is in service aboard surface ships, submarines, and shore facilities [BL-6].

The over-the-side torpedo system is a surface ship launched underwater weapon system. Example of the surface ship launched torpedo is MK 46 which is a deep diving, high speed torpedo fitted with active/passive acoustic homing [BL-7].

B.6 COMMAND AND DECISION SYSTEM

The command and decision system provides the commander of the task force and ship necessary tactical information obtained from the force's organic sensors and from intelligence sources outside the force and process a complete picture of the local operations situation. There are two levels of the command and decision systems: the task force command and decision system and the ship's command decision system. The task force command and decision is concerned with command and control of task force units for the accomplishment of the mission assigned by the fleet, and the detailed and usually local direction and control of maneuver necessary to accomplish mission or tasks assigned [PA-1]. The ship's command and decision system is concerned with the acquisition of sensor information from organic sensors[1] and from external to the task force via FLTSATCOM channels, Link 11, Link 16, and other EXCOMM systems and to issue command and control directions to the weapon subsystems shown on the right side of Figure 19.1.

B.7 WEAPON CONTROL SYSTEM

The weapon control system is a guidance and control system, which receives targeting and attitude data from either shipboard system or a third party for over-the-horizon operations and provides en route guidance to the missile. Currently, the Harpoon weapon system uses the AN/SWG-1A(V) Harpoon control system [BL-10].

The electro-optical sighting system is a naval fire control system for directing gunfire. The system uses a forward looking infra-red (FLIR) sensor and the electro-optical unit, and daylight television system. Currently the Seafire Naval Fire Control System is used aboard ships equipped with MK 86 Gunfire Control System [BL-11].

B.8 WEAPON SYSTEM

The shipboard weapon system consists of: (a) shipboard surface-to-surface weapons for anti-ship and shore bombardment, (b) shipboard surface-to-air weapons for air defense, and (c) guns. These weapon systems deliver weapons (missiles and guns) and perform target engagement functions by aiming and releasing weapons. Some of the representative shipboard weapon systems are described below.

- The Tomahawk weapon system is a long-range (up to 2500 km) cruise missile system. The sea launched cruise missile (SLCM) AN/BGM-109 Tomahawk missile is capable of being launched from surface ships and submerged submarines.

---

[1]Sensors that are a part of the ship's resources.

The Tomahawk SLCM is capable of carrying out both land attack and anti-ship attack using a Terrain Contour Matching (TERCOM) system [BL-12].

- The Harpoon weapon system is an all-weather anti-ship, medium range (90 km) tactical cruise missile system for use against surfaced submarines, patrol craft, destroyers, and larger vessels. The AN/RGM-84A shipborne surface-to-surface missile is capable of being ship- or submarine-launched [BL-10]. The air-launched version is the AN/AGM-84 airborne air-to-surface missile.

- The Standard Missile-2 (SM-2) (AN/RIM-66A and AN/RIM-67A) is a surface-to-air ship launched missile for defense against attacking aircraft and anti-ship missile. The SM-2 is deployed on surface ships. The SM-2 has an advanced guidance system (mid-course and terminal guidance) [BL-4; OL-1; WI-1].

- The MK-15 Phalanx close-in weapon system is a terminal defense anti-ship missile, using M61A1 Gatling gun. It combines on single mount fire-control radars and a six-barrel Gatling gun [BL-13; NL-1].

- The Vertical Launch System (VLS) is an improved missile launching system for several missiles such as the Standard Missile-2 (SM-2) and the Tomahawk SLCM. Currently the MK 41 vertical launch system is installed in surface ships.

B.9 TELEMETRY SYSTEM

The shipboard sensor data link transfers tactical and imagery data from airborne platforms to the ship. The shipboard sensor data link shown in Figure B.1 is the AN/SRQ-4 LAMPS [SN-1; BL-8]. Another shipboard sensor data link is the AN/USQ-123(V) Common High Band Data Link [PS-1].

B.10 EXTERIOR COMMUNICATIONS (EXCOMM) SYSTEM

The shipboard EXCOMM system provides communications services from the ship to other ships and to shore facilities. Shipboard EXCOMM systems shown in Figure B.1 are Link 11, Link 4A, Link 16, Radios, and OTCIXS. As discussed in previous chapters of this book, other shipboard EXCOMM systems include radios in ELF/VLF/LF, MF, HF, VHF/UHF LOS, UHF SATCOM, SHF SATCOM, and EHF SATCOM frequency bands.

B.11 INTERIOR COMMUNICATIONS (INTERCOM) SYSTEM

The shipboard INTERCOM system provides control and switching of the interior voice and data terminals and the exterior communications system. Typical shipboard INTERCOM systems are the MCS-2000 INTERCOM System, the AN/SRC-47 Flight Deck Communications System, the AN/WTC-2 Sound Power Telephone System, and the LS-474/U Shipboard Announcing System as described in Chapter 18.

B.12 METEOROLOGICAL SYSTEM

The shipboard meteorological system provides a means to receive continuous up-to-date weather information. The weather information can be received via Defense Meteorological Satellites (DMS) and HF transmission. Shipboard meteorological systems are the Tactical Environmental Support System (TESS) with the AN/SMQ-11 satellite receiver, and the AN/URC-93 HF transceiver.

B.13 COMBAT TRAINING SYSTEM

The AN/SSQ-91(V) Combat Simulation Test System (CSTS) is the organic and training system for the LHD-1 class. Other combat training systems are the Aegis Combat Training System (ACTS) for Aegis ships (the CG-47 and DDG-51 classes) and the AN/SQQ-89 ASW Onboard Training (OBT) system for surface ships (the CVN, CG-47, DDG-51, CGN-36, DDG-993, DD-963, FFG-7, LCC-19, and LHA-1 classes).

B.14 REFERENCES

[BL–1] Blake, B., ed., "AN/SQS-53/26 Sonars," *Jane's Underwater Warfare Systems 1989–1990*, Jane's Information Group, Alexandria, Va. 22314-1651, 1989, p. 95.

[BL–2] Blake, B., ed., "AN/SQS-35, AN/SQS-36, AN/SQS-38 Sonars," *Jane's Underwater Warfare Systems 1989–1990*, Jane's Information Group, Alexandria, Va. 22314-1651, 1989, p. 94.

[BL–3] Blake, B., ed., "AN/UPX-23(V) Shipborne IFF System," *Jane's Military Weapon Systems 1987–88*, Jane's Information Group, Alexandria, Va. 22314-1651, 1987, pp. 692–693.

[BL–4] Blake, B., ed., "Standard (RIM-66A and RIM-67A) Surface-to-Air Missile," *Jane's Military Weapon Systems 1987–88*, Jane's Information Group, Alexandria, Va. 22314-1651, 1987, pp. 510–511.

[BL–5] Blake, B., ed., "AN/SLQ-25 Countermeasures," *Jane's Underwater Warfare Systems 1989–1990*, Jane's Information Group, Alexandria, Va. 22314-1651, 1989, p. 155.

[BL–6] Blake, B., ed., "AN/WQC-2A Communication Set," *Jane's Underwater Warfare Systems 1989–1990*, Jane's Information Group, Alexandria, Va. 22314-1651, 1989, p. 165.

[BL–7] Blake, B., ed., "MK 46 Torpedo," *Jane's Underwater Warfare Systems 1989–1990*, Jane's Information Group, Alexandria, Va. 22314-1651, 1989, p. 18.

[BL–8] Blake, B., ed., "LAMPS MK III," *Jane's Underwater Warfare Systems 1989–1990*, Jane's Information Group, Alexandria, Va. 22314-1651, 1989, pp. 124–125.

[BL–9] Blake, B., ed., "AN/SLQ-32 Shipboard EW Equipment, *Jane's Weapon Systems 1987–1988*, Jane's Information Group, Alexandria, Va. 22314-1651, 1989, p. 722.

[BL–10] Blake, B., ed., "Harpoon Shipborne Surface-to-surface Missile (RGM-84A)," *Jane's Weapon Systems 1987–1988*, Jane's Information Group, Alexandria, Va. 22314-1651, 1989, pp. 487–488.

[BL–11] Blake, B., ed., "Gun Fire Control System MK 86," *Jane's Weapon Systems 1987–1988*, Jane's Information Group, Alexandria, Va. 22314-1651, 1989, pp. 468–469.

[BL–12] Blake, B., ed., "Tomahawk Sea-Launched Cruise Missile (BGM-109)," *Jane's Weapon Systems 1987–1988*, Jane's Information Group, Alexandria, Va. 22314-1651, 1989, pp. 486–487.

[BL–13] Blake, B., ed., "MK 15/16 Phalanx Close-in Weapon System," *Jane's Weapon Systems 1987–1988*, Jane's Information Group, Alexandria, Va. 22314-1651, 1989, pp. 508–509.

[KI–1] Kiely, D. G., *Naval Electronic Warfare*, Pergammon-Brassey's International Defence Publishers, McLean, Va. 22102, 1988.

[LA–1] Law, Preston, "AN/SLQ-32," *Shipboard Antennas*, Artech House, Inc., Dedham, Mass. 02026, 1984, pp. 439–442.

[NL–1] "Phalanx Close-in Weapons System," *Sea Power*, Navy League of the U. S., Arlington, Va. 22201-3308, January 1993, p. 182.

[NL–2] Upgrades Planned for ASW Trainer, *Sea Power*, Navy League of the U. S., Arlington, Va. 22201-3308, May 1991, pp. 44–46.

[OL–1] Oliver, M. L. and W. N. Sweet, "Standard Missile: the Common Denominator," *Johns Hopkins University APL Technical Digest*, October-December, 1981, Laurel, Md. 20707, pp. 283–288.

[PA–1] Pakenham, W.T.T., *Naval Command and Control*, Pergammon-Brassey's International Defence Publishers, McLean, Va. 22102, 1989.

[PH–1] Phillips, C. C., "Aegis: Advanced Multi-Function Array Radar," *Johns Hopkins University APL Technical Digest*, October–December, 1981, Laurel, Md. 20707, pp. 246–249.

[PS–1] *Paramax Data Sheet on Common High Bandwidth Data Link-Ship Terminal AN/USQ-123(V)* , Paramax Systems Corporation, Salt Lake City, Utah 84116, 1993.

[SN–1] *AN/SRQ-4 Radio Terminal Set*, Technical Manual, EE185-AA-OMI-0120, Space and Naval Warfare Systems Command, Washington, D. C. 20363-5100, September 1, 1984.

[UR–1] Urick, R. J., *Principles of Underwater Sound*, 3rd Ed., McGraw-Hill, New York, N.Y., 1983, pp. 7–11.

[WI–1] Witte, R. W. and R. L. McDonald, "Standard Missile: Guidance Development," *Johns Hopkins University APL Technical Digest*, October–December, 1981, Laurel, Md. 20707, pp. 289–298.

Appendix C

Joint Electronics Type Designation System

First Letter INSTALLATION	Second Letter EQUIPMENT	Third Letter PURPOSE
A - Piloted Aircraft	A - Infrared	B - Bombing
B - Submarine	C - Carrier	C - Communications
D - Pilotless Carrier	D - Radiac	D - Direction finding/reconnaissance
F - Ground, Fixed	G - Telegraphic-Teletype	E - Ejection/release
G - Ground, General	I - Interphone	G - Fire control
K - Amphibious	J - Electromechanical	H - Record–Reproduce
M - Ground, Mobile	K - Telemetering	K - Computing
P - Pack or Portable	L - Countermeasures	M - Maintenance/test
S - Water	M - Meteorological	N - Navigation aids
T - Ground Transportable	N - Sound in Air	Q - Special type
U - General Utility	P - Radar	R - Receiving
V - Ground, Vehicular	Q - Sonar	S - Detect/range/bearing
W - Water, Surface and Underwater	R - Radio	T - Transmitting
Z - Pilotted and Pilotless Airborne Vehicular Combination	S - Special Type	W - Remote control
	T - Telephone (wire)	X - Identification and recognition
	V - Visible light	Y - Surveillance and control
	W - Armament	
	X - Facsimile or TV	
	Y - Data Processing	

This table is in accordance with MIL-STD-196.

Example of AN/URC-93	
AN/	= Standard Prefix
U	= General Utility
R	= Radio
C	= Communications
93	= Design Number

Index

A

AAW (*see also* Anti-air warfare)
  area defense, 24, 25
  communications, 42, 43
  connectivity, 24
AAW weapon coordination circuit, 42, 45
Acoustic communications, 478
ACP (*see* Allied communications policy)
ACP-121, 261
ACP-127, 261
Advanced narrowband digital voice
    terminal, 58, 61, 177, 256, 276
AEW (*see* Airborne early warning)
AEW control circuit, 42, 45, 46
AF (*see* Audio frequency)
AFSAT (*see* Air Force satellite
    communications)
AFSCN (*see* Air Force satellite control
    network)
AI (*see* Articulation index)
Air coordination circuit, 42, 45
Air Force satellite communications, 331
Air Force satellite control network, 302

Air homing control net, 42, 46
Air operations communications, 42, 46
Air reporting net, 42, 45
Air strike control net, 42, 46
Air tactical data system, 36
Airborne early warning, 46
Airborne warning and control system, 36, 37
Aircraft-to-aircraft connectivity, 37
Aircraft-to-ship connectivity, 37
Aircraft-to-shore connectivity, 37
Aircraft-to-submarine connectivity, 37
AJ (*see* Anti-jamming)
ALE (*see* Automatic link establishment)
Allied communications policy, 62
Alternate routing, 83, 87
American National Standard Institute, 431
American standard computer information
    interchange code, 279
Amphibious task force, 30
Amphibious warfare, 29, 30, 422
Amphibious warfare, connectivity, 30, 43,
  46
AMW (*see* Amphibious warfare)

AN/ARC-101, 201
AN/ARC-159A, 201
AN/ARC-164, 108, 220
AN/ARC-171, 220
AN/ARC-182, 50, 80, 201, 211–12, 220
AN/ARC-201, 201
AN/ARC-210, 108
AN/ART-52, 149
AN/BRA-6, 132
AN/BRA-23, 129
AN/BRA-34, 129
AN/BRR-3, 147
AN/BRR-6, 127
AN/BRR-8, 127
AN/FRC-64, 108
AN/FRR-21, 147
AN/FRT-3, 114
AN/FRT-31, 114
AN/FRT-64, 114
AN/FRT-67, 114
AN/FRT-87, 114
AN/FSC-79, 302
AN/GRC-171, 220
AN/GRR-23, 6, 207
AN/GRR-1986C, 207
AN/GRT-21(V)3, 210, 207
AN/PRC-77, 208
AN/PRC-114, 414
AN/PRC-117, 414
AN/PRC-119, 208
AN/PSC-2, 210
AN/SLQ-32, 256
AN/SQQ-33, 388
AN/SRQ-4 48, 223–24
AN/SSQ-88, 389–92
AN/SSR-1, 53, 271, 272, 275
AN/SYQ-7, 408–11
AN/TRQ-35, 163, 394
AN/TRQ-42, 163, 394, 395
AN/TSQ-129, 214
AN/UGC-6, 278
AN/UGC-48, 278
AN/UGC-77, 278
AN/UGC-136AX, 279
AN/UGC-143A, 277, 278
AN/URA-38, 173
AN/URC-93, 201, 214
AN/URC-94, 80
AN/URC-109, 108, 169, 177, 178, 376
AN/URQ-23, 392–93
AN/URR-69, 80
AN/URR-76, 147, 149

AN/URR-103, 80
AN/URT-23, 169, 171, 172–74, 193–95, 376
AN/USC-28, 108
AN/USC-38, 108, 327, 334–36
AN/USC-43, 276–77
AN/USQ-36, 169
AN/USQ-59, 169
AN/USQ-63, 169
AN/USQ-74, 169
AN/USQ-76, 169
AN/USQ-79, 169
AN/USQ-83, 169, 171, 176–77
AN/USQ-92, 169
AN/USQ-123, 48, 221, 222
AN/UYK-44, 150
AN/VRC-12, 208
AN/VRC-40, 212–13
AN/VRC-46, 207, 208
AN/VRC-49, 208
AN/VRC-88, 201, 208
AN/VRC-90, 108, 208, 209
AN/VRC-91, 208
AN/VRC-92, 208
AN/WLR-2, 256
AN/WRR-3, 171, 212–14
AN/WRR-7, 149
AN/WSC-3, 50, 53, 80, 212, 220, 241, 271–74
AN/WSC-5, 240, 260
AN/WSC-6, 108, 240, 312–13, 315, 317, 321
AN/WTC-2, 414, 417
ANDVT (*see* Advanced narrowband digital voice terminal)
Announcing system, 413, 417–18
ANSI (*see* American National Standard Institute)
ANSI standard X3.139, 431, 436
Antenna:
 agile, 328
 AN/WSC-6, 315
 array, 72
 beamwidth, 237
 broadband, 179, 356
 buoyant cable, 127, 129, 131, 132
 characteristics of VLF receiver, 138
 coupler, 115, 129, 132, 225–27, 353, 354
 dipole, 72, 129, 183, 348, 356
 directional, 71, 368
 discage, 225, 356, 358
 discone, 179, 356, 357
 earth coverage, 309, 328
 effective area, 71

efficiency, 157, 237
EW, 344
fan type, 177–80, 185
gain, 71, 72, 237, 239
GPS receiver, 129
height, 79
helical, 358
HF, 179
HF/MF whip, 129
impedance matching, 132
integration, 368, 369
light weight communications, 186
log-periodic array, 179, 180, 225, 358
long wire, 113, 180, 183
long-wave receiver, 114, 115, 126, 128
long-wave transmitter, 115, 126
loop, 113, 128, 129, 132
LORAN-C receiver, 129
mast-mounted, multifunction, 127, 128, 130, 132
MILSTAR, 337
monopole, 179, 180, 355
multicoupler, 395
narrow beam coverage, 309
near vertical incident sky wave, 186–88
omnidirectional, 71, 72, 368
parabolic reflector, 352, 359
parameters, 237
placement, 341, 368
quarter-wave, 158
radar, 344
radiation pattern, 183, 184
reflector, 237, 313
retrofitting, 368, 369
RF distribution, 395–402
separation, 374
shipboard communications (*table*), 349
steerable, 328
TACAMO trailing, 126
towed buoy, 127, 132
towel bar, 187
trailing wire, 55, 56
tuner, 158
twin fan, 178
twin whip, 179, 181, 182
whip, 113, 179, 180, 225, 348, 355
Yagi-Uda, 198
Antenna size, HF, 157, 158
Anti-air warfare, 23, 253, 422
Anti-jamming, 89, 155, 158, 216, 325, 344
HF, 158, 177
link performance, 91, 93, 97

processing gain, 323
protection, 76, 325
systems (*table*), 108
Anti-submarine warfare, 26, 27, 422
connectivity, 26
Anti-submarine warfare operations center, 11, 28, 35
Anti-surface warfare, 25, 27, 422
connectivity, 26
Antipodal point, 117
Application layer, 435, 440, 442
ARQ (*see* Automatic repeat request)
Articulation index, 81
AS-390, 225, 358
AS-1729, 225
AS-1735, 225
AS-2410, 274
AS-2537A, 358
AS-2629B, 132
AS-2802, 356
AS-2815, 272
AS-2865, 357
AS-2876, 358
AS-3018, 275, 358
AS-3275, 223–24
AS-3399, 315, 318
ASCII (*see* American standard computer information interchange code)
ASK (*see* Modulation, amplitude-shift keying)
ASuW (*see also* Anti-surface warfare)
communications, 42, 45
connectivity, 26
contact and reporting circuit, 42, 45
coordination circuit, 42, 45
ASW (*see also* Anti-submarine warfare)
air coordination circuit, 42, 45
communications, 42, 44, 45
connectivity, 27
ASW control net, 42, 44
ASW helicopter coordination circuit, 42, 45
ASWOC (*see* Anti-submarine warfare operations center)
ATDS (*see* Air tactical data system)
ATF (*see* Amphibious task force)
Atlantic ocean region, 290, 293, 319
Atmospheric attenuation, 233
Atmospheric noise (*see* Noise, atmospheric)
Atmospherics, 73, 363
Audio frequency, 406
Audio signal, 391

Auroral effect, 122
AUTODIN (*see* Automatic digital network)
Automatic digital network, 12, 20, 62, 64, 263
Automatic link establishment, 54, 155, 174, 189–95
Automatic repeat request, 64
Automatic secure voice communications system, 12
Automatic voice network, 12
AUTOSEVOCOM (*see* Automatic secure voice communications system)
AUTOVON (*see* Automatic voice network)
Availability, 70, 86
AWACS (*see* Airborne warning and control system)

B

Back-door EMI, 365
Back-door intermodulation, 370
Band-filtering factor, 137
Bandspread, 90, 103
Bandspread factor, 106
Bandwidth:
    ASK, 139, 142
    channel, 89
    effectiveness, 82, 141
    efficiency, 82, 145
    FSK, 139, 142
    MSK, 143
    occupancy, 143
    PSK, 143
Barreling, 75
Baseband communications switching (*table*), 403
Battle group, 23, 158
    communications, 40
Baudot code, 279
BCS (*see* Black communications switch)
Beachmaster coordination net, 43, 46
BER (*see* Bit error rate)
Beyond line-of-sight, 13, 51
Bit energy to noise power density ratio (E_B/N_0), 81, 240
Bit error rate, 80
BIT (*see* Built-in-test)
Black communications switch, 391–92, 402–5
Black signal, 19
Black switch matrix, 392
Blockage, 368
BLOS (*see* Beyond line of sight)

Boltzmann's constant, 76, 79, 157, 234
Bonding, 366
Brewster angle, 180
Bridge-to-bridge circuit, 51
Broadband jammer, 99
Broadcast control authority, 265–66
Built-in-test, 391–92

C

C-band, 289
C2 (*see* Command and control)
C3 (*see* Command, control and communications)
C3I (*see* Command, control, communications, and intelligence)
C3I data flow, 252
CA-1100, 400–401
CAINS (*see* Carrier, aircraft inertial navigation system)
CAP (*see* Combat, aircraft patrol)
Carrier aircraft inertial navigation system, 219
Carrier coordination circuit, 42, 46
Carrier intelligence center, 254, 256
Carrier-to-noise ratio (C/N_0), 77, 240
Cassegrainian feed, 359
CCC (*see* CINC command complex)
CCITT (*see* Consultative Committee of International Telephone and Telegraph)
CCOW (*see* Channel control order wire)
CCSC (*see* Combat cryptological support console)
CDMA (*see* Code division multiple access)
CELP (*see* Code excited linear prediction)
Central Radio Propagation Laboratory, 162, 163
Channel capacity, 89
Channel control order wire, 280
Channel usage, 141
Channel vocoder, 60
CHBDL (*see* Common high bandwidth data link)
Chip, 105, 108, 109
Chirpsounder®, 163–66, 394
CIC (*see* Combat information center)
CINC (*see* Commander-In-Chief)
CINC command complex, 454
Circuit, definition of, 5
Circuit quality monitoring, 386
Circuit switching, 20
Circuits on a navy platform (*figure*), 254
CLF (*see* Commander, Landing Force), 30

Clocking, 321
Clutter, 90, 91, 94
Coastal earth station, 241, 243, 289–90, 293, 294
Coaxial cable, 429, 447
Code:
  continental Morse, 134, 139
  Fire, 75
  five-unit, 134
  Hamming, 75
  international Morse, 133
  Morse, 133
  pseudo-noise, 105, 109
  seven-unit, 134
Code division multiple access, 104, 107, 109
Code excited linear predictive coding, 60–62, 276
Coding, 1/2 rate, 317
Coding gain, 317
Colocation interference, 372
Combat aircraft patrol, 23, 37, 45
Combat cryptological support console, 256
Combat information center, 37, 254, 256, 446
Combat training system, 481
Command and control, 64, 428
Command control, and communications, 30
Command control, communications, and intelligence, 31, 36, 252
Command and control traffic, 321
Command and decision system, 422, 476, 479
Commander, Landing Force, 30
Commander-In-Chief, 11
Commander's ship-to-shore circuit, 52, 54
Commercial-off-the-shelf, 302
Common high bandwidth data link, 48, 49, 168, 221–23
Common user digital information exchange subsystem, 53, 252, 255, 257, 261, 410
Communication support system, 156, 211, 302, 447, 449–53
Communications antenna, 341
Communications plan, 448, 453
Communications security, 208, 210, 381
  equipment, 210
  equipment area, 387
  operations, 18
Compact VLF, 146, 149, 150
COMPLAN (*see* Communications plan)
Computer-to-computer data transmission, 83
COMSEC (*see* Communications security)
Conductivity, 115
Consolidated Space Operations Center, 297
Constant envelop signal, 101

Consultative Committee of International Telephone and Telegraph, 421
Continuous wave, 50, 147
Control circuit patchpanel, 400
Control ship coordination net, 42, 43, 47
Copernicus architecture, 10, 454
COSAM, 370
Cosite interference, 209, 377
Coupler, 132
Coupling, 365, 368
Coverage, 70, 71
Crosslinking, 328
Crossmodulation, 364
Cryptographic auxiliary unit, 411
CSOC (*see* Consolidated Space Operations Center)
CSS (*see* Communication support system)
CU-1396/BRA-16, 132
CU-1441/BRR, 132
CU-1559, 226–27
CU-1772/SRA-56, 396–97
CU-1774/SRA-57, 397–98
CU-1776/SRA-58, 397–98
CU-1789, 398–99
CU-2364/BRR, 132
CUDIXS (*see* Common user digital information exchange subsystem)
CV-3333, 58
CV-3591, 278
CVIC (*see* Carrier intelligence center)
CVSD (*see* Modulation, continuously variable sloped delta modulation)
CW (*see* Continuous wave)

D

D layer, 118, 122, 125, 126
DAMA (*see also* Demand assigned multiple access)
  frame format, 280
  subscriber terminal, 285
Data, 60
Data link, 43, 48
Data link layer, 433
Data rate, 141
  ELF, 83, 150
  HF, 158
  VLF, 83, 141
Data signal, 391
Data switching, 408–13
Data terminal set, 169, 170
Data transfer switch, 408–9

Date time group, 64
Day-night changes, 122, 126
DCAOC (*see* Defense communications
    agency operation center)
DCS (*see* Defense communications system)
DCT (*see* Digital communications terminal)
DDN (*see* Defense digital network)
DDSP (*see* Distributed digital signal
    processor)
DECAL, 370
Defense Communications Agency, 421
Defense communications agency operation
    center, 318–19
Defense communications system, 11, 20, 64
Defense data network, 12, 64, 252
Defense Information Systems Agency, 421,
    436
Defense satellite communications system, 12,
    51, 52, 109, 230, 232, 305–22
Defense secure network 1, 2, 3, 12, 64
Defense switched network, 12
Delay-locked loop, 106
Demand assigned multiple access, 53, 249,
    262, 273, 280–86, 303
Demodulation, 7
Demodulation processing gain, 89
Demultiplexing, 7
DF (*see* Direction finding)
Dielectric constant, 115, 117
Dielectric materials, 116
Digital communications terminal, 210
Digital time division command/response mul-
    tiplex, 211
Direct sequence, 105, 108, 109
Direct wave, 116, 203
Direction finding, 43
DISA (*see* Defense Information Systems
    Agency)
Distress communications, 43, 50
Distributed digital signal processor, 150
Distribution, RF, 396
Diurnal change, 122
Diversity, 83
    frequency, 75, 204–5
    space, 205
Doppler shift, 321
Downlink path loss, 92
Downtime, 84
DS (*see* Direct sequence)
DSB (*see* Modulation, double sideband)
DSBAM (*see* Modulation, double-side-band
    amplitude)

DSCS (*see* Defense satellite communications
    system)
    control, 318
    footprint (*figure*), 308
    frequency plan, 311–12
    ship terminal, 312
    shore terminals, 306
DSCS II, III, 108, 109
DSCS III satellite characteristics, (*table*), 309
DSCS satellite control facility, 310
DSN (*see* Defense switched network)
DSNET (*see* Defense secure network) 1, 2, 3
DTG (*see* Date time group)
DTS (*see* Data terminal set)
Duct, 113

E

E layer, 159, 161
E3 (*see* Electromagnetic, environmental
    effects)
EAM (*see* Emergency action message)
Early warning, 422
Earth coverage, 310
Earth's conductivity, 117
Earth's magnetic field, 119
ECCM (*see* Electronic counter-
    countermeasures)
ECM (*see* Electronic countermeasures)
Effective isotropic radiated power, 72, 91,
    240, 241, 242, 312–13
Effective reflection height, 118
EHF (*see* Extremely high frequency)
EHF SATCOM, 4, 56, 246, 323–39
EIA (*see* Electronic Industries Association)
EIRP (*see* Effective isotropic radiated power)
EK-070, 147
Electric breakdown, 116
Electric field propagation mode, 119
Electric field strength, 73, 117, 203
Electromagnetic:
    compatibility, 360, 362, 365
    environmental effects, 362
    interference, 360, 362, 365
    pulse, 151, 362
    radiation, 362
    radiation hazard, 313, 362
Electron density, 122
Electronic combat, 32
Electronic counter-countermeasures, 89, 208,
    210, 214, 220, 344, 362
Electronic countermeasures, 29, 45, 344

Electronic Industries Association, 421
Electronic intelligence, 29
Electronic support measures, 45, 344, 422
Electronic warfare, 344–46, 478
     antennas, 341, 344–46
Electronic warfare control system, 450
Electronic warfare system, 422, 476, 478
ELF (*see also* Extremely low frequency)
     communications system, 55, 56
     propagation, 117
     seawater penetration, 56, 150
     transmitter power, 150
ELF submarine communications system
     (Sanguine), 56, 112, 150
ELINT (*see* Electronic intelligence)
Elliptical orbit, 231
ELOS (*see* Extended line-of-sight)
EMC (*see* Electromagnetic compatibility)
EMC advisor board, 365, 366, 382
EMC analysis, 369
EMCAB (*see* EMC advisor board)
EMCON (*see* Emission control)
Emergency action message, 310
EMI (*see also* Electromagnetic interference)
     assessment, 365, 366
     control method, 374
     filtering, 381
     matrix (*table*), 367
     source, 363–65
Emission control, 24
Emission, RF, 368
EMP (*see* Electromagnetic pulse)
Epoch, 216, 217
Equivalent noise temperature, 73, 236
Equivalent operating temperature, 74
Error:
     correction, 314
     curve, 81
     detection and correction, 64, 81
     rate, 78, 81, 83, 98
ESM (*see* Electronic support measures)
European Space Agency, 290
EW (*see* Early warning)
EW (*see* Electronic warfare)
EW communications, 42, 45
EW coordination circuit, 42, 45
EXCOMM (*see* Exterior communications)
Extended line-of-sight communications, 13
Exterior communications, 58, 255, 425,
     449–51, 476, 480
Extremely high frequency, 2, 4, 107, 323
Extremely low frequency, 3, 111, 112

F

F layer, 159, 161
Facsimile, 60, 64, 83
Facsimile transmission, 65
Fading, 75, 155, 158, 160, 161
Failure, 84
FBM (*see* Fleet ballistic missile)
FCC (*see* Fleet command center)
FDCS (*see* Flight deck communications
     system)
FDDI (*see* Fiber distributed data interface)
FDM (*see* Frequency division multiplex)
FDMA (*see* Frequency division multiple
     access)
FED-STD-1045, 191, 193
FED-STD-1046, 290
FED-STD-1047, 193
FED-STD-1048, 193
Federal information processing standard-146,
     436
FEP (*see* Fleet satellite, EHF package)
FH (*see* Frequency hopping)
FH system, 107
Fiber distributed data interface, 429–32, 448
Fiber optic, 440, 442
Fiber optics in surface ships (*table*), 432
Fighter air direction circuit, 42, 46
Fighter control circuit, 42, 46
File transfer protocol, 436
Filter, cosine, raised cosine, square, 136, 137,
     144
Filtering, 381
Filtering index, 136, 142
FIPS (*see* Federal information processing
     standard)
Flag data display system, 450
Fleet ballistic missile submarine, 151,
Fleet broadcast, 34, 35, 62
Fleet command center, 11
Fleet LF multichannel broadcast, 112
Fleet ocean surveillance information center,
     252, 263
Fleet primary ship-to-shore circuit, 52, 54
Fleet satellite, 230, 246, 249, 302
     broadcast, 108, 247, 257, 260, 330
     broadcast subsystem, 252, 253, 255, 260,
        261, 328
     characteristics (*table*), 249
     communications, 51–53, 108, 109, 245–50
     communications circuits (*figure*), 254
     communications control, 287

Fleet satellite (*cont.*)
  communications end-to-end subsystem configuration, 258
  communications information transmission systems (*table*), 257
  communications secure voice, 54
  coverage (*figure*), 247
  EHF package, 246, 324, 328
  frequency plan, 246, 247
  information transmission system (*table*), 257
Fleet tactical circuits (FLETAC), 44
Flight deck communications system, 413–15
FLTBROADCAST (*see* Fleet broadcast; Fleet satellite broadcast)
FLTSAT (*see* Fleet satellite)
FLTSAT information exchange subsystems, 52, 53
FLTSATCOM (*see* Fleet satellite communications)
FM (*see* Modulation, frequency)
Footprint, 231, 308
FOSIC (*see* Fleet ocean surveillance information center)
Fractional power occupancy bandwidth, 143
Frame, 103, 216, 217, 281
Frequence optimum de travail (FOT), 162
Frequency:
  allocation, 199, 362
  assignment, 362
  band designations, (*figure*), 206
  diversity, 75, 158
  domain, 103
  standard, 392, 400
Frequency and wavelength of long-wave, 112
Frequency assignment, microwave (*table*), 206
Frequency division multiple access, 103
Frequency division multiplexing, 109, 320–21, 332
Frequency hopping, 107, 108, 207, 208, 325, 330
Frequency hopping filter, 378–79
Frequency regions, shipboard communications, 3
Front-door EMI, 365
Front-end noise (*see* Noise, front-end)
FSK (*see* Modulation, frequency-shift keying)
Full-period termination, 54

G

G/T (receiver figure of merit), 240, 241–43
G-band, 50, 48
Gain, 71, 89, 90, 156, 234, 237, 239

Gapfiller satellite, 245
Gaussian noise (*see* Noise, gaussian)
GENSER (general service), 257
Geomagnetic constant, 126
Geomagnetic field, 125, 126
Geostationary satellite, 231
Geosynchronous orbit, 231
Global connectivity, 155
Global information exchange subsystem, 34, 454
Global positioning system, 129, 256, 342, 441
GLOBIXS (*see* Global information exchange subsystem)
GMF (*see* Ground mobile force terminal)
GMF modem, 108
GOSIP (*see* Government open system interconnection profile)
Government open system interconnection profile, 432, 436–37
GPS (*see* Global positioning system)
Great circle path, 125
Ground conductivity, 116, 117, 125, 126
Ground constant, 126
Ground mobile force terminal, 317
Ground wave, 115, 116, 155, 168
  field strength, 117
Grounding, 366, 380
Guard time, 220, 281

H

H-200 chestset, 415
H-202 headset, 415
H-203 handset, 415
Hamming code (*see* Code, Hamming)
Handshaking, 190
Harbor communications, 43, 51
Harmonic frequency, 364
HAVE QUICK II, 48, 50, 108, 109, 198, 220–21
Header, OSI protocol, 433
Helicopter control circuit, 42, 46
HEMP (*see* high altitude electromagnetic pulse)
HERF, 362
HERO, 362
HERP, 362
HESSA (*see* High efficiency solid state amplifier)
HF (*see also* High frequency)
  anti-jamming, 108, 158
  anti-jamming communciations system, 375, 376

band, 154
communications, 51, 52, 54, 155, 156
design consideration, 158
fading, 75, 160, 161, 166
long-haul circuit, 154
operating frequency, 161, 163
operational limitations, 158
performance, 154
propagation, 159, 161
propagation forecast, 163
radio, 394
receiver, 171
sounding, 163, 394
transceiver, 174
transmitter, 172
HF improvement program, 155
HF shipboard antenna, 180
HF shipboard antennas, comparison (*table*), 180
HF-80, 174–76
HHR (*see* High hopping rate)
HICOM net, 52, 55
High altitude electromagnetic pulse, 158
High efficiency solid state amplifier, 116
High frequency, 3, 17, 54, 72, 154–95
High hopping rate, 333
High speed fleet broadcast, 302–3
High speed TTY, 83
High-level radiation, 368
High-power amplifier, 335
Hop rate, 108
Hostile interference, 94
Hot spare, 87
HPA (*see* High-power amplifier)
HSFB (*see* High speed fleet broadcast)
Hull grounding, 380
HY-2, 61

I

I-carrier (*see* In-phase carrier)
ICS/SCAN (*see* Integrated communications system/shipboard communications area network)
ICS-3 (*see* Integrated communication system-3)
ICW (*see* Interrupted continuous wave)
Identification, friend or foe, 49, 216, 422, 477
antennas, 341, 342–44, 346–47
IEEE (*see* Institute of Electrical and Electronic Engineers)
IEEE 802.5, 429–31, 436
IFF (*see* Identification, friend or foe)
IM product, 370, 372, 373

Impedance matching, 115, 227, 353
Impulse noise (*see* Noise, impulse)
In-phase carrier, 135, 140, 141, 150
Index of refraction, 162
Indian ocean region, 290, 293, 319
Inertial navigation system, 256
Information bits duration, 82
Information exchange subsystem, 51, 53
Initial entry, 218, 219
INMARSAT (*see also* International maritime satellite)
characteristics (*table*), 292
coverage (*figure*), 292
services, 289
INS (*see* Inertial navigation system)
Institute of Electrical and Electronic Engineers, 421
Integrated communications system/shipboard communication area network, 447–48
Integrated communication system-3, 108, 171, 177
Integrated undersea surveillance system, 11, 35, 252
Intelligibility of spoken language, 82
INTELSAT, 290
INTERCOM (*see* Interior communications system)
Interconnecting group, 279, 408, 411–13
Interference, 92
Interference sources, 370
Interfering transmitter, 370
Interior communications system, 385, 389, 407–8, 422, 476, 480
Intermodulation, 364
hull-generated, 364
products, 372
signal, 100
International maritime satellite, 288–94
International Standards Organization, 421, 432
International Telecommunications Union, 205, 290
Internet protocol, 433, 436, 449
Interrupted continuous wave, 133, 146
Ionosphere, 112, 116, 117, 233
Ionospheric:
fluctuation, 155
layers, 159–62
propagation, 394
reflection, 118, 155, 159, 160
sounding, 155–56
variation (*table*), 161
IP (*see* Internet protocol)

ISB (*see* Modulation, independent sideband)

ISO (*see* International Standards Organization)

ISO and military protocols, comparison (*table*), 436

ITU (*see* International Telecommunications Union)

IUSS (*see* Integrated undersea surveillance system)

IXS (*see* Information exchange subsystem)

J

Jam-resistant satellite communications, 305
Jam-resistant secure communications, 52
Jammer-to-signal ratio, 95
Jamming, 89–91, 94
    power, 91–94
JANAP 128(J), 62, 257, 261
JCS (*see* Joint Chiefs of Staff)
JINTACCS (*see* Joint interoperability of command and control system)
Joint Chiefs of Staff, 11, 305
Joint interoperability of command and control system, 257, 260, 261
Joint operational tactical system, 256, 450
Joint tactical information distribution system, 48, 198, 216–19, 256
JOTS (*see* Joint operational tactical system)
JRSC (*see* Jam-resistant secure communications)
JTIDS (*see also* Joint tactical information distribution system)
    class 2 terminal, 108
    net participant group (*table*), 219

K

Keying waveform, 137
KG-84, 302, 276, 279
Ku-band, 48, 49, 205–6, 221–23
KW-46, 302
KY-57, 62
KYV-5, 278

L

L-band, 289
LAMPS (*see* Light airborne multi-purpose system)
LAN (*see* Local area network)
Land/launch control net, 42, 46
Landing force, 47
    command net, 42, 47

gunfire support net, 43, 47
intelligence net, 43, 47
logistics net, 43, 47
reconnaissance net, 43, 47
Latency, 430, 431
LDMX (*see* Local digital message exchange)
LDR (*see* Low data rate)
Leakage, 364
LEASAT, 245, 247–51
    characteristics (*table*), 250
    control, 287–88
    coverage (*figure*), 250
    frequency plan, 250, 251
    operations control center, 288
LF (*see* Landing force; Low frequency)
LF multichannel broadcast, 112
LFSR (*see* Linear feedback shift register)
Life cycle EMC, 382
Light airborne multi-purpose system, 28, 36, 37, 49, 223
Light airborne multi-purpose system data link, 48, 49, 168, 223–25
Lightweight suite, 439, 448
LINCAL, 370
Line replaceable unit, 86
Line-of-sight communications, 13, 198, 200, 202
Linear feedback shift register, 105
Linear predictive coding-10, 60, 61, 276
Link 1, 48, 50, 168, 256
Link 11, 48, 49, 158, 168, 170, 426, 446
Link 11 Improvement (LEI), 155
Link 11 message format, 426–27
Link 14, 48, 50, 446
Link 16, 36, 48, 49, 70, 168, 253, 256
Link 4A, 48, 49, 218–20, 446
Link:
    availability, 82, 84
    budget, 239–43
    capacity, 70, 82
    definition of, 5
    design equation, 78
    layer, 433, 440, 441
    loss, 91
    margin, 72
    quality, 70, 80
    quality analysis, 190
LLTV (*see* Low light television)
LNA (*see* Low noise amplifier)
Local area network, 422, 424, 429, 431, 432, 438
Local digital message exchange, 20, 64, 447

Logistic delay, 84
Long distance propagation, 117
Long range navigation (LORAN-C), 112, 132, 342
Long-haul circuit, 155, 185
Long-haul communications, 40, 51, 52, 54, 83, 154
Long-wave, 111
    modulation system, 133
    propagation, 112, 115–25
    radio, 112, 113
    receiving antennas, 115, 126
    receiving equipment, 145
    signal, 113, 116
    signal field strength, 116
    transmission, 113–15
    transmitting antennas, 126
Long-wave propagation capability, 125
LORAN-C (*see* Long range navigation)
LOS (*see* Line-of-sight)
Low data rate (MILSTAR), 324, 325, 329–32
Low frequency, 3, 111, 112
Low light television, 68
Low noise amplifier, 312–13, 337
Low probability of detection, 324, 344
Low probability of interception, 38, 158, 214, 344
Low speed TTY, 83
Lower sideband (LSB), 166
Lower usable frequency (LUF), 162
LPC-10 (*see* Linear predictive coding-10)
LPD (*see* Low probability of detection)
LPI (*see* Low probability of interception)
LRU (*see* Line replaceable unit)
LS-474, 417
LST command net, 42, 47
LWCA (*see* Antenna light weight communications)
LWPC (*see* Long-wave propagation capability)
L_x band, 168

M

Magnetic field propagation mode, 119
Main communications center, 254
Maintainability, 85
Maintenance action, 84
Man-on-the-move radio, 414
Man-pack, 208
Manchester II waveform, 442
MARECS, 290–92

Marine radiotelephone, 51
MARISAT, 245, 290–92
Maritime patrol aircraft, 28
Mark and space, 141
Master control station, 280, 283
Matching networks, 158
Maximum length sequence, 105
Maximum observed frequency (MOF), 165
Maximum usable frequency, 157, 162
MBA (*see* Multibeam antenna)
MCE (*see* MILSTAR control element)
MCM (*see* Mine countermeasures communications)
MCM TG command net, 50
MCM TG reporting net, 51
MCM TG tactical net, 50
MCS-2000, 407–8
MD-900, 271
MD-942, 296
MD-1030/A, 314, 316–18
MDR (*see* Medium data rate)
Mean logistic delay time, 85, 86, 87,
Mean time between failure, 85, 86
Mean time to repair, 85–87
Medium data rate (MILSTAR), 324, 325, 331–32
Medium frequency, 3
MEECN (*see* Minimum essential emergency communications network)
Message:
    priority, 84
    routing, 84
    switching, 12
    traffic, 260
Meteor burst, 202
Meteorological system, 422, 476, 481
Meteorology and telemetry antennas, 341, 347
MF (*see* Medium frequency)
MIDL (*see* Modular interoperable data link)
MIL-STD-188-110, 107, 334
MIL-STD-188-135, 202
MIL-STD-188-141A, 174, 189
MIL-STD-188-203-1, 49
MIL-STD-188-203-3, 49
MIL-STD-188C, 177
MIL-STD-461, 380
MIL-STD-1397, 169
MIL-STD-1553B, 174, 193, 210–11
MIL-STD-1553B bus, 422, 424, 442–43
MIL-STD-1582, 107, 334
MIL-STD-1777, 436
MIL-STD-1778, 436

MIL-STD-1780, 436
MIL-STD-1781, 436
MIL-STD-1782, 436
Military (DoD) protocol standards (*table*), 436
Military grid reference, 214
Military strategic and tactical relay satellite, 56, 108, 230, 246, 324–39
MILSTAR (*see also* Military strategic and tactical relay satellite)
    base band capabilities (*table*), 327
    connectivity, 330
    terminal, 335
    waveform, 334
MILSTAR operations center, 338–39
Mine countermeasures communications, 43, 50
Mini-DAMA equipment, 286
Minimum essential communications, 14
Minimum essential emergency communications network, 15
MIST (*see* Modular interoperable surface terminal)
MLDT (*see* Mean logistic delay time)
MOC (*see* MILSTAR operations center)
Modular interoperable data link, 221–23
    surface terminal, 221
Modulation, 7
    AM suppressed carrier, 167
    amplitude, 82
    amplitude-shift keying, 134, 135, 139, 140, 143, 144
    binary PSK, 141, 144, 145
    coherent FSK, 81
    coherent PSK, 81, 93, 99
    continuously variable sloped delta, 60, 61, 62, 276
    CW keying, 133
    differential PSK, 314, 330
    double sideband, 166, 167, 171
    duo-binary PSK, 169
    frequency, 82
    frequency-shift keying, 134, 139, 140, 143, 144
    ICW keying, 134
    independent sideband, 167, 171
    minimum-shift keying, 135, 140, 141, 143–47
    multilevel phase-shift keying, 143
    multiple frequency-shift keying, 333
    narrowband FSK, 133
    narrowband PSK, 133
    offset quadrature phase-shift keying, 141, 143
    on-off keying, 134
    phase-shift keying, 133, 136, 140, 143, 144, 314
    quadrature phase-shift keying, 143, 145, 212
    rate, 136
    shaped AM, 133
    single side-band, 154, 155, 166–68, 171
    sinusoidal FSK, 144, 145
    suppressed carrier, 167
Modulation index, 139
Modulation method comparison, 145
Monitoring, 385–89
MPA (*see* Maritime patrol aircraft)
MPA reporting and control circuit, 42, 46
MSK (*see* Minimum-shift keying)
MTBF (*see* Mean time between failure)
MTTR (*see* Mean time to repair)
Multibeam antenna, 310
Multibeam antenna with nulling, 108
Multichannel HF radio, 370, 374, 375
Multicoupler, 128, 185, 225, 354, 385, 396
Multimedia networking, 448–49
Multipath, 74, 75, 155, 158, 201, 202–4
Multiplexer, DAMA, 281
Multiplexing, 7
MX-512P, 169
MX-1986C/SRC, 207

N

Narrowband preselector filter, 374, 375
National command authority, 33
National Institute for Standards and Technology, 436
National Television Standards Committee, 68
Naval communications:
    connectivity, 7
    history, 7
    radio spectrum, 2
Naval communications processing and routing system, 17, 64, 260, 263, 447
Naval communications station, 13, 17–20, 36, 54, 269
Naval computer and telecommunications area master station, 13, 16–20, 36, 54, 306
Naval digital voice systems, 61
Naval gunfire air spot net, 43, 47
Naval gunfire control circuit, 43, 47
Naval gunfire ground spot net, 43, 47
Naval gunfire support communications, 43, 47

Naval message, 63
Naval modular automated communications
    system, 257, 260–63, 408–11, 446–47
Naval shipboard communication services,
    characteristics (*table*), 60
Naval shipboard EXCOMM connectivity, 34
Naval Space Command, 286, 287
Naval tactical data system, 158, 168, 169,
    424, 428, 446
Naval telecommunications:
  automation programs, 445
  center, 18
  system, 12
  system architecture, 447
  system, area, 16
Naval Telecommunications Command, 286
Naval Telecommunications Command
    operation center, 286–87
Naval warfare, 22
NAVCOMPARS (*see* Naval communications
    processing and routing system)
NAVCOMSTA (*see* Naval communications
    station)
Navigation, 422
  antennas, 342
  receiver, 132
  satellite, 342
  system, 422, 476, 478
NAVMACS (*see* Naval modular automated
    communications system)
Navy digital system interconnect standard
    (MIL-STD-1397A), 422, 424,
Navy EHF satellite program, 324, 328, 334,
    336
Navy enhanced terminal, 303
Navy satellite control station, 297
Navy standard teleprinter terminal, 279
NCA (*see* National command authority)
NCTAMS (*see* Naval computer and
    telecommunications area master station)
Near-vertical incident skywave, 186–87, 188
NEC (*see* Numerical electromagnetic code)
NESP (*see* Navy EHF satellite program)
Net control station, 169, 428
Net management, 218, 219
Network layer, 433, 440, 441
Network participant group, 218–19
NGFC (*see* Naval gunfire control circuit)
Noise, 73–76, 123, 234
  atmospheric, 115, 123, 155, 157
  background, 73
  cosmic, 123

  figure, 79
  front-end, 79, 363
  galactic and solar, 73, 79, 363
  gaussian, 75
  human-made, 123, 363
  impulse, 75
  level, receiver, 155
  operating temperature, 233, 235
  power, 76, 234
  power density, 73, 76
  quantization, 75
  sources, 234
  temperature, 234–36
  thermal, 73
Normal-through, 395
NPG (*see* Network participant group)
NSCS (*see* Navy satellite control station)
NTCOC (*see* Naval Telecommunications
    Command operation center)
NTDS (*see* Naval tactical data system)
NTDS interface, 169, 424
NTS (*see* Naval telecommunications system)
NTSC (*see* National Television Standards
    Committee)
Numerical electromagnetic code, 369
NVIS (*see* near-vertical incident skywave)

O

Ocean surveillance information system, 11,
    252, 263
OE-176, 129
OE-207, 129
OE-305, 127
OE-315, 132
Officer-in-tactical command information
    exchange subsystem, 252, 257, 263–65,
    411
OM-55, 108
OMEGA, 112, 132, 342
ON-143(V)6, 264, 265, 280, 411–13
On-call circuit, 54
One-sided bandwidth, 82
One-way announcing system, 417–18
OOK (*see* Modulation, on-off keying)
Open system interconnection, 432, 448
Operating temperature, 79
Operational availability (A_0), 84
Operational requirement, 88
OPNAVINST 2410.4, 362
Optical fiber, 429, 432
Optimum frequency shift, 135

Optimum usable frequency (OUF), 162
Optimum working frequency (OWF), 162, 163
OQPSK (*see* Modulation, offset quadrature
    phase-shift keying)
OR (*see* Operational requirement)
OR-209, 223–24
Orbital parameter, 231
Order wire, 318
ORESTES, 260–63, 271
Organic sensor, 25
OSI (*see* Open system interconnection)
OSI model, 194, 433–36
OSIS (*see* Ocean surveillance information
    system)
OTCIXS (*see* Officer-in-tactical command in-
    formation exchange subsystem)
OTH (*see* Over-the-horizon)
OTH-T gold format, 257, 268
Over-the-horizon, 29
    targeting, 253, 263, 267

P

Pacific ocean region, 290, 293, 319
Packet switching, 12
Parabolic reflector, 72
Parallel A_0, 86
Parasitic oscillation, 364
Patching, 19, 115, 385–89
Patching and testing, 386
Patchpanel:
    audio frequency, 406–7
    HF, 396, 398
Path loss, 72, 115, 239
PB (*see* Phonetically balanced)
PECAL, 370
Periscope depth, 122
Phase-locked loop, 83
Phonetically balanced, 81
Physical layer, 433, 440, 441
Picket station, 169
Pitch, 61
PLA (*see* Plain language address)
Placement, HF antennas, 368
Plain language address, 62
Plain text, 385
PLRS (*see* Position location reporting system)
PN (*see* Pseudo-noise code)
PN-sequence, 105
Polar orbit, 231
Polarization:
    horizontal, 187

    vertical, 115
Polling, 168, 190, 192
Position location reporting system, 214–16
Power balancing, 100
Power budget for satellite links, (*tables*),
    241–43
Power law device, 100
Precedence level, 62, 264
Preemption, 64
Presentation layer, 434, 440, 442
Primary ship-to-shore (NATO) circuit, 52, 54
PRITAC (*see* TG tactical circuit)
Processing gain, 78, 90, 93, 104, 107
Propagating mode, 118, 119
Propagation, 116, 233
    prediction, 123, 162–66
    VHF/UHF, 202
Prototype ocean surveillance system, 450
PSK (*see* Modulation, phase-shift keying)
Pulse width, 136

Q

Q-carrier (*see* Quadrature carrier)
QMCS (*see* Quality monitoring control
    system)
QMN (*see* Quasi minimum noise)
QMS (*see* Quality monitoring system)
QPSK (*see* Modulation, quadrature phase-
    shift keying)
Quadrature carrier, 135, 141, 150
Quality monitoring control system, 389–92
Quality monitoring system, 252
Quality monitoring and test, 388
Quantization:
    noise, 75
    step, 75
Quasi-minimum noise, 372, 373
QUICKSAT, 317–18

R

R-1051, 169, 171
R-2368, 147, 169, 171
R800, 147
Radar, 422, 476
Radar antennas, 342
Radar electronic warfare, 31
RADHAZ (*see* Electromagnetic radiation haz-
    ard)
Radio frequency interference, 281
Radio link dropout, 162
Radio TTY, 44

Radios, VHF/UHF (*table*), 201
Range, 157
Range equation, 72, 78, 156, 239–43
Range, VHF/UHF, 201
Ratio of symbol energy to the noise power
    density (E_B/N_0) 78
Ratio of the desired signal to the interference,
    93
RATT (*see* Radio TTY)
RCCOW (*see* Return channel control order
    wire)
Reaction time, 70, 84
Receiver antenna matching, 113
Receiver front-end, 200
Receiver noise temperature, 79
Receiver performance, VLF, 149
Receiver sensitivity, 78, 79, 80, 115, 129, 145,
    201
    long wave receiver, 128
Receiver synchronization, 411
Record message, 60, 62, 83
Record traffic, 321
Red communications switch, 402, 404–6
Red signal, 19
Red-black isolation, 381, 411
Redundancy, 83, 86
Redundancy coding, 75
Reflected wave, 159, 162
Reliability, 85
Response time, 84
Return channel control order wire, 281
RF:
    distribution system, 395–402
    emission, 368
    frequency standard patchpanel, 401–2
    impedance matching, 395, 397
    patchpanel, 386, 394, 398–402
    signal, 391
    signal distribution, 385
    switch, 385, 394
    system design, 370, 371
RF and baseband area, 388
RI (*see* Routing indicator)
Roll-call mode, 169, 428
Routing indicator, 62
RS-232, 210, 433
Rusty bolt effect, 364, 370, 380

S

S-band, 246
SA-2112, 6, 8, 391–92, 402–06

SA-2561, 401–2
SA-4176, 408
SAFENET (*see* Survivable adaptable fiber
    embedded network)
SAFENET characteristics (*table*), 438
SAFENET I, II, 425, 436–42, 448
SAFENET lightweight suite, 439, 441–42,
    448
SAFENET OSI suite, 440–41
Sanguine, 55, 56, 112, 150
SAR (*see* Search and rescue)
SAS (*see* Single audio system)
SATCOM (*see* Satellite communications)
Satellite antenna gain, 91
Satellite communications, 51, 83, 109
Satellite orbit, 230–32
SB-3648, 400
SB-4249, 398–99
SB-4268, 406–7
SB-4269, 406–7
Scanning, 191
Screen tactical net, 42, 44
SCT (*see* Single channel transponder)
Sea-following buoy, 128
Search and rescue, 37, 50
Search and rescue net, 50
Seawater:
    attenuation, 113, 120, 121
    penetration, 115, 120
Secure voice:
    equipment, 276
    subsystem, 252, 255, 257, 268–70, 312, 411
SECVOX (*see* Secure voice subsystem)
Selective calling, 190
Self-steering filter, 374
SELSCAN, 174
SEMCAC (*see* Ship EMC analysis for
    communications)
SEMCIP (*see* Shipboard EMC improvement
    program)
Serial A_0, 86
Services (CSS), 451, 453
SESS (*see* Special signal exploitation space)
Session layer, 434, 440, 441
SEW (*see* Space and electronics warfare)
SHF (Super high frequency)
SHF SATCOM (*see* SHF satellite
    communications)
SHF satellite, 247
    communications, 4, 51, 52, 230, 305–22,
    328
Shielding, 366, 380

Ship alteration, 382
Ship earth station, 241, 243, 289–90, 293
Ship EMC analysis for communciations, 382
Ship-to-aircraft connectivity, 36, 198
Ship-to-ship connectivity, 36, 198
Ship-to-shore connectivity, 35, 198
Ship-to-submarine connectivity, 36
Shipboard:
    antenna couplers (*table*), 354
    C3I system, 253
    combat systems, 341, 422, 423, 477
    communications antennas, 348–59
    communications protocol, 421–43
    exterior communications monitor and
       control system, 388
    LAN, 422, 429–32, 448
    signal interfaces, 423
    technical control facility, 392
    transmission cable, 429
Shipboard communications, frequency regions
    (*figure*), 3
Shipboard EMC improvement program, 365,
    382
Shore-to-aircraft connectivity, 35
Shore-to-ship connectivity, 35
Shore-to-shore connectivity, 34
Shore-to-submarine connectivity, 35
Short haul communications, 40, 83
SI (*see* Special intelligence)
Sideband splatter, 364
Signal:
    duration, 93
    element, 108
    isolation, 380
    management, 32
    suppression, 100
Signal energy ratio (E_B/N_0), 78, 94, 96
Signal plus noise plus distortion to noise, 79, 80
Signal-to-interference ratio, 93
Signal-to-noise ratio, 75–77, 78, 90, 115, 157,
    200, 239
Simple mail transfer protocol, 436
Simultaneous operation, 158
SINAD (*see* Signal plus noise plus distortion
    to noise)
SINCGARS (*see* Single channel ground to air
    radio system)
Single audio system, 208, 210, 273, 403, 417
Single channel ground to air radio system, 80,
    108, 207–11, 377–78
Single channel mode, 208
Single channel transponder, 310

Single integrated operations plan, 15, 310
Single mode propagation prediction, 124
SIOP (*see* Single integrated operational
    plan)
Skip zone, 159, 186
Sky wave, 115, 116, 155, 168
    interference, 122
Skynet, 306
Small-signal suppression, 92, 94, 100
Snell's law, 162
Sound-powered telephone system, 413–17
Sounding, 162, 190, 192
Space and electronics warfare, 31, 32
    connectivity, 31, 32
Space loss, 239
Special intelligence, 252, 260, 261
Special signal exploitation space, 254, 256
Spread-spectrum, 199, 207, 261, 324, 344
    communications, 73, 76, 104, 105, 306
SSBAM (*see* Modulation, single side-band
    amplitude)
SSBN, 35
SSECMS (*see* Surface ship exterior communi-
    cations monitor and control system)
SSIXS (*see* Submarine satellite information
    exchange subsystem)
SSN, 35
STANAG 3050, 147
STANAG 3838, 422, 443
STANAG 4253, 334
Standby, 86
Station-keeping, 231
STDL (*see* Submarine tactical data link)
Strategic communications, 13
Strategic submarine communications, 31, 33,
    41
Streamliner, 260
Strike warfare, 28
    connectivity, 28, 29
STU-III, 62
STW (*see* Strike warfare)
Submarine communications, 43, 48, 111, 116
Submarine coordination circuit, 43, 48
Submarine operations authority, 17, 37, 48,
    263
Submarine operations and distress net, 37, 43,
    48
Submarine satellite information exchange sub-
    system, 55, 56, 257, 265–67, 411
Submarine tactical data link, 36
Submarine-to-aircraft connectivity, 38
Submarine-to-ship connectivity, 38

Submarine-to-shore connectivity, 37
Submarine-to-submarine connectivity, 38
SUBOPAUTH (*see* Submarine operations authority)
Super high frequency, 4, 52, 53
Supportability, 84, 85
Suppression factor, 101
Surface ship exterior communications monitor and control system, 388, 405–6
Surface wave, 115, 116
SURTASS (*see* Surveillance towed array sensor system)
SURTASS acoustic data relay, 52
Surveillance system, 31
Surveillance towed array sensor system, 28, 52
Survey, EMC, 382
Survivable adaptable fiber embedded network, 448, 438
Switch matrix, 392
Symbol energy to noise ratio (*see* Signal energy ratio)
Symbol rate, 82, 99
Syracuse, 306
System noise, 75, 76

T

TA-970, 406, 417–19
TACAMO (Take-charge-and-move-out) 33, 55, 112, 149, 151
TACAN (*see* Tactical air navigation)
TACC (*see* Tactical air control center)
TACINTEL (*see* Tactical intelligence information exchange subsystem)
TACSATCOM (*see* Tactical satellite communications)
Tactical air command net, 42, 46
Tactical air control center, 30
Tactical air navigation, 342
Tactical air traffic control circuit, 42, 46
Tactical command center, 454
Tactical communications, 13, 41, 42
Tactical data information exchange subsystem, 257, 267, 268, 411, 454
Tactical data link, 48
Tactical data processor, 428
Tactical digital link-A, 48, 49, 168, 169, 426–27, 446
Tactical digital link-B, 48
Tactical digital link-C, 48, 49, 220
Tactical digital link-J, 48, 49, 168, 218
Tactical environmental support system, 256

Tactical flag command center, 28, 254, 256
Tactical frequency management system, 163, 394
Tactical group communications, 42, 44
Tactical intelligence information exchange subsystem, 257, 260, 261, 267
Tactical receive equipment, 256, 267, 268
Tactical satellite communications, 108
TADIL-A (*see* Tactical digital link-A)
TADIL-B (*see* Tactical digital link-B)
TADIL-C (*see* Tactical digital link-C)
TADIL-J (*see* Tactical digital link-J)
TADIXS (*see* Tactical data information exchange subsystem)
Tally-ho report, 45
TCC (*see* Tactical command center)
TCP (*see* Transport control protocol)
TCP/IP, 433
TD-1063, 272
TD-1271B/U, 303, 281, 282, 285
TDM (*see* Time division multiplexing)
TDMA (*see* Time division multiple access)
TE-mode, 119
Technical control, 17, 18, 20, 385–94
Telemetry system, 422, 476, 480
Telemetry, tracking, and command, 252, 295, 297
TELENET protocol, 436, 438
Telephone set, 413, 417–19
Teletypewriter, 58, 252, 278, 312
	circuits, 321
	subsystem, 271
TEMPEST, 381
Terrestrial waveguide, 118
TESS (*see* Tactical environmental support system)
Testing, 385–89
TFCC (*see* Tactical flag command center)
TFMS (*see* Tactical frequency management system)
TG broadcast, 42, 44
TG command net, 44
TG Orestes, 42, 44, 62
TG SI (Orestes) circuit, 42, 44
TG tactical circuit, PRITAC, 36, 42, 44
Thermal noise (*see* Noise, thermal)
TIDET (*see* Topside integration design engineering team)
Time-bandwidth product, 78, 93, 104, 144
Time division multiple access, 103, 215–16, 217, 280, 321, 330

Time division multiplexing, 109, 261, 321, 332, 392, 426
Time domain, 103
Time-hopping, 108
Time slot, 103, 216, 217
Time to establish communications, 87
TM-mode, 119
Token ring, 429, 430, 440, 438, 442
Top-side antenna placement, 129, 130
Topside integration design engineering team, 367
Topside integration, EMC, 366
Tracking, satellite 232
Training system, 422, 476, 481
Transceivers, AN/WSC-3, (*table*), 273
TRANSEC (*see* Transmission security)
Transmission, 7
    bandwidth, 143
    security, 210, 336
    symbol duration, 80, 208, 210
Transmitter antenna gain, 92
Transmitter efficiency, 126
Transport control protocol, 433, 436
Transport layer, 433, 440, 441
Traveling wave tube amplifier, 100, 101, 330
TRC-251, 147
TRE (*see* Tactical receive equipment)
Tropospheric scatter, 202
Trouble-shooting and testing, 398
TT&C (*see* Telemetry, tracking and command)
TTY (*see* Teletypewriter)
Tuner, 132
Tuners and couplers, VLF, 132
Twisted pair cable, 429
Two-way announcing system, 417
TWT (*see* Traveling wave tube amplifier)
Typical signal and noise values (*table*), 94

U

U.S. message text format, 257, 261
UHF (*see also* Ultra high frequency)
    communications, 198–201
    fleet satellite broadcast, 52, 53
    FLTSATCOM, 56
    LOS, 83
    satellite communications, 3, 53, 230, 245–50
UHF follow-on satellite, 294–302
UHF SATCOM (*see* UHF satellite communications)
UHF SATCOM control, 286–88

Ultra high frequency, 3, 198–28
Underwater system, 422, 476, 478
Unified networking technology, 448
UNT (*see* Unified networking technology)
Uptime, 84
USMTF (*see* U.S. message text format)

V

Verdin, 55, 112, 149, 151
Very high frequency, 3, 198–228
Very low frequency, 3, 111–50
VF (*see* Voice frequency)
VFCT (*see* Voice frequency carrier telegraph)
VHF (*see also* Very high frequency)
    communications, 198–201
    EMI control, 377
Victim, 365
    receiver, 370
Video, 66
    imagery, 58, 60, 68, 83
    service, characteristics (*table*), 68
VINSON, 62, 256, 273
VLF (*see also* Very low frequency)
    bandwidth effectiveness, 141
    channel usage, 141
    circuit, 33
    ground wave, 116
    message injection, 151
    message rate, 138
    modulation rate, 138
    propagation, 117
    receiver circuit, 148
    transmitter, 114, 115
    wavelength, 118
VLF/LF:
    antenna coupler system, 132
    digital communications equipment, 146
    propagation waveguide, 119
VOCODER, 61, 268, 276, 413
Voice, 60, 61, 83
Voice frequency, 111
Voice frequency carrier telegraph, 177
Voice intelligibility, 81
Voltage standing wave ratio, 179, 227, 397
VSWR (*see* Voltage standing wave ratio)

W

Waterfront corrective action program, 382
Waveform, 82, 107
    filtering, 136
Waveguide mode, 117

Wavelength, 112, 368
WCAP (*see* Waterfront corrective action
    program)
Weak signal suppression, 101
Weapon control system, 422, 476, 479
Weapon system, 422, 476, 479
WT product, 144, 240

X

X-band, 205–6, 221, 223, 305
XPRESS transfer protocol, 441

Z

Zulu time, 64

About the Authors

John C. Kim received his Ph.D. and M.S. degrees in Electrical Engineering from Michigan State University. He has 25 years of experience in Naval Communications. Since joining TRW in 1969, he has held a number of positions in communications engineering. Currently, he is a principal communications systems designer of Navy and Air Force combat training systems. As the manager of advanced communications systems, he has conducted company sponsored R&D projects in shipboard multimedia networks, the simulation of an automated shipboard HF broadcast system, and the development of an LPD radio. He was deputy manager of the Navy communications project, in which he was engaged in the automation of shore technical control facilities and the radio room design of the SSN-21 class submarine. He was also manager of the Royal Saudi Navy communications project. He designed an HF digital communications (AN/USC-32) for the U.S. Navy, which was installed on SURTASS, and FFG class ships, as well as on P-3C aircraft.

Eugen I. Muehldorf received his M.S. degree in Electrical Engineering and a Doctor of Science degree from the Technical University, Vienna, Austria. His career spans computer and communications system design and analysis. He has published over 50 papers and two books, and holds several patents in his areas of expertise. He has received several awards, including three IBM invention achievement awards. He has served as an advisor to the Austrian Academy of Science. He is currently a senior technical staff member at

TRW, leading a group contributing to the development of the FAA's voice communications switching system for the National Air Space system. He also contributes solutions to the communications architecture for the U.S. Navy. For over 10 years, he has contributed to the development of U.S. Naval communications with emphasis on satellite communications, data transmission, voice communications, and command and control message transmission. He has contributed to the design and testing of the satellite communications system for the U.S. Navy's SURTASS ships.